T0135708

Machine Learning of Information Extraction Procedures

- An ILP Approach -

Bernd Thomas

Bibliografische Information Der Deutschen Bibliothek

Die Deutsche Bibliothek verzeichnet diese Publikation in der Deutschen Nationalbibliografie; detaillierte bibliografische Daten sind im Internet über http://dnb.ddb.de abrufbar.

ISBN 3-8325-0791-4

Logos Verlag Berlin
Comeniushof, Gubener Str. 47,
10243 Berlin
Tel.: +49 030 42 85 10 90
Fax: +49 030 42 85 10 92
INTERNET: http://www.logos-verlag.de

Machine Learning of Information Extraction Procedures
- An ILP Approach -

von

Dipl. Inform. Bernd Thomas

Vom Promotionsausschuß des Fachbereichs Informatik der Universität Koblenz-Landau zur Verleihung des akademischen Grades Doktor der Naturwissenschaften (Dr. rer. nat.) genehmigte Dissertation.

Vorsitzender des Promotionsausschusses:
Prof. Dr. Jürgen Ebert

Vorsitzender der Promotionskommission:
Prof. Dr. Jürgen Ebert

Berichterstatter:
Prof. Dr. Ulrich Furbach, Dr. Nicholas Kushmerick

Tag der wissenschaftlichen Aussprache: 13. Dezember 2004

Acknowledgments

"... You can't always get what you want
But if you try sometimes well you might find
You get what you need ..."
The Rolling Stones

It takes usually more than just the author to finish a thesis. It is the same in this case. This work is the product of two research projects I was involved in and the eminent atmosphere at the *Artificial Intelligence Research Group* at the University of Koblenz-Landau lead by Ulrich Furbach. The group provided the backbone for fruitfull discussions, motivating ideas and critical expert opinions for developing the presented approach. Besides being very grateful to all of the group members I would like to thank especially the following members and ex-members of the group, since they all have contributed to the formation of this thesis, (in no particular order): Peter Baumgartner, Margret Gross-Hardt, Bernhard Beckert, Jürgen Dix, Frieder Stolzenburg. Nevertheless, there are a few persons that deserve my special thanks, excuses to those I forgot to mention here. First of all I would like to thank Ulrich Furbach, he was a great mentor to me and gave me the greatest possible support for my research. I would like to thank Nicholas Kushmerick for his work as reviewer and the inspirations given by his work on *wrapper induction*. Big thanks to Gerd Beuster working together with him in the *LExIKON* and the *IWIA – Intelligent Web Information Agents* projects was a great pleasure. I am very grateful to Oliver Obst as a colleague and friend he gave me moral and scientific support not only during "Long-Jogs". Christian Wolff did a great job in supporting me in tedious programming tasks and was always prepared for discussions on new ideas. Additional thanks to the proof-readers especially to Jan Murray and Thomas Kleemann.

Finally, the greatest encouragement was given to me by my *Family*: my parents Helli and Achim Thomas, my sister Jutta, relatives Jörg & Nico, my best friend Thorsten Möhring and my partner Cora Arendt. Whenever I thought I was doomed it was my *Family* and Cora who gave me strength to endure.

Koblenz, Mai 2004

Abstract

Automatic fact retrieval from text documents is becoming one of the key technologies for the *Information Age.* One category of *Intelligent Information Systems* aims at supporting the user in search and retrieval of precious information from data resources like *intranets* or the *World Wide Web* containing billions of *web pages* and linked documents. Until now, most of the existing systems are restricted to document retrieval tasks and only a few hand tailored systems exist allowing the user to query and retrieve *facts* from the vast amount of online information available.

In the last decade several approaches have been developed in the *Information Extraction* (*IE*) research area that are able to automatically construct (*learn*) extraction procedures, so called *wrappers.* *Wrappers* allow documents to be interpreted and accessed like relational databases. They form one of the core components in future *Intelligent Information Systems,* since they allow the user to query, compare and combine information from various textual information sources.

This thesis presents an *Logic Programming* and *Inductive Logic Programming* (*ILP*) framework for *supervised learning* of *wrappers* from positive examples only. In contrast to existing systems that adapt some methods from the *Artificial Intelligence* subfield of *Inductive Logic Programming* the here presented *machine learning* approach follows a pure logical *bottom-up learning* approach under a new *IE-ILP semantics.* The presented learning approach for *multi-slot* extraction programs is independent of the chosen wrapper model and document view.

Three classes of *Inductive Logic Programming* algorithms are presented, two *one step learning* algorithms, a set of iterative learning algorithms, and one algorithm combining *clustering techniques* with an iterative *ILP* algorithm.

Several extraction tasks are investigated and a formal definition of wrapper classes is given. Based on these wrapper classes three *wrapper models* are presented using two different document representations, a sequential token and a *DOM* related representation.

The introduced learning algorithms and wrapper models are evaluated on standard test cases and they are compared with related methods and *machine learning based information extraction systems.* For some of the *single-slot* extraction tasks the implemented methods yield better results than the best state-of-the-art systems. Learned wrappers for *multi-slot* extraction tasks show promising competitive quality scores in comparison to the leading extraction systems.

Contents

Part I.

Introduction

1. Preamble

1.1. Motivation

Finding the *right* information in the constantly growing vast amount of electronic documents becomes one of the key technologies for the *Information Age.* Currently the greatest pool of varying information sources is the *World Wide Web.* It offers billions of web pages containing information, links to documents, data files, and other web pages. Although numerous successful techniques and systems have been developed in the area of *Information Retrieval* [Baeza-Yates and Ribiero-Neto, 1999] that support the user in finding web pages in this pool of information the automatic retrieval of facts from the web still remains a big challenge.

Fact retrieval methods can support the user in information acquisition tasks from online documents and can free him from tedious time consuming *surfing*, document selection, reading and manual filtering of relevant data. Hence, these *information extraction* methods build a core component for future *Intelligent Information Systems* [Klusch *et al.*, 2003]. Of course we can manually handcraft such procedures for a quite small number of documents of certain web sites, but the general task apparently requires techniques for an automatic construction to *wrap* new documents fast and easily. Consequently a construction technique is required building extraction procedures (*wrappers*) which are capable of extracting relevant information from unseen documents. For instance, some online documents change rapidly in its content and structure and therefore *wrappers* must be general enough to cover slight changes. Or the construction methods have to be general and flexible enough to handle these changes efficiently such that they can build new wrappers or can adapt existing ones. So, what is needed for the automatic construction of wrappers are general techniques that are not bound to or tailored for certain web sites and documents. Instead they have to have the ability to build automatically wrappers given only a subset of documents of an arbitrary document class (e.g. web pages containing product offers, list of restaurants, etc.).

Note that we are not interested in developing user interactive techniques, that assist the user in assembling wrappers. In this thesis, we present and discuss methods that automatically construct wrappers solely based on input consisting of example extractions, the facts (text fragments) occurring in a given document. Based on these inputs, the system *learns* a wrapper capable to extract all the relevant data from the given documents and similar future ones. Thus the overall goal is to develop techniques to induce knowledge about how the requested information is represented with respect to a certain class of documents. Therefore the induced knowledge builds the basis for the automatic wrapper construction process. If the *learned* or *induced wrapper* is general enough but not overly general it can be applied to similar documents like those used for *learning.* Figure 1.1 illustrates this process of learning information extraction procedures and its application.

With such a *wrapper induction method* [Kushmerick *et al.*, 1997] at hand a manifold of applications can be implemented. For instance, the extracted data can be used to populate databases, to monitor online information sources (e.g. stock rates), to compare content from different online vendors (e.g. price comparisons), or to build large online encyclopediae. But in general these methods are not restricted to online documents (e.g. HTML documents) only. Depending on representational and conceptual issues the learning algorithms are also applicable to natural language text. Obviously, a technology to automatically construct information extraction programs is one of the core components in future intelligent information systems.

1.1.1. Information Extraction and Machine Learning

Beyond dispute the idea of *Information Extraction*(IE) [Cowie and Lehnert, 1996] is rooted within the area of *natural language processing (NLP)*. But *IE* is different from the general research on *text understanding* in *NLP*, because *IE* focuses on predefined concept sets in a specific domain while ignoring other textual given information. In contrast to *text understanding* the general *IE* task always requires a clearly defined information need and it aims on mapping text fragments on slots, tuples or concept fillers.

Especially with the emerging need for automatic text processing tools due to the growing number of electronic documents and online resources an increasing demand for information extraction methods exist. This methods should be capable of handling various document formats and not well formed natural language text. Substantial work in the more conventionally oriented *IE* community mostly based on linguistic *NLP* methods can be found in the proceedings of the DARPA *Message Understanding Conferences (MUC)* [Def, 1992; Def, 1993; Def, 1996]. In Chapter 10 of this thesis various machine learning based information extraction systems are discussed, which are not so closely related to *NLP* based approaches.

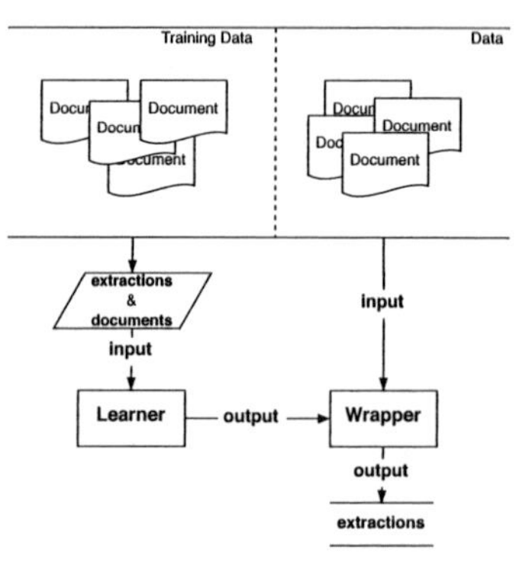

Figure 1.1.: wrapper induction

In the late 90's stronger tendencies in the *IE* community emerged using *Machine Learning* (ML) [Mitchell, 1997] techniques to automatically construct wrappers. Vice versa researchers from the machine learning area became more interested in applying their methods to this application domain. Associated with the chosen machine learning techniques also new wrapper representations and implementation techniques appeared differing from the conventionally linguistic phrase patterns or linguistic rule based approaches. Following the idea of online information agents the focus was also set on non natural language text, messy text like newsgroup postings, and web pages. Consequently, several approaches completely omit the use of linguistic information and solely use document views and representations making use of semi-structuring text elements (e.g. *HTML* tags) to identify relevant text fragments. Hence, current state-of-the-art *Machine Learning* based *IE* systems differ strongly in their used learning techniques, wrapper implementations and document representations. The majority of them can be separated into two classes: one using *finite state automata* based wrapping techniques and a second implementing and learning wrappers as *relational rules*. Accordingly, most of the systems use *Grammatical Inference* techniques [Murphy, 1996; Parekh and Honavar, 1998] respectively strongly influenced ones or they are based on *Inductive Logic Programming* (ILP) [Lavrac and Dzeroski, 1994; Bergadano and Gunetti, 1996; Muggleton, 1991; Muggleton and Raedt, 1994] or closely related ideas for relational rule learning. An overview of machine learning based *IE* systems is given in Section 10.1 of this work.

Combining the two areas of *Machine Learning* and *IE* was for instance researched in projects like the *WebKB* [Craven *et al.*, 2000]. Within this project text classification and information extraction techniques were developed and combined with the aim to automatically construct knowledge bases from the World Wide Web. Besides successful non commercial systems like *Cora* [McCallum *et al.*, 1999], which is a domain-specific search engine over computer science research papers, also more or less successful commercial systems and companies have emerged out of this research community. For instance, the start-up company *WhizBang!*[1] and one of their systems *flipdog.com*, which is an online database of job openings directly extracted from company web sites[2], uses *ML based IE* techniques. The relatively new research area of *Text* and *Web Mining* [Kosala and Blockeel, 2000] can be considered to be a cross-disciplinary field which includes among other techniques like retrieval, analysis and knowledge acquisition the mentioned *Information Extraction* methods. Independently of how *IE* is understood in the *mining* context either as a pre-processing step or as the overall goal it is apparent that it is one of the key technologies opening the door to more intelligent information and knowledge acquisition systems for the World Wide Web.

1.1.2. ILP and Relational Rule Based Wrappers

Existing *Machine Learning* based *IE* approaches can be separated into two classes. The major class uses *finite state automata* related wrappers and therefore *Grammatical Inference* [Parekh and Honavar, 1998; Carraso and Oncina, 1994] related learning techniques. The second minor class uses *relational rules* for information extraction which are automatically constructed by adopted *Inductive Logic Programming* techniques [Bergadano and Gunetti, 1996; Lavrac and Dzeroski, 1994].

Relational rules have the advantage that they are more or less human "readable" and "understandable". Another advantage of relational rule based wrapper methods is the closeness to *logical Knowledge Representation* techniques [Brewka, 1996]. By these means additional knowledge can be quite conveniently incorporated into a logical rule induction process, as for example semantic knowledge derived from ontologies. Additionally the logical approach to wrapper induction provides the possibility to simply combine or replace different wrapper languages and document views by appropriate predicates without having to modify the learning technique itself. Hence, wrappers can be described by logical rules on different representational levels. For instance, some predicates used within a wrapper rule describe pre and postfix character sequences of relevant text fragments, whereas other predicates describe relational properties of the text fragment regarding neighbor nodes in a *HTML* parse tree.

Although there are some quite successful existing systems inspired by *ILP* techniques most of them do not use a pure *Logic Programming* and *ILP framework* for wrapper induction. Most approaches are based on variants of the *FOIL* top-down rule induction algorithm by [Quinlan, 1990], like *SRV* [Freitag, 1998] or WL^2 [Cohen *et al.*, 2002]. This is somewhat astonishing since learning in a bottom-up manner seems to be the more promising method in this domain, because the hypothesis space is extremely large due to the complexity of possible rule instances. A bottom-up method inducing rules by generalization of most specific example description rules reduces the search for good rule literals significantly. [Califf, 1998] investigated and presented

---

[1] *WhizBang!* has been acquired by Inxight Software in late 2002.

[2] The wrapper techniques used by the *WhizBang Labs Wrapper Learner* are discussed in Section 10.1.

in her thesis an *ILP* based bottom-up approach to induce single slot wrappers for natural text. But similar like other *ILP* related approaches no unifying *ILP* framework independently of the wrapper language and document view is presented. Instead tailored generalization (learning) operators with a very strong linguistic background using additional semantic knowledge are introduced.

1.2. Purpose & Claims

Almost no state-of-the-art wrapper induction system uses a pure *Logic Programming* and *ILP* approach. But there are several systems that use successfully closely related *ILP* approaches. Obviously the question arises if a domain independent wrapper induction system purely based on *Logic Programming* and *bottom-up ILP* methods using standard operators can power an *IE* system competitive to existing state-of-the-art approaches. Investigations and results on these topics contribute to a better understanding of standard bottom-up based inductive logic programming methods for automatic wrapper induction and serves as a starting point for refinements in this hybrid research field.

Because of these reasons, the goal of this thesis is to demonstrate that an *ILP* framework for automatic wrapper construction can be developed such that the following claims hold:

Claim No.1: The presented *ILP* learning framework is independent of the document view and wrapper model. Different document views and wrapper models can be conveniently represented by means of logic programming and integrated into the wrapper learning process.

Claim No.2: A pure *LP* and *bottom-up ILP* framework based on the *least general generalization of clauses* represents an adequate learning technique for single and multi-slot wrappers from positive examples only.

The purpose of this thesis is to show that a *Logic Programming* and *Inductive Logic Programming* framework based on a standard bottom-up learning technique, namely the *least general generalization* of *logic program clauses* [Plotkin, 1970], can be successfully used to learn single and multi-slot wrappers from positive examples only.

1.3. Thesis Organization

The thesis is divided into three major parts *Introduction*, *Inductive Logic Programming Based Wrapper Learning* and *Results & Discussion*. The reader is advised to read these parts consecutively since each chapter relies more or less on the previous ones.

In Chapter 2 of Part I the basic notions, some basics of *Machine Learning* and two logic programming based document representations are established. The concept of a *wrapper* is extensively studied in Chapter 3. The chapter starts with a formal classification of several different wrapper types which have been studied or at least mentioned in *ML based IE* literature. The set up wrapper classes are independent of the *ILP* and *LP* paradigm and thus can serve as a formal basis for other *IE* researchers. Succeeding Sections 3.2, 3.3 and 3.4 impose evaluation metrics for wrapper learning, three different wrapper representation models and finally two wrapper languages. Section 3.5 briefly sketches how the proposed wrapper models can be implemented by means of logic programming. The first part of the thesis concludes with a short discussion on observable properties of the introduced wrapper classes.

Part II starts with Chapter 5 giving a short introduction into the field of *Inductive Logic Programming* (Section 5.1) and standard *ILP* learning methods (Section 5.3). In Section 5.2 missing properties of existing *ILP semantics* in the context of wrapper induction are disucssed. This leads to the definition of a new *ILP semantics*, namely the *IE-ILP setting*. The chapter concludes with the Sections 5.4 and 5.5 which determine how the afore defined wrapper models are represented in this *IE-ILP setting*.

In Chapter 6 the first of three bottom-up *ILP* learning algorithms for wrapper induction is presented. The *OSL* algorithm is the most simple one of the introduced algorithms. In Chapter 7 the basic *OSL* concept is extended to yield an iterative learning algorithm, the *basic-BFOIL* algorithm. The third algorithm *cluster-BFOIL* is a hybrid algorithm combining *ILP* methods and *Clustering* techniques to overcome some of *basic-BFOIL*'s weak spots. It is defined in Chapter 8. In each of these chapters properties and observations regarding the efficiency and quality of learned wrappers and the according algorithms are exemplified.

In Part III the quality of the overall approach is experimentally evaluated and compared to existing approaches. Chapter 10 includes an overview and comparison of related machine learning based information extraction systems, experimental results and comparison to other system results on standard test cases. The experimental results are critically discussed. Chapter 11 briefly summarizes and discusses *ILP* related work. Chapter 12 concludes this thesis.

2. Preliminaries

This chapter starts with the introduction of the basic notations that are used throughout the thesis. In Section 2.2, a very brief introduction to the area of machine learning in the context of automatic construction of extraction procedures (wrappers) is given. The chapter concludes with Section 2.3 presenting two document representation formalisms suitable for logic programming and inductive logic programming based learning of wrappers, respectively.

2.1. Notations

2.1.1. Text and Wrappers

One of the basic objects involved in an information extraction task is a *document*. In general a document is any $D \in \Sigma^*$ where Σ is an alphabet. If we do not want to allow the empty document we restrict D to be an element of $\Sigma^+ = \Sigma^* - \{\epsilon\}$ where ϵ denotes the empty word. The length of a document is given by $|D|$. In practice the notion of a *word* determines a text or string occurring in a document. In theory the notion of a *word* is identical to that of the previously introduced notion of a *document*.

For information extraction tasks it is helpful to define the set of words occurring in a document denoted by $W(D)$ as $W(D) = \{w|D = xwy \text{ with } x, w, y \in \Sigma^*\}$. Here xw is the concatenation of x and w. If the empty word is to be excluded from the set of words regarding D we denote this by $W^+(D) = W(D) \setminus \{\epsilon\}$.

For a word w we say that p is a *prefix*, i an *infix* and s a *suffix* if w is the concatenation of p,i and s, written $w = pis$. For determining the greatest common parts of two words w_1 and w_2 the *longest infix*, *prefix* and *suffix* are helpful concepts.

We say $\text{suf}(w_1, w_2)$ returns the *longest common suffix* s of w_1 and w_2 if $w_1 = us$, $w_2 = vs$, and u and v have no non-empty common suffix. The *longest common prefix* denoted by $\text{pre}(w_1, w_2)$ is given by p if $w_1 = pu$, $w_2 = pv$, and u and v have no non-empty common prefix. Consequently, the *longest common infix* is given by $\text{inf}(w_1, w_2) = i$ if $w_1 = ris$, $w_2 = tiu$, and s and u have no non-empty common prefix and r and s have no non-empty common suffix.

In the sequel the notion of *text* or *text fragment* denotes a word in D with a specific start position s and end position e. By $t^D_{s,e} = w$ we refer to the word $w \in W(D)$ starting at s and ending at e if $D = xwy$, $|x| = s$ and $|xw| = e$.

One common way in the context of information extraction is to interpret a document as a *sequence* of words. Normally, special characters contained in Σ are used to determine word boundaries. In practice these are white space characters like *blank*, *punctuation marks*, or *newline*. Formally, there is a (sequence) function $Seq : \Sigma^* \to Z$ with $Z = \bigcup_{0 \leq n \leq \infty} (\Sigma^*)^n$ returning for a given document D a sequence of words of $W(D)$. By $Seq(D) =< w_1, w_2, \ldots, w_n >$ we denote a sequence of words of D such that $D = w_1 w_2 \ldots w_n$. The empty sequence is denoted by $<>$. If S is a sequence with $S.i$ we refer to the i-th element w_i of S. A *subsequence* of a sequence S starting with the i-th element and ending with the j-th element of S is denoted by $S.i.j$. The concatenation of two sequences $S =< w_1, \ldots, w_n >$ and $S' =< u_1, \ldots, u_m >$ is written as $S \circ S' =< w_1, \ldots, w_n, u_1, \ldots, u_m >$. If the context is clear we leave out the concatenation operator and write SS' for $S \circ S'$. By $length(S)$ we denote the length of a sequence given by $\sum_{i=1}^{n} |S.i|$ and by $|S|$ we denote the number of words in S.

Now that the basic notations for describing documents, words and sequences are introduced the concept of a *wrapper* is defined.

Definition 2.1.1 (Wrapper Universe) *Given an alphabet* Σ *the* wrapper universe $\mathcal{U}$ *is defined by* $\mathcal{U} = \bigcup_{n>0} U_n$ *with* $U_n = \{(D, x) | D \in \Sigma^* \text{ and } x \in (W(D))^n \text{ with } n \in \mathbb{N}\}$. *We call* x an extraction tuple *from* D. □

Usually a wrapper is introduced as a mapping from a document D onto a set of words from D or a set of vectors of words from D. This represents the intuitive meaning of wrappers that a wrapper is some sort of extraction procedure. In contrast to the usual definition a declarative (relational) definition is used throughout this thesis. In the following a wrapper is defined as a set of tuples consisting of documents D and vectors of words from D. One reason for this relational representation is the easier integration into a logic programming framework (Chapter 5) and a easier set oriented view on the different wrapper concepts like *target wrapper, learned wrapper, wrapper model, partial wrapper model* and *wrapper classes* (Chapter 3).

Definition 2.1.2 (Wrapper) *Given an* $n \in \mathbb{N}$, *an alphabet* Σ *and wrapper universe* $\mathcal{U}$, *we call* $\mathcal{W}$ *an* n-slot wrapper *with* n *the number of slots iff* $\mathcal{W} \subset U_n$ *with* $U_n \in \mathcal{U}$.*In the sequel* $\mathcal{W}$ *denotes a wrapper. For* $e \in \mathcal{W}$ *with* $e = (D, x)$ *we denote by* $e.D$ *the document* D *and by* $e.i$ *the* i*-th component of* x. *The components of* x *are called* slots *and the value of* $x.i$ slot filler. *If for every* $e = (D, x) \in \mathcal{W}$ *exists a total function* $o : \{1, \ldots, n\} \rightarrow \{1, \ldots, n\}$ *such that* $D = u\, e.o(1)\, v\, e.o(2) \ldots w\, e.o(n)\, z$ *we call the function* o *the* slot filler occurrence order *of* e. *If no such mapping exists for every* $e \in \mathcal{W}$ *then* o *is undefined.* □

Note that for an equivalent functional interpretation of a wrapper $\mathcal{W}$ as defined in Definition 2.1.2 we assume a wrapper function W_f such that for all extractions $x_1, x_2, \ldots, x_n$ from D with (D, x_1), (D, x_2), $\ldots$, $(D, x_m) \in \mathcal{W}$ it holds that $W_f(D) = \{x_1, x_2, \ldots, x_m\}$.

Example 2.1.1 (Wrapper) *Assume a document* D *is given containing names as shown in Figure 2.1. The task is to extract tuples of first and last names.*

As can be seen, the order in which first and last names occur within the document vary, but the order of the first and lastname slots has to be fixed. This is important, because a fixed semantics for the slots of $\mathcal{W}$ *is required to provide a basis for automatic processing of extraction results (e.g. for using extraction tuples to populate databases). According to Definition 2.1.2 a wrapper* $\mathcal{W}$ *for this extraction task is given by* $\mathcal{W} = \{(D, < \text{John}, \text{McMurphy} >), (D, < \text{Kiefer}, \text{ Sutherland} >)\}$. *Hence, the* slot filler occurrence order *for the first tuple* $e_1 : (D, < \text{John}, \text{McMurphy} >)$ *is given by* $o = \{(1,1)(2,2)\}$ *and for the second extraction tuple* $e_2 : (D, < \text{Kiefer}, \text{ Sutherland} >)\}$ *by* $o = \{(1,2),(2,1)\}$. *So, By* $e_2.o(1)$ *we refer to the slot filler Sutherland that occurs before all other slot fillers of* e_2 *in the document* D. *Let us assume another wrapper* $\mathcal{W}$ *for the document shown in Figure 2.2.* $\mathcal{W}$ *constains extractions* $e_1 : (D, < \text{56070 Koblenz}, \text{Koblenz/Germany} >)$ *and* $e_2 : (D, < \text{58097 Hagen}, \text{Hagen/Germany} >)$. *Obviously for this wrapper we cannot define a slot filler occurrence order neither for* e_1 *nor for* e_2 *(according to Definition 2.1.2), because*

```
...John...McMurphy...
...Sutherland,Kiefer...
```

Figure 2.1.: document

the slots of both extractions overlap. In fact we could extend Definition 2.1.2 such that the order function is also defined on overlapping and included slots, but since in the remainder of this thesis we do not need the occurrence order for this class of wrappers we omit a more sophisticated definition. ⌟

Note that if o is defined it is given for every extraction tuple and that a unique mapping o cannot be calculated solely from w_i of Σ^* if there are multiple occurrences of w_i in D. Example 2.1.1 illustrates that each slot of a wrapper is associated with a certain concept or semantics. Informally, we can say that each slot is related to a concept C and that each instance of C can be described by some word from Σ^*. In this sense, an extraction tuple x is an *intended extraction* from a document regarding a wrapper $\mathcal{W}$ if each slot filler of $x.i$ is an instance of the concept associated with the i-th slot.

```
... 56070 Koblenz/Germany...
... 58097 Hagen/Germany...
```

Figure 2.2.: document

For the later discussion on wrapper classes in Section 3.1 it is helpful to assume that potential slot fillers occur only once within a document. For instance, assume that in the document from Example 2.1.1 an additional last name *Sutherland* occurs before the first name *John.* In this case the wrapper from Example 2.1.1 does not uniquely determines the relevant extraction text. Hence it is not clear which of the two words *Sutherland* is meant, because the *slot filler occurrence order* only defines a order among the slot filler occurrences and does not determine the exact position within the document. In general, there is no need to take this *unique occurence assumption*, since the wrapper definition and function o can be extended to use absolute position information. On the other hand, this would result in more complex wrapper class representations given in Section 3.1.

Definition 2.1.3 (Unique Occurrence Assumption) *Given an n-slot wrapper $\mathcal{W}$ for all $e \in \mathcal{W}$ with $e.D = m_0\, e.o(1)\, m_1 \ldots m_{n-1}\, e.o(n)\, m_n$, there exist only unique $m_0, m_1, \ldots, m_n \in W(D)$.*

A very common pre-processing step in information extraction is the *tokenization* of documents. Roughly speaking, a token represents a word occurring in a document as a list of attribute value pairs. The most simple token representation consists solely of one attribute classifying the word according a certain class. For instance, a word $w = 1969$ can be represented by the token $(type, num)$. Tokens with one attribute are often denoted by the value of the only attribute, e.g. num. In the remainder of this thesis tokens possess more than one attribute and are represented as a tuple of attribute value pairs in the form of $((a_1, v_1), \ldots, (a_n, v_n))$ where $a_i \neq a_j$ for $i \neq j$ and a_i are attributes and v_i are arbitrary values. By $tok(w)$ we denote the tokenized representation of a word w. The value of an attribute can be any word from a given alphabet. So, formally there is a *tokenization function* $\mathcal{T} : \Sigma^+ \rightarrow (A \times V)^n$ with A an arbitrary set of words representing *attribute names* and V an arbitrary set of words representing attribute *values*. A tuple of attribute value pairs $((a_1, v_1), \ldots, (a_n, v_n))$ is called *token* and by $t.a$ we denote the value of attribute a of token t. Throughout this thesis we will also use a *Prolog* term notation of tokens given by $token([a_1 = v_1, \ldots, a_n = v_n])$.

A *tokenized document* is obtained in two steps. First, a document D is transformed into a sequence by $Seq(D) = S$. Second, each word in the sequence S is replaced by its token. The tokenized version of a document D is denoted by $Tok(S) =< t_1, \ldots, t_n >$ iff $S =< w_1, \ldots, w_n >$ and $tok(w_i) = t_i$ with $i = 1, \ldots, n$. If it is clear from the context what sequence function and what tokenization function is used $Tok(D)$ is used to denote the tokenized document D.

Actually, this notion of a *token* is more or less related to the notion of *feature structures* [Carpenter, 1991; Carpenter, 1992; Carpenter, 1993; Smolka and Treinen, 1994; Smolka, 1988; Aït-Kaci *et al.*, 1997]. In general *feature structures* are able to represent more complex structures, because they allow hierarchical attribute value structures. This means, attribute values are allowed to consist of feature structures and they can contain references to other attribute values.

2.1.2. Logic Programming

In the following some of the basic logic programming notions used in this thesis are introduced. The definitions are taken from [Lloyd, 1987] and build the basis for existing *Prolog* systems [Kowalski, 1974] and thus also for almost all inductive logic programming methods. Model theoretic definitions (i.e. Tarski semantics, Herbrand Interpretation and Model) and discussions are deliberately not given, because they are only needed in Section 5.1, which is intended for the reader who is familiar with the field of logic programming. A comprehensive explanation on this topic is also found in [Lloyd, 1987]. In this work the commonly used syntax for logic programs is used. The reader familiar with predicate logic and logic programming may skip this part and is referred to Section 2.1.3.

The *first-order alphabet* from which logic programs are built consists of seven classes of symbols: *variables* (starting with an upper case letter), *constants* (starting with an lower case letter), *function symbols* (starting with a lower case letter), *predicate symbols* (starting with a lower case letter), *connectives* ($\wedge, \vee, \neg, \leftarrow, \leftrightarrow$), *quantifiers* ($\forall, \exists$), and *punctuation symbols* (,.). We assume the usual meaning for connectives and quantifiers as used in [Lloyd, 1987].

A *term* is defined inductively as follows: A variable is a *term*. A constant is a *term*. If f is a n-ary function symbol and $t_1, \ldots, t_n$ are *terms*, then $f(t_1, \ldots, t_n)$ is a *term*. A *ground term* is a term not containing variables.

The set of all *well formed formulae* is defined inductively as follows:

1. If p is an n-ary predicate symbol and $t_1, \ldots, t_n$ are terms, then $p(t_1, \ldots, t_n)$ is a formula called *atom*.

2. If F and G are formulae, then $(\neg F)$, $(F \vee G)$, $(F \vee G)$, $(F \rightarrow G)$ and $(F \leftrightarrow G)$.

3. If F is a formula and x a variable, then $(\forall x F)$ and $(\exists x F)$ are formulae.

A *first-order language* is the set of all well formed formulae. A *literal* is an atom or the negation of an atom. A *positive literal* is an atom. A *negative literal* is the *negation* of an atom. The implication $F \rightarrow G$ is also written as $G \leftarrow F$ and G is called the consequent.

The scope of an quantifier $\forall x$ (resp. $\exists x$) in $\forall x F$ (resp. $\exists x F$) is F. A variable is *bound* if it is immediately following a quantifier or is occurring within a quantifier's scope. Otherwise a variable is *free*. A *closed formula* is a formula with no free variables. The *universal closure* of F, which is the closed formula obtained by adding universal quantifiers for every free variable in F, is denoted by $\forall x(F)$.

Definition 2.1.4 (Clause) *A clause is a formula of the form:* $\forall x_1 \ldots \forall x_s (L_1 \vee \ldots \vee L_m)$ *where each* L_i *is a literal and* $x_1, \ldots, x_s$ *are all variables occurring in* $L_1 \vee \ldots \vee L_m$. □

Clauses of the form $\forall x_1 \ldots \forall x_s (H_1 \vee \ldots \vee H_m \vee \neg B_1 \vee \ldots \vee \neg B_i)$ are written as a rule $H_1 \vee \ldots \vee H_m \leftarrow B_1, \ldots, B_i$ or in *clause normal form* as a set $\{H_1, \ldots, H_m, \neg B_1, \ldots, \neg B_i\}$.

Definition 2.1.5 (Definite Program Clause, Unit Clause, Definite Program)
A definite program clause *is a clause of the form* $H \leftarrow B_1, \ldots, B_n$ *which contains precisely one literal in its consequent.* H *is called the* head *and* $B_1, \ldots, B_n$ *the* body *of the program clause. A* unit clause *is a clause of the form:* $H \leftarrow$ *that is the definite program clause with an empty body. A* definite program *is a finite set of definite program clauses.* □

Definition 2.1.6 (Predicate Definition, Goal, Horn Clause) *In a definite program the set of all program clauses with the same predicate symbol p in the head is called the* definition *of p. A* goal *(query) is a clause of the form* $\leftarrow B_1, \ldots, B_n$ *that is, a clause which has an empty consequent. Each* B_i *(i=1,n) is called a* subgoal *of the goal. A* Horn clause *is either a definite program clause or a definite goal.* □

Definition 2.1.7 (Program Clause, Normal Program, Normal Goal)
A program clause *is a clause of the form* $H \leftarrow B_1, \ldots, B_n$ *where* H *is an atom and* B_i *is a positive or negative literal for* $i = 1, \ldots, n$*. A* normal program *is a set of program clauses. A* normal goal *is a clause of the form* $\leftarrow B_1, \ldots, B_n$ *where* $B_i(i = 1, n)$ *are literals.* □

Example 2.1.2 (Definite Logic Program)

Figure 2.3 shows a definite logic program implementing a toy-like knowledge base. Stating that a bear and a chicken both are animals; that a chicken has wings; a bear has claws and that something that is an animal and has wings is a bird. We expect that a query like $\leftarrow bird(X)$ *asking "What is a bird?" given to this logic program results in an answer* $bird(chicken)$. ┘

$$
\begin{aligned}
&animal(bear) \leftarrow \\
&animal(chicken) \leftarrow \\
&has(chicken, wings) \leftarrow \\
&has(bear, claws) \leftarrow \\
&bird(X) \leftarrow animal(X), \\
&\qquad\qquad has(X, wings)
\end{aligned}
$$

Figure 2.3.: definite program

In the remainder of the thesis program clauses are also written in standard *Prolog* syntax. The syntax differs in that the symbol `:-` is used instead of the $\leftarrow$ connective; for unit clauses the $\leftarrow$ connective is omitted, and every program clause is terminated with a period.

After having briefly introduced the syntax of logic programs it has to be clarified how to compute answers for queries like the one given in Example 2.1.2. In the following a standard proof procedure the *SLD-resolution* and the related concepts of *unification* and *answer computation* are very briefly introduced.

Definition 2.1.8 (Substitution)
A substitution θ *is a finite set of the form* $\{v_1/t_1, \ldots, v_k/t_k\}$, *where each* v_i *is a variable, each* t_i *is a term distinct from* v_i *and the variables* $v_1, \ldots, v_n$ *are distinct. Each element* v_i/t_i *is called* binding *for* v_i. θ *is called a* ground substitution *if the* t_i *are all ground terms.* θ *is called a* variable-pure *substitution if the* t_i *are all variables. If* $\theta = \{u_1/s_1, \ldots, u_m/s_m\}$ *and* $\sigma = \{v_1/t_1, \ldots, v_m/t_n\}$ *are substitutions then the composition* $\theta\sigma$ *of* θ *and* σ *is the substitution obtained from* $\{u_1/s_1\sigma, \ldots, u_m/s_m\sigma, v_1/t_1, \ldots, v_m/t_n\}$ *by deleting any bindings* $u_i/s_i\sigma$ *for which* $u_i = s_i\sigma$ *and any bindings* v_j/t_j *for which* $v_j \in \{u_1, \ldots, u_m\}$. □

Definition 2.1.9 (Expression, Instance,Variants) *An* expression *is either a term, a literal or a conjunction or disjunction of literals. Let* $\theta = \{v_1/t_n, \ldots, v_n/t_n\}$ *be a substitution and* E *be an expression. Then* $E\theta$, *the instance of* E *by* θ *is the expression obtained from* E *by simultaneously replacing each occurrence of the variables* v_i *in* E *by the term* t_i *for* $i = 1, \ldots, n$. *If* $E\theta$ *is ground, then* $E\theta$ *is called* ground instance *of* E. *If* $S = \{E_1, \ldots, E_n\}$ *is a finite set of expressions then* $S\theta$ *denotes the set* $\{E_1\theta, \ldots, E_n\theta\}$. *Expressions* F *and* E *are* variants *if there exist substitutions* θ *and* σ *such that* $E = F\theta$ *and* $F = E\sigma$. □

Definition 2.1.10 (Unification, Most General Unifier) *Given* $S = \{l_1, \ldots, l_n\}$ *a finite set of terms and atoms. A substitution* θ *is called a* unifier *for* S *if* $S\theta$ *is a singleton. A unifier* θ *is called* most general unifier *(mgu) for* S *if, for each unifier* σ *of* S, *there exists a substitution* γ *such that* $\sigma = \theta\gamma$. □

Definition 2.1.11 (Answer, Correct Answer) *Let* P *be a definite program and* G *be a definite goal* $\leftarrow B_1, \ldots, B_n$. *An* answer *for* $P \cup \{G\}$ *is a substitution* θ *for variables in* G. *We say* θ *is a* correct answer *for* $P \cup \{G\}$ *if* $\forall((B_1 \wedge \ldots \wedge B_n)\theta)$ *is a logical consequence of* P. □

We say that a formula F is a *logical consequence* of a formula S written as $S \models F$, if every interpretation I that is a model for S is also a model for F. If a formula F is not satisfiable (e.g. there is no model for F) we write $F \models \square$ and call F inconsistent. For a formal definition of the notions *interpretation, model, Herbrand model* and *Herbrand interpretation* see [Lloyd,

1987]. Unless we do not explicitly state any other proof procedure it is assumed that the most common proof procedure, namely the *SLD-Resolution* resp. *SLDNF-Resolution* on which *Prolog* systems are based, is used.

Definition 2.1.12 (SLD-Derivation) *Let G be $\leftarrow A_1, \ldots, A_m, \ldots, A_k$ and C be $A \leftarrow B_1, \ldots, B_q$. Then G' is* derived *from G and C using mgu θ if the following hold:*

- *A_m is an atom, called the* selected *atom, in G.*
- *θ is an mgu of $\{A_m, A\}$.*
- *G' is the goal $\leftarrow (A_1, \ldots, A_{m-1}, B_1, \ldots, B_q, A_{m+1}, \ldots, A_k)\theta$.*

Let P be a definite program and G a definite goal. An SLD-Derivation *of $P \cup \{G\}$ consists of a (finite or infinite) sequence $G_0 = G, G_1, \ldots$ of goals, a sequence $C_1, C_2, \ldots$ of variants of program clauses of P and a sequence $\theta_1, \theta_2, \ldots$ of mgu's such that each G_{i+1} is derived from G_i and C_{i+1} using θ_{i+1}.* □

Definition 2.1.13 (SLD-Refutation) *A* SLD-Refutation *of $P \cup \{G\}$ is a finite SLD-Derivation $G_0, G_1, \ldots, G_n$ of $P \cup \{G\}$ which has the empty clause □ as the last goal in the derivation. If $G_n = \square$, we say the refutation has length n.* □

For normal logics programs where negation in the body of a program clause is allowed SLD-Resolution has to be extended. Therefore the *closed world assumption* (CWA) [Reiter, 1978] is taken, which states that something that is not a logical consequence of a program is considered to be false. Now, assume a rule with a body literal $\neg A$. The proof obligation is to show that A does not hold. If we can show in finitely many *SLD-Resolution* steps that there does not exist a SLD-tree containing a success branch for the query $\leftarrow A$, then we can conclude under the *CWA* that $\neg A$ holds. For a more detailed discussion on *SLDNF-Resolution* see [Lloyd, 1987].

Definition 2.1.14 (Computed Answer) *Let P be a definite program and G a definite goal. A* computed answer *θ for $P \cup \{G\}$ is the substitution obtained by restricting the composition $\theta_1 \ldots \theta_n$ to the variables of G, where $\theta_1, \ldots, \theta_n$ is the sequence of mgu's used in an SLD-Refutation of $P \cup \{G\}$.* □

By $P \vdash_{SLD} G$ we denote that there is a *SLD-Refutation* of $P \cup \{G\}$. We say G is derivable or follows from P under SLD-Refutation. By $P \vdash_{SLD} G\theta$ we denote that θ is a computed answer for $P \cup \{G\}$. If it is clear from the context which proof procedure is used we omit the subscript $\vdash_{SLD}$ and write $\vdash$.

Example 2.1.3 (SLD-Refutation) *Assume we want to compute an answer for the query $\leftarrow bird(X)$ from the definite program P given in Figure 2.3 using* SLD-Resolution. *So, what we have to do is to show that there is an* SLD-Refutation *for $P \cup \{bird(X)\}$ by finding a sequence of* SLD-Derivations *such that we derive the empty clause □. The computed answer θ for X is then given by the mgu's used in the derivation steps. Figure 2.4 shows a* SLD-Refutation *for the query $\leftarrow bird(X)$ and the computed answer.* ┘

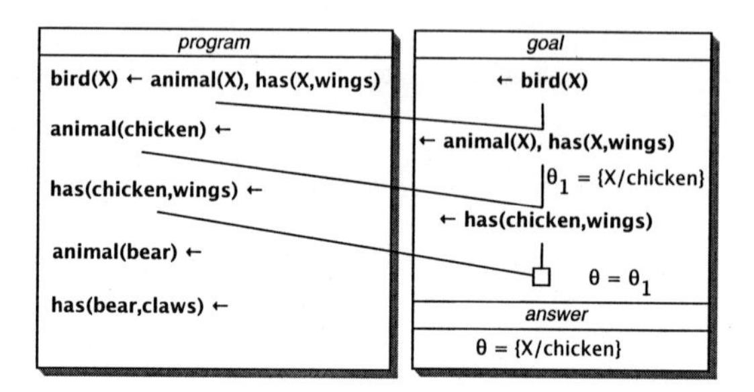

Figure 2.4.: SLD-Refutation

2.1.3. Least General Generalization

Roughly speaking, *Unification* tries to equalize terms by finding suitable substitutions. A similar operation like *Unification* that tries to equalize terms is the *least general generalization (lgg)* respectively *θ-Subsumption.* In contrast to *Unification* it works in an opposite manner by replacing terms with variables. This concept can be successfully used for generalization and therefore as learning operator. Plotkin [Plotkin, 1970] has been one of the first, who studied the usage of *lgg* operations for machine learning.

Definition 2.1.15 (θ-Subsumption) *Let c and c' be program clauses. Clause c θ-subsumes c' ($c \preceq c'$) if there exists a substitution θ, such that $c\theta \subseteq c'$ [Plotkin, 1970]. Two clauses c and d are θ-subsumption equivalent if $c \preceq d$ and $d \preceq c$. A clause is* reduced *if it is not θ-subsumption equivalent to any proper subset of itself.* □

The following definitions 2.1.16 and 2.1.17 are taken from [Lavrac and Dzeroski, 1994].

Definition 2.1.16 (LGG of Terms) *Given two terms t_1 and t_2 the least general generalization of t_1 and t_2 written $lgg(t_1, t_2)$ is defined as follows:*

- $lgg(t, t) = t$,
- $lgg(f(s1, \ldots, s_n), f(t_1, \ldots, t_n)) = f(lgg(s_1, t_1), \ldots, lgg(s_n, t_n))$,
- $lgg(f(s1, \ldots, s_m), g(t_1, \ldots, t_n)) = V$ *where* $f \neq g$, *and* V *is a variable which represents* $lgg(f(s_1, \ldots, s_m), g(t_1, \ldots, t_n))$.
- $lgg(s, t) = V$ *where* $s \neq t$ *and at least one of* s *and* t *is a variable; in this case,* V *is a variable which represents* $lgg(s, t)$.

□

Note, that for all possible occurrences of the *lgg* of subterms the same variable has to be used. For instance, $lgg(q(a, a), q(b, b)) = q(X, X)$ and not $q(X, Y)$.

Definition 2.1.17 (LGG of Atoms and Literals) *Given two literals l_1 and l_2 the least general generalization of l_1 and l_2 written $lgg(l_1, l_2)$ is defined as follows:*

- *if* l_1 *and* l_2 *are atoms, then* $lgg(l_1, l_2)$ *is defined as follows:*
  - $lgg(p(s_1, \ldots, s_n), p(t_1, \ldots, t_n)) = p(lgg(s_1, t_1), \ldots, lgg(s_n, t_n))$
  - $lgg(p(s_1, \ldots, s_m), q(t_1, \ldots, t_n))$ *is undefined if* $p \neq q$.
- *if both* l_1 *and* l_2 *are negative literals,* $l_1 = \neg a_1$ *and* $l_2 = \neg a_2$*, then* $lgg(l_1, l_2) = \neg lgg(a_1, a_2)$
- *if* l_1 *is a positive and* l_2 *is a negative literal, or vice versa,* $lgg(l_1, l_2)$ *is undefined*

□

Definition 2.1.18 (LGG of Clauses) *The least general generalization (lgg) of two reduced clauses* c *and* c'*, denoted by* $clause\_lgg(c, c')$*, is the least upper bound of* c *and* c' *in the* θ*-subsumption lattice. [Lavrac and Dzeroski, 1994]* □

Definition 2.1.19 (Compute LGG of Clauses) *Let* c *and* d *be clauses. To compute the clause lgg* C' *of* c *and* d *compute the set* $C = \{l | lgg(l_1, l_2) = l \text{ with } l_1 \in c \wedge l_2 \in d\}$ *and let* C' *be the set of clauses obtained after removing redundant literals from* C *such that* $\forall c' \in C' : c' \preceq c \wedge c' \preceq d$*. A literal* l *in a clause* C *is called redundant if there exists* $C - \{l'\}$ *is* θ *subsumption equivalent to* C*. By* $clause\_lgg(c, d) = C'$ *the clause lgg of clause* c *and* d *is denoted.* □

Note, as for the *lgg of terms* for the computation of a *clause lgg* also the the same variables have to be used for the occurrences of *lgg of terms* and *subterms*.

Theorem 2.1.1 ($\preceq$ logically entails $\models$) *Given* c *and* c' *program clauses. If* $c \preceq c'$ *then* c *logically entails* c'*,* $c \models c'$*.*

A proof of the subsumption theorem can be found in [Kowalski, 1970].

2.2. Machine Learning

When talking about Machine Learning and especially about learning extraction procedures (wrappers) for information extraction (IE) it is necessary to clarify the *learning task* from a theoretical machine learning perspective. In this thesis *supervised* learning of wrappers will be investigated. Roughly speaking, supervised in contrast to unsupervised learning implies that a learner is trained on pre-classified examples. Whereas unsupervised learning methods learn from unclassified examples and thus try to learn a hypothesis for partitioning data into classes. In supervised learning the task is either to learn a hypothesis which is capable of describing, characterizing or classifying future data according to their correct class or relationship membership.

Supervised learning of wrappers can be referred to as *concept learning*. Following the idea of [Bratko, 1989] learning a concept C means to learn to recognize objects in C.

Abstractly, a concept C is a subset of objects O. For instance in the IE context O is the *wrapper universe* (Definition 2.1.1). Consequently with $C \subset O$ a class of extractions and documents is defined. This is identical to the definition of a wrapper and the discussed wrapper semantics in Section 2.1.1.

Learning a wrapper according to this interpretation means to be able to determine if an arbitrary element from O belongs to the concept C. For example, assume a wrapper $C \subset O$

is given that contains product description extractions from amazon.com web pages. Given an arbitrary amazon web page D we simply check if tuples $(D', x) \in C$ exist with $D = D'$. If this is the case x represents an intended address extraction from D. A wrapper is obviously not correctly covering the concept of an amazon.com wrapper, if not all product offers for any amazon.com web page are contained in C. Nevertheless C is still a wrapper according to Definition 2.1.2. On the other hand, if C provides for every amazon web page all possible product descriptions and no false offers, we call C a *target wrapper*, because it perfectly models the concept of the intended amazon product wrapper. In the remainder of this section we assume, if not stated otherwise, wrappers to be *target wrappers.*

Formally, a concept can be defined as a boolean function, the *target concept* $c : O \rightarrow \{0, 1\}$. In general, instances $x \in O$ for which $c(x) = 1$ are called *positive examples* and those for which $c(x) = 0$ holds are *negative examples. Training examples* are represented as ordered pairs $< x, c(x) >$ where x is an instance and $c(x)$ its corresponding concept value. This is the standard definition used in many machine learning text books ([Mitchell, 1997]).

In general a learner has to solve the problem of finding a *hypothesis* that characterizes the target concept as good as possible. This means, the learner has to search through a space of hypotheses and has to choose the one that fits all the presented training examples best. We denote the set of all possible hypotheses by $\mathcal{H}$. The overall goal is to find an $h \in \mathcal{H}$ such that $\forall x \in O : h(x) = c(x)$.

Since the only information presented to the learner concerning the learning of C are some training examples, the best result that can be assured is that the concept is learned correctly concerning the training examples. This leads to the fundamental *assumption of inductive learning*: any found hypothesis to approximate the target concept well over a sufficiently large training set will also approximate the target concept well over unseen examples (taken from [Mitchell, 1997]).

It is important to note that the best hypothesis on the training data does not necessarily fits best on unseen examples. Although the learned hypothesis perfectly fits on the training data it can produce poor results on unseen data. Reasons for this can be that the training set is to small to learn a sufficient general hypothesis, or that most of the training examples have identical characteristics which are not necessarily representative for the target concept, or that the learning algorithm biases the hypothesis construction on training set specific features. This process is also known as *overfitting.* Consequently sometimes a hypothesis having a larger error on the training data can have a smaller error on the unseen data than the best hypothesis on the training data. Evaluation methods like *cross-validation* can help to select the best non overfitting hypothesis.

Usually objects and concepts do not have to be described by the same language, instead objects like documents and text tuples are described by an *object language* and concepts like the wrappers by a *concept language.* Furthermore, a concept can be described extensionally or intensionally. Where extensionally means to list all instance descriptions of the concept and intensionally to state the instances by descriptions in a certain concept description language in a compact and clear manner. In general such intensional descriptions are often given in form of rules providing information about significant features typical for all instances of the concept.

Choosing a concept description language directly defines the hypothesis space that is potentially searched by the learner. Thus the question arises if the chosen concept description language really suffice in the sense that it is expressive enough to describe the target concept? On the other hand, if the concept language is too rich and the hypothesis space becomes too large then learning might become too hard. Almost all inductive learning methods are confronted with this problem and therefore make a priori assumptions regarding the target concept. This is known as the *inductive bias.* In general, a learning system's search for a good hypothesis can be constrained according to the search in the hypothesis space (*search bias*) or the hypothesis space itself can be constrained (*language bias*). For instance, one inductive bias for the learning of wrappers might be to define the hypothesis language consisting of propositional logic rules. Another one might use restricted first-order predicate logic rules with two body literals. Consequently, questions arise about the expressiveness, learnability of concepts with respect to the size of the hypothesis space or how many training examples are needed depending on the hypothesis space. Work on the impact of inductive bias on learning is discussed in detail in [Baxter, 2000; Haussler, 1988; Utgoff, 1986].

Definition 2.2.1 (Positive Example) *Given an n-slot target wrapper $\mathcal{W}$ we call e^+ a positive example iff $e^+ \in \mathcal{W}$. A set of positive examples E^+ is defined as $E^+ \subseteq \mathcal{W}$.* □

Definition 2.2.2 (Negative Example) *Given an n-slot target wrapper $\mathcal{W}$ we call $e^- = (D, e)$ a negative example iff $e \in \{(D, w)|w \in W(D)^n\} \setminus \mathcal{W}$. By E^- we denote a set of negative examples.* □

Note, that here in the context of learning wrappers the definition of negative examples is determined regarding a specific document D. A negative example must consist of words occuring in D, instead of defining E^- to be a subset of $\mathcal{U} \setminus \mathcal{W}$.

Definition 2.2.3 (Exhaustive Set of Examples) *Let E^+ be a set of positive examples regarding a target wrapper $\mathcal{W}$. Further let $\mathcal{D}(E^+)$ be the set of documents occurring in E^+ defined as $\mathcal{D}(E^+) = \{D|(D, x) \in E^+\}$ and $\mathcal{X}_{E^+}(D)$ be the set of all extractions x from a document D contained in E^+ defined as $\mathcal{X}_{E^+}(D) = \{x|(D, x) \in E^+\}$. Further let $\mathcal{X}_{\mathcal{W}}(D)$ be the set of all extractions from a document D contained in $\mathcal{W}$. We say a positive example set E^+ is exhaustively enumerated regarding a target wrapper $\mathcal{W}$ iff $\forall D \in \mathcal{D}(E^+) : \mathcal{X}_{E^+}(D) = \mathcal{X}_{\mathcal{W}}(D)$.*
□

This definition simply states that if one extraction from a document D is in E^+ then all other intended extractions from document D regarding a target wrapper W have to be in E^+ otherwise it is no exhaustive set of examples.

Similar to tokenizing a document, the tokenized version of an example $e = (D, x) \in E$ is obtained by replacing every slot filler w_i given by $e.i = w_i$ with $tok(w_i)$. The tokenization of an example set E is denoted by $Tok(E)$ where every $e \in E$ is tokenized.

One basic property one expects from a learned hypothesis is *consistency.* In this thesis the notion consistency means that at least all presented training examples are classified correctly by the learned hypothesis according to the target concept.

Definition 2.2.4 (Consistent Hypothesis) *Given a set of training examples E, a hypothesis h and target concept c. The hypothesis h is consistent iff $h(x) = c(x)$ for all $(x, c(x)) \in E$.*
□

From a theoretical point of view a target wrapper is a perfect wrapper providing all possible and only correct extractions. Hence positive and negative training examples as defined in Definition 2.2.1 and 2.2.2 are also true positive and true negative examples, because they depend on the concept of a target wrapper. Nevertheless, in practice it might happen, due to some circumstances, that some positive examples are incorrectly presented as negative examples or vice versa. In this case the general question arises if the used learning technique is robust enough to treat the noisy training data in an appropriate way such that the learned hypothesis construction is not biased by these false examples. In this thesis the focus is set on learning from noise free data, which is also implicitly assumed in Definition 2.2.4.

Definition 2.2.5 (Complete Hypothesis) *Given a hypothesis h, target concept c and X the set of all objects for which $c(x) = 1$. The hypothesis h is complete iff $h(x) = c(x)$ for all $x \in X$.* □

Until now, the wrapper semantics is defined in a declarative manner that allows to integrate it nicely into the concept learning theory. But obviously, a wrapper follows a different functionality than the abstract definition of its target function $c : \mathcal{U} \rightarrow \{0, 1\}$. In a broader sense the target function classifies a presented pair of document and text tuples. Nevertheless, it is obvious that if a target concept is learned it can be used for extraction, albeit in a very brute force and infeasible manner. Someone just has to present all possible text tuples of a given document to the learned target concept. Those text tuples classified with 1 are good extractions. Evidently, this is not a suitable approach for practical applications, but it serves as a starting basis for combining information extraction with the inductive logic programming paradigm for automatic wrapper construction in Part II.

2.3. Document Representation

Several different document representations depending on various wrapper methods have been proposed in the *IE* literature. Many of these approaches enrich the original document in a pre-processing step by adding additional semantic information at the level of word meaning. In most cases this semantic information consists of linguistic knowledge about morphological or grammatical properties of words and phrases [Ciravegna, 2000; Freitag and McCallum, 2000]. But also additional semantic information derived from ontologies regarding the meaning of words appearing in a document [Califf, 1998; Soderland, 1997] or structural properties regarding the syntactic structure have been used within some approaches [Muslea *et al.*, 1999; Soderland, 1999].

The basic idea followed in this thesis is to use as little as possible additional semantic information. Hence, we do not want to incorporate large dictionaries or additional knowledge sources into the pre-processing of documents. This is motivated by the aim to be as flexible as possible for any type of document and for any document content without spending much time for setting up domain specific knowledge bases. Nevertheless, the chosen representation should offer the possibility to incorporate such additional information in cases it is needed or inalienable due to practical application issues. In this thesis the main focus is on semi-structured documents, especially *HTML* documents. Although the focus is on learning wrappers for this special type of markup documents the presented wrapper induction approach is easily adaptable to other markup languages than *HTML* and non semi-structured documents.

Since we discuss logic programming based techniques for the learning of wrappers in this thesis a document representation is needed, which can be used efficiently in terms of logic programming. Therefore, a transformation is required mapping markup documents (e.g. *SGML* [Goldfarb, 1994], *XML*, *HTML*) into a suitable representation. Such a representation should be *structure preserving*. The grammatical, syntactical features and the markup based structural properties of the document must be preserved. Furthermore, the document representation should be easy to integrate with the fundamental logic programming paradigm. This means that a representation is used which makes it easy to access document elements by means of unification [Knight, 1989], and not for example by substring operations. Document elements are interpreted as logical terms following the logic programming paradigm e.g. unit facts, clauses or lists of terms. A third requirement is the removal of redundant text elements under the assumption that removing text elements does not change the documents semantics. This sort of filtering is helpful to minimize the size of the document representation.

2.3.1. Preprocessing

The introduced *tokenization* of documents based on *sequencing* and mapping text parts onto lists of attribute-value pairs allows for a very convenient way to access and to describe certain text properties. Moreover, if we use a term notation of these attribute-value pairs, smart access and query techniques by means of logic and logic programming are provided (see Section 3.5). Sequencing and tokenization, which are usually implemented by means of lexical analyzation tools [GNU, 1995; GNU, 1997], build the first step in preprocessing documents. The tokenization serves as basis for the approaches presented in Section 2.3.2 and 2.3.3. In the following, the sequencing and tokenization of documents (Section 2.1.1) is discussed in broader detail.

Starting with sequencing a document, the most common approach is to interpret certain symbols like *blank*, *tabulator* and *newline* as word[1] delimiters. These symbols (*white-spaces*) together with rules for the recognition of hypertext tags form the basic sequence function for parsing hypertext documents. Once a sequence function has been determined a document is transformed into a sequence of texts according to this function. Throughout this thesis an intuitive sequence function that builds sequences splitting the document in texts containing *tags*, *white-spaces*, *punctuation marks*, *special symbols*, *arithmetic operators*, *integers*, *floats*, *dates* and *words* is used. More general, this is the common way documents are read and understood by humans, interpreting markup tags as annotating text blocks and all other text parts belonging to natural language text, as is depicted in Figure 2.5.

Next, the role of tokenization and how it is used to incorporate additional information into the document representation is explained in broader detail. In contrast to existing approaches that learn text patterns (e.g. regular expressions) for information extraction on basis of letters solely, the use of tokens allows to enrich information associated with text parts in the document. Therefore, tokenization based approaches provide more information for the subsequent learning of wrappers. The question arising is, what useful additional information can be associated with certain text parts in a document? And how can this additional information, which will be represented as tokens to describe and replace the original text, build a promising basis for logic based machine learning techniques? The answer strongly depends on the document type. As we will investigate hypertext documents it makes sense to map the attribute-value pairs of hypertext tags one-to-one to token attribute-value pairs. Note, this is independent from the

[1] Note that we are now talking about "natural" words and not all words from Σ^*.

Document

<b>Day 19: US Troops storm central Baghdad</b>

Sequence

< '<b>', 'Day','', '19',':',' ','US',' ','Troops',' ','storm',' ','central',' ','Baghdad','</b>' **>**

Figure 2.5.: example sequence of a partial *HTML* document

domain or context modeled by the hypertext document. Secondly, for the *normal text*, in case of *HTML* pages this is the rendered text appearing on the *document's surface* displayed by the web-browser (except the special characters which also have to be represented by *HTML*-tags), a lexical analyzation is used, introducing types and certain features. Figure 2.6 shows an example tokenization of the texts from Figure 2.5. At this point and in the remaining part of the thesis it is not important which sort of attributes are introduced by the tokenization process, since none of the presented approaches will use techniques that try to learn from the semantic meanings of these attributes. More precisely, neither the meaning of the attribute's name nor the meaning of its value bias the learning. An example for an attribute semantic based approach would be the following: Assume tokens can have attributes `type` and `txt`. Further the learning algorithm (learner) would recognize that in all of its presented learning examples there are three successive tokens with `(type=int,txt=030)`, `(type=op, txt=-)` and `(type=int,txt=675610)`. A reasonable conclusion would be to say that the examples contain telephone numbers, since it is quite common to represent telephone numbers by a local area code separated by a minus symbol followed by a number. Thus the learner generalizes based on the attribute types (names), its values (e.g. `030` is the local area code for Berlin) and some common world knowledge about telephone numbers. This example is in fact very attractive and describes a desirable more elaborated information extraction task somewhat related to *text understanding* research topics.

But the crux is, when does the learner know which world knowledge to apply and where does it comes from and how do we represent it? This obviously leads to the research areas of *Knowledge Representation* and *Automated Deduction*. Without any doubt these two areas have shown huge successful steps in the last decades and there are a lot of movements especially to integrate knowledge representation techniques (e.g. *Description Logics*) with web contents as in the *Semantic Web Project* [W3C, 2004]. Nevertheless, the pitfall of modeling more than small domains still remains. As with the integration of *computational linguistics* these *knowledge representational* methods can be used as sensible refinement and extension of the basic approaches presented in this thesis.

The last step in the preprocessing phase is the removal of unnecessary tokens, the filtering. In general white-spaces are removed since the word boundaries are now determined by the tokens itself. But other filters depending on the document type and the IE-task are conceivable. Technically the token filtering is a trivial process of sorting out tokens from the tokenized sequence according to predefined attributes. Figure 2.6 shows the token sequence after removing

Document

<b>Day 19: US troops storm central Baghdad</b>

Tokenization

< [ttype = 'html', value = '<b>', tag = 'b', s_pos = '0', e_pos = '2'],
[ttype = 'word', value = 'Day', first = 'upper', s_pos = '3', e_pos = '5'],
[ttype = 'ws', value = ' ', s_pos = '6', e_pos = '6'],
[ttype = 'int', value = '19', s_pos = '7', e_pos = '8'],
[ttype = 'punct', value = ':', s_pos = '9', e_pos = '9'],
[ttype = 'ws', value = ' ', s_pos = '10', e_pos = '10'],
[ttype = 'word', value = 'US', first = 'upper', s_pos = '11', e_pos = '12'],
[ttype = 'ws', value = ' ', s_pos = '13', e_pos = '13'],
[ttype = 'word', value = 'Troops', first = 'upper', s_pos = '14', e_pos = '19'],
[ttype = 'ws', value = ' ', s_pos = '20', e_pos = '20'],
[ttype = 'word', value = 'storm', first = 'lower', s_pos = '21', e_pos = '25'],
[ttype = 'ws', value = ' ', s_pos = '26', e_pos = '26'],
[ttype = 'word', value = 'central', first = 'lower', s_pos = '27', e_pos = '33'],
[ttype = 'ws', value = ' ', s_pos = '34', e_pos = '34'],
[ttype = 'word', value = 'Baghdad', first = 'upper', s_pos = '35', e_pos = '41'],
[ttype = 'html_end', tag = 'b', value = '</b>', s_pos = '46', e_pos = '49'] >

Filtering

< [ttype = 'html', value = '<b>', tag = 'b', s_pos = '0', e_pos = '2'],
[ttype = 'word', value = 'Day', first = 'upper', s_pos = '3', e_pos = '5'],
[ttype = 'int', value = '19', s_pos = '7', e_pos = '8'],
[ttype = 'punct', value = ':', s_pos = '9', e_pos = '9'],
[ttype = 'word', value = 'US', first = 'upper', s_pos = '11', e_pos = '12'],
[ttype = 'word', value = 'Troops', first = 'upper', s_pos = '14', e_pos = '19'],
[ttype = 'word', value = 'storm', first = 'lower', s_pos = '21', e_pos = '25'],
[ttype = 'word', value = 'central', first = 'lower', s_pos = '27', e_pos = '33'],
[ttype = 'word', value = 'Baghdad', first = 'upper', s_pos = '35', e_pos = '41'],
[ttype = 'html_end', tag = 'b', value = '</b>', s_pos = '46', e_pos = '49'] >

Figure 2.6.: example tokenization of a partial *HTML* document and filtered token sequence all tokens of type `ws` (white-space).

It is worth pointing out that the overall extraction results of learned wrappers considerably depend on the selected sequence and corresponding tokenization function, but the introduced learning methods and techniques in Part II are not affected in their operational function by this choice. This is important, since it allows to adapt the presented approaches to a wide variety of different document types (e.g. *XML*) by simply modifying the sequence and tokenization function. The reader is referred to Chapter 10 for observations concerning the expected quality of learned wrappers depending on the number and type of attributes introduced by the tokenization function.

2.3.2. Attribute-Value Representation

After a brief motivation and discussion of some aspects of sequence and tokenization functions this section describes a representation tokenized documents as logic programs. The basic idea is to represent a tokenized version of a document D as a set of *unit clauses* such that each token in $Tok(D)$ is one unique unit clause. Uniqueness is ensured by encoding the token's sequence position in $Tok(D)$ as additional argument into its unit clause representation. This preserves the order of tokens as they occur in the original document and it also allows, depending on the tokenization function $\mathcal{T}$ and the resulting attribute values, to reconstruct the original document. Additionally, each unit clause is extended by an argument serving as reference to the original document. This allows us to represent several documents within one logic program by avoiding clashes of identical unit clauses stemming from tokens of different documents. Formally the logical attribute-value representation of an arbitrary document is determined by Definition 2.3.1.

Definition 2.3.1 (Attribute-Value Representation) *Given a document D, a unique identifier D_{ID} for D, the* logical attribute-value representation *of D wrt. a* sequence function *Seq and* tokenization function *$\mathcal{T}$ is defined as* $AV(D) = \{token(D_{ID}, i, t) | t = Tok(D).i \wedge i = 1 \ldots |Tok(D)|\}$. □

Some remarks concerning Definition 2.3.1. The tokenized document D (a sequence of tokens) is given by $Tok(D)$. By $Tok(D).i$ we select the i-th sequence element of $Tok(D)$.

To transform a document into a representation as logic program its *AV-representation* is interpreted to form a set of unit clauses. Finally, arbitrary documents can be represented by means of the *AV-representation* as a logic program consisting of unit clauses. Figure 2.7 illustrates the document from Figure 2.5 represented as logic program (*Prolog* syntax) based on the *AV-representation*.

2.3.3. Document Object Model Based Representation

For the rest of this chapter we assume the reader to be familiar with the *document object model* (*DOM*) [Dom, 2000] of *XML* documents. One obvious shortcoming of the *AV-representation* is that it ignores the structural information inherently given by the *XML* or *HTML* tags. These annotations (tags) define text properties on two possible levels: the text layout level and the semantic level. The annotations at the semantic level in most cases give further information about the meaning of the text. The annotations at the layout level in general influence the visual appearance of the text. But independently of their different meanings

AV-representation

```
token(0,1,[ttype='html', value = '<b>', tag = 'b', s_pos = '0', e_pos = '2']).
token(0,2,[ttype = 'word', value = 'Day', first = 'upper', s_pos = '3', e_pos = '5']).
token(0,3,[ttype = 'int', value = '19', s_pos = '7', e_pos = '8']).
token(0,4,[ttype = 'punct', value = ':', s_pos = '9', e_pos = '9']).
token(0,5,[ttype = 'word', value = 'US', first = 'upper', s_pos = '11', e_pos = '12']).
token(0,6,[ttype = 'word', value = 'Troops', first = 'upper', s_pos = '14', e_pos = '19']).
token(0,7,[ttype = 'word', value = 'storm', first = 'lower', s_pos = '21', e_pos = '25']).
token(0,8,[ttype = 'word', value = 'central', first = 'lower', s_pos = '27', e_pos = '33']).
token(0,9,[ttype = 'word', value = 'Baghdad', first = 'upper', s_pos = '35', e_pos = '41']).
token(0,10,[ttype = 'html_end', tag = 'b', value = '</b>', s_pos = '46', e_pos = '49']).
```

Figure 2.7.: example *AV-representation* in *Prolog* syntax

(layout and semantics) they both define structural environments which are very helpful for wrapper learning.

Annotations defining the text layout, as for example in *HTML* the tags for paragraphs (`<p>`), tables (`<table>`) or list environments (`<ul>`), can give important hints for discovering similarities among extraction examples. For instance, imagine all extraction examples are occurring within table columns, but they are embedded in larger text parts within these columns. If a learner focuses on investigating the surrounding text parts of the examples to learn extraction rules, it probably will not recognize the structural information that all examples occur within a table environment. Because each text example is preceded and succeeded by text not including any table tags.

Therefore extending the learner by the ability to recognize structural contexts in which the learning examples occur most likely leads to better learning results. Thus, a plausible attempt is to change the document representation from a linear document representation, as the *AV-representation*, to a tree-structured hierarchical representation. Such a representation easily allows to retrieve the environment in which a text part is embedded by simply traversing the tree structure.

In order to capture and model these syntactical and hierarchical aspects of *HTML* and *XML* documents, we define the concept of *TDOM-trees* (token based document object model). This concept is strongly related to that of a *DOM*-tree. There are only few minor differences between these two models, a *TDOM* is defined for any document type. This includes documents which do not contain any tags at all, but also well formed *XML* documents. The essential idea is, that every text in a document that forms a syntactical correct tag according to the *XML* definition plus the known exceptions of *HTML* tags are identified and parsed as inner nodes of a *DOM* resp. *TDOM-tree.* Based on this idea every document is transformed into a *DOM* based representation. In more detail, a node in a *TDOM-tree* consists of four features: a document reference D_{ID}, a node identifier n_{ID}, the corresponding token t describing the text occurring in the document denoted by the node and a sequence of child node identifiers $< ch_1, \dots, ch_n >$ where the child nodes sequence position determines its actual child number. A *TDOM node* is represented as an atom $node(D_{ID}, n_{ID}, t, [ch_1, ch_2, \dots, ch_n])$. Figure 2.9 shows a *TDOM node.*

This representation differs not too much from the *AV-representation* of a document and thus depicts the basic idea for representing *TDOM-trees* as sets of unit clauses. Similar to encoding the linear occurrence of tokens in the *AV-representation* by enumerated token numbers (Definition 2.3.1) a *TDOM* node also has a unique identification argument determining its position in the *TDOM*. Such *node identifiers* have two functions: 1) Every node in the tree can be referenced by a unique identifier. 2) Every *node identifier* determines the exact position of the node within the *TDOM-tree* in terms of relational information.

There is an important difference between simply enumerating all nodes in a pre-order traversal of a tree to determine a unique node identifier, or using a more sophisticated identifier expressing the node's position in terms of relational information like *the second child node of the first child of the root node.*

```
<html>
<head><title>Example</title></head>
<body>
<h1>Example</h1>
<ul>
<li> A simple <b>example</b>
<li> of a TDOM
</ul>
</body></html>
```

Figure 2.8.: simple *HTML* document

Assume two *HTML* documents having an identical layout but differ in their number of words, e.g. two automatically produced web pages presenting products in an online catalogue. In these web-pages the price information always is given in the second column of the first table appearing in the document. Since product detail descriptions vary in their lengths it is obvious that simply enumerating the nodes to determine node identifiers results in two different identifiers for the node describing the price information. By choosing the proposed relational representation the price information nodes from both pages would be the same. This simple example demonstrates how much influence a representational issue may have on the later bias for a learning technique. The introduced *node identifiers* offer essential information about the occurrence of example texts with respect to the document structure.

More concrete, *node identifiers* are terms representing a path from the root node to a node in the *TDOM-tree.* For example, the term $child(child(child(0,1),0),3)$ refers to the fourth child of the first child of the second child of the root node in the *TDOM.* Here the root node is an abstract node with number 0. For better readability we notate node identifiers as lists in the following way `[1,0,3]`. Hence a *node identifier* is used to assign a unique term to each node in a *TDOM.* In fact, this representation is a slight variant of the *Dewey-Notation* [Scott, 1998] which is a library classification scheme very common in librarianship. In Figure 2.9 a simplified *TDOM-tree* (only one node atom is displayed) for the small *HTML* example from Figure 2.8 is given.

TDOM Construction

The essential part needed for the construction of *TDOM-trees* is a function defining how to parse and to construct the tree structure associated with the markup tags occurring within the document. In general, three trivial rules concerning the types of tags can be defined to obtain a tree construction method. There are two different types of tags: those introducing environments, which are determined by *start* and *end tags*, and those introducing *empty environments,* which consist of single tags. The tree is then defined as follows:

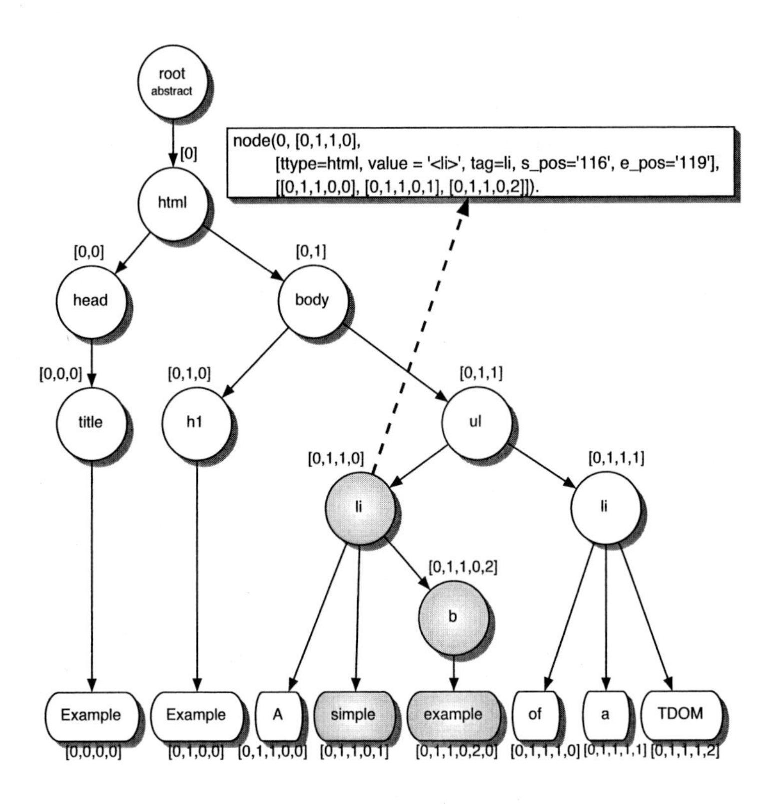

Figure 2.9.: simplified *TDOM-tree*, node and span ([0,1,1,0],1,2)

- If the current token to be processed is a *start tag*, create a new child node. Set the new node to be the new child node of the current node. Let the new node be the current node.

- If the current token to be processed is an *end tag*, set the new current node to be the father node of the current node.

- For all other tokens to be processed including *empty environment tags*, create a new child node for the current node.

This algorithm is not covering any exceptions and should just roughly give a sketch of the basic idea. There are numerous parser tools [Xerces, 2003] for *XML* and *XHTML* available, which can easily be extended to be used for the construction of *TDOM-trees*.

A further difference between the *TDOM* concept and the standard DOM is the treatment of text nodes. A leaf node in a DOM-tree represents text appearing at the "surface" of the hypertext document. For example, a whole paragraph consisting solely of ordinary text and no tags, is associated with one leaf node in a DOM-tree. In many cases, this representation is not detailed enough for IE tasks. For instance, if the information to be extracted is just a part of this paragraph, then it is more helpful to modify the concept of a DOM-tree such that a leaf node in a DOM-tree becomes many leaf nodes in a *TDOM-tree*. Each of these nodes represent one token from the text. We summarize the concepts introduced so far by following definitions.

Definition 2.3.2 (Node Identifier) *Given a tree T where each node $n \in T$ has a unique node number. Let $ch : \mathbb{N}_0 \times \mathbb{N}_0 \rightarrow \mathbb{N}_0$ be a partial function selecting for a given node number n the i-th child node of node n. A* node identifier *is defined as a sequence of concatenations of ch starting with the root node of T. Node identifiers are written as sequences according to the scheme: $ch((...ch(ch(0, c_0), c_1), ...), c_n) \Rightarrow < c_0, c_1, \ldots, c_n >$. Since node identifiers are sequences we alternatively notate them as* Prolog *lists.* □

Definition 2.3.3 (Order of Node Identifiers) *A node identifier n_i is smaller than a node identifier n_j written $n_i < n_j$ iff n_i is a subsequence of n_j with $n_j.1.|n_i| = n_i$ (n_i is a prefix of n_j) or $\exists x \in \mathbb{N}_0 : n_j.x > n_i.x$ and $\forall y \in \mathbb{N}_0 : y < x$ it holds that $n_j.y = n_i.x$ where $n_i.n$ denotes the n-th element (child number) of the node identifier. Two node identifiers n_i and n_j are equal if they have the same length and $n_i \not< n_j$ and $n_j \not< n_i$.* □

Definition 2.3.4 (TDOM representation) *Given a document D with identifier D_{ID}, an* AV-representation *$AV(D) = \{token(D_{ID}, 0, t_0), token(D_{ID}, 1, t_1), \ldots\}$ of D and a TDOM tree T build from tokens $t_0, t_1, \ldots$ from $AV(D)$. Let N be the set of node identifiers of T. The TDOM representation of D is given by $TD(D) = \{node(D_{ID}, n, t, ch) | n \in N$ and t the token associated with the node identifier n and ch the sequence of child node identifiers of node $n\}$.*
□

As for the *AV-representation* the *TDOM-representation* of a document D is interpreted as a logic program, where every element in $TD(D)$ is an atom and builds a unit clause. Figure 2.10 shows the *TDOM* unit clause set representing the example document of Figure 2.8.

Node identifiers have nice properties for wrapper-learning. Similar to expressions in the XPATH language [XPa, 1999] node identifier expressions can be used to refer to more than

TDOM-representation

```
node(0,root,[],[root-0]).
node(0,root-0,token([ttype,html),(value,'<html>'),(tag,html),(s_pos,'0'),(e_pos,'5']),[root-0-0,root-0-1]).
node(0,root-0-0,token([ttype,html),(value,'<head>'),(tag,head),(s_pos,'7'),(e_pos,'12']),[root-0-0-0]).
node(0,root-0-0-0,token([ttype,html),(value,'<title>'),(tag,title),(s_pos,'14'),(e_pos,'20']),[root-0-0-0-0]).
node(0,root-0-0-0-0,token([ttype,word),(value,'Example'),(first,upper),(s_pos,'21'),(e_pos,'27']),[]).
node(0,root-0-1,token([ttype,html),(value,'<body>'),(tag,body),(s_pos,'59'),(e_pos,'64']),[root-0-1-0,root-0-1-1]).
node(0,root-0-1-0,token([ttype,html),(value,'<h1>'),(tag,h1),(s_pos,'66'),(e_pos,'69']),[root-0-1-0-0]).
node(0,root-0-1-0-0,token([ttype,word),(value,'Example'),(first,upper),(s_pos,'70'),(e_pos,'76']),[]).
node(0,root-0-1-1,token([ttype,html),(value,'<ul>'),(tag,ul),(s_pos,'88'),(e_pos,'91']),[root-0-1-1-0,root-0-1-1-1]).
node(0,root-0-1-1-0,token([ttype,html),(value,'<li>'),(tag,li),
                (s_pos,'93'),(e_pos,'96']),[root-0-1-1-0-0,root-0-1-1-0-1,root-0-1-1-0-2]).
node(0,root-0-1-1-0-0,token([ttype,word),(value,'A'),(first,upper),(s_pos,'97'),(e_pos,'97']),[]).
node(0,root-0-1-1-0-1,token([ttype,word),(value,simple),(first,lower),(s_pos,'99'),(e_pos,'104']),[]).
node(0,root-0-1-1-0-2,token([ttype,html),(value,'<b>'),(tag,b),(s_pos,'106'),(e_pos,'108']),[root-0-1-1-0-2-0]).
node(0,root-0-1-1-0-2-0,token([ttype,word),(value,example),(first,lower),(s_pos,'109'),(e_pos,'115']),[]).
node(0,root-0-1-1-1,token([ttype,html),(value,'<li>'),(tag,li),
                (s_pos,'125'),(e_pos,'128']),[root-0-1-1-1-0,root-0-1-1-1-1,root-0-1-1-1-2]).
node(0,root-0-1-1-1-0,token([ttype,word),(value,(of)),(first,lower),(s_pos,'129'),(e_pos,'130']),[]).
node(0,root-0-1-1-1-1,token([ttype,word),(value,a),(first,lower),(s_pos,'132'),(e_pos,'132']),[]).
node(0,root-0-1-1-1-2,token([ttype,word),(value,'TDOM'),(first,upper),(s_pos,'134'),(e_pos,'137']),[]).
```

Figure 2.10.: *TDOM-representation* of the example *HTML* document

one node by the use of variables. The node identifier `[0,1,1,X]` refers to every child node of the <ul> environment of Figure 2.9. For example, the term `[X,3]` refers to all third child nodes of children of the root node. It is important to point out, that variables can only be substituted by $n \in \mathbb{N}_0$ and not by partial node identifier expression like `[0,1]`. It should be mentioned that constraints can be introduced by more than one occurrence of the same variable (e.g. `[0,X,2,X,0]`) and that such pattern variables are not treated disjunctively.

In comparison to the XPATH query language, these expressions can only be expressed by means of iterative programming language constructs like for-loops and thus are not as elegant and compact and easy to handle.

The presented notation makes it easy to generalize on node identifiers by means of *lgg* operations. Assume two text examples are located in the node `[0,1,1,0,0]` and `[0,1,1,1,0]`. A reasonable first step in learning an extraction rule is the assumption that all nodes described by the generalized node identifier `[0,1,1,X,0]` are good extractions.

One further essential concept of *TDOM* based document and example representation is that of a *span*. Informally a span determines a subtree in a *TDOM-tree*. We pick up the idea mentioned by [Cohen *et al.*, 2002] where a span is defined as a triple consisting of a node identifier n and a left and right delimiter l,r. Delimiters determine the left and right boundaries of an interval of child nodes contained in a span. For example the span `([0,1,1,0],1,2)` of the example *TDOM* depicted in Figure 2.9 refers to the sequence of node identifiers `<[0,1,1,0,1], [0,1,1,0,2], [0,1,1,0,2,0]>`.

Definition 2.3.5 (Span) *Given a tree T and corresponding node identifiers N. A* span *S is determined by a triple (n, l, r) with $n \in N$ and $l, r \in \mathbb{N}_0$ such that S is the sequence of all reachable descendant nodes obtained in a left depth first traversal starting at the i-th child node of node identifier n with $i = l..r$. The depth first traversal ensures the left to right* *order of the surface text of a document under the assumption that T has been constructed analogically.* □

For the description of example texts by determining their position within a *TDOM-tree*, we introduce the concept of a *minimal text span*.

Definition 2.3.6 (Minimal Span) *Given a TDOM T for document D and a text t occurring in D. A minimal span MS for t is the shortest span including all nodes (node identifiers) representing text t. Analogously, we denote with $MS(D,t) = (n,l,r)$ the function calculating the minimal span (n,l,r) for t in T of D.* □

Example 2.3.1 (Minimal Span) *Let t be a text fragment with $t =$ `simple example` from the document shown in Figure 2.9. Let S_1 be a span with $([0,1,1],0,1)$ and S_2 be determined by $([0,1,1,0],1,2)$. Clearly both S_1 and S_2 contain t but $|S_1| > |S_2|$ and therefore S_2 is the only existing minimal example span of t with respect to the example* TDOM *because: $\neg\exists S' : |S'| < |S_2|$ where S' is a span including t.* ⌟

3. Information Extraction Wrappers

The research field of Information Extraction by its origins is strongly motivated by practical tasks. Since one major aim of this thesis is to discuss heuristics for the extraction of information from hypertext documents, we present an overview of several wrapper classes motivated by practical extraction tasks in Section 3.1. Since it is necessary to evaluate wrappers concerning their extraction quality and reliability, several properties used to define measures for evaluation are introduced in Section 3.2. In Section 3.3 the three wrapper models used throughout this work are introduced. For these models in Section 3.4 two wrapper languages are presented. Finally, Section 3.5 depicts three possible implementation techniques for the advocated wrapper models.

3.1. Wrapper Classes

In the last decade numerous researchers in the area of machine learning based information extraction (ML4IE) have tested their automatic wrapper construction algorithms on different types of documents [Muslea, 1998]. Such tasks a wrapper has to fulfill, range from a simple telephone number extraction to a more sophisticated task like the extraction of nested structures as sometimes given in product descriptions with varying number of attributes and order. Such extractions may contain different structural layout elements like for instance tables and itemize environments, which require to discover specific relational dependencies among text parts in a document. Due to the nature of the documents and the different extraction tasks several interesting problems regarding the occurrence or structural position of slot fillers within documents are observed. Independent of the used representation of wrappers and applied learning algorithms, the overall goal of almost all approaches can be abstractly described to consist of automatically constructing an extraction procedure that computes extraction tuples covering a given wrapper and associated wrapper semantics as discussed in Section 2.1.1. In the remainder of the thesis the results provided by a wrapper procedure have to be a tuple of fixed arity with a fixed semantics for each slot.

Note that the underlying assumption is that in general there is more than one extraction tuple contained in one document. If this assumption is dropped the task of multi-slot extraction would become simply the task to identify fillers for n single slots independently of each other and discussions about varying slot semantics could be left out.

In practice relevant text information in the source code of a web page can be represented in many different ways such that it is very easy or very hard to construct an extraction procedure meeting the intended wrapper semantics. For instance, someone wants to extract from a web page of the local cafeteria a set of tuples `<weekday,meal>`. If for example there is more than one meal offered at one day, this information may be given as a list `monday: chicken madras, wienerschnitzel or pan cake` or it might be represented as `monday:chicken madras, monday:wienerschnitzel, monday:pan cake`.

Consequently the question arises if it is easier to construct a wrapper for one representation than for another. And even more, are there representations for which no suitable automatic wrapper construction algorithms can be found? So far the definition of a wrapper (Defintion 2.1.2) is a declarative one setting up no constraints on slot fillers regarding their occurrence in a document or their relationships to each other. Given the following observations it is reasonable to adapt the general wrapper definition according to such insights drawn from practical observations. From these observations reasonable wrapper classes are defined which in turn will help to build more efficient automatic wrapper construction algorithms.

Most of the observations are based on the assumption almost all existing ML4IE approaches have in common. It is the concept of *delimiters*. Delimiters underlie the assumption that they are unique text sequences, which surround the *slot filler text* or attributes that are to be extracted. They are some sort of fundamental concept to define wrapper procedures and this concept can be applied to other more structural and hierarchical document representations as we will show in Section 3.3.3.

This section defines a set of abstract wrapper classes which are derived by observations from practical application domains and various research publications. Most of these observations are taken from [Kushmerick and Thomas, 2003; Kushmerick, 2000] where they are informally stated. The presented classes can be used for exhaustive investigations on the learnability of wrappers, since it defines a formal basis. Figure 3.12 summarizes the wrapper classes introduced in the following sections.

For the following discussion the reader is reminded of Definition 2.1.3. There we took the basic assumption that for each extraction tuple e of a wrapper $\mathcal{W}$ only one unique set of words $m_0, m_1, \dots, m_n \subset W(D)$ exists such that $e.D = m_0 e.o(1) m_1 \dots e.o(n) m_n$. This assumption helps to define several wrapper classes in a more convenient way. Dropping this quite strong assumption would require to use distinct information about the extraction text $e.i$ of each tuple to uniquely determine its position within a document. Without the explicit knowledge about the relation of slot fillers regarding their position within a document some observations consequently cannot be made. Therefore either the unique occurrence assumption has to be made or for each extraction tuple argument absolute text positions have to be supplied. Since choosing the absolute position alternative requires the adaptation of several standard definitions (e.g. word concatenation with respect to sub words and their absolute position) and thus for ease of notation, readability and comprehension the unique occurrence assumption is taken throughout this section.

3.1.1. Linear Wrappers

Though each extraction of a wrapper $(D, (w_1, \dots, w_n)) \in \mathcal{W}$ has a fixed semantics and thus an order on its slot arguments w_i it cannot be assumed that such an ordering also holds regarding the occurrences of $w_1, \dots, w_n$ in a document D.

Definition 3.1.1 (Linear Wrapper) *An n-slot wrapper $\mathcal{W}$ is a linear wrapper iff for all $x \in \mathcal{W} : x.D = m_0\ x.1\ m_1\ x.2\ m_2\ \dots m_{n-1}\ x.n\ m_n$ with $m_i \in W(D)$. Alternatively, for all $x \in \mathcal{W}$ and o it holds that $o(i) = i$. A wrapper where o is undefined or $\exists i \in 1, \dots, n : o(i) \neq i$ is called* non-linear wrapper. □

x ∈ W: x = (D,<A,B,C>) o = {(1,1), (2,2), (3,3)}
__ : arbitrary text
D: A B C
m0 x.1 m1 x.2 m2 x.3 m3

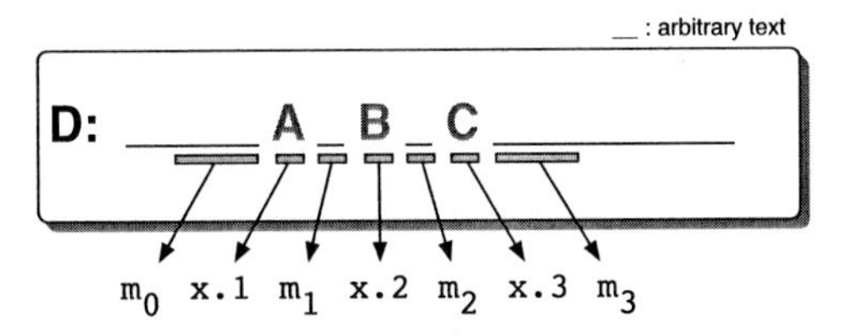

Figure 3.1.: linear wrapper example

x,y ∈ W: x = (D,<A,B,C>) o = {(1,1), (2,3), (3,2)}
y = (D,<E,F,G>) o = {(1,1), (2,2), (3,3)}
__ : arbitrary text
D: A C B
E F G
m0 x.1 m1 x.3 m2 x.2 m3
x.o(1) x.o(2) x.o(3)

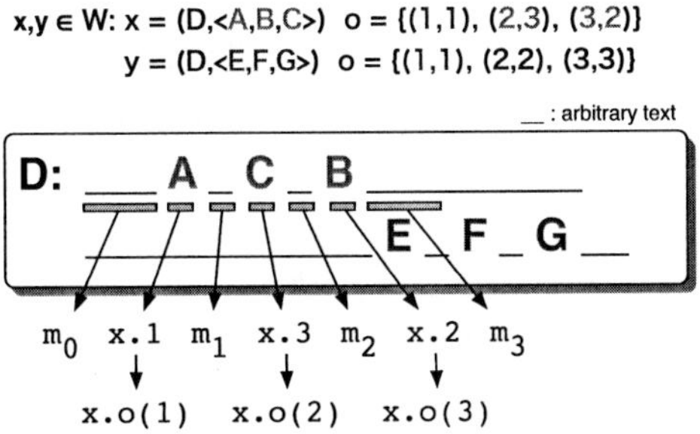

Figure 3.2.: non-linear wrapper example

3.1.2. Empty Slot Wrappers

In many extraction tasks some slot fillers are missing in a document but the significant delimiters for these slots are present. For instance, a wrapper extracting product IDs must be able to handle empty slot fillers. Assume that normally a product ID occurs between the `</b>` and `<p>` tag but there are also parts in the document like `<b>ID:</b><p>`. Since we don't restrict a wrapper to one document the general task is to learn wrappers to extract many information from several documents. So, the following definitions consider this aspect that certain properties of tuples are observable among extraction tuples taken from different documents.

Definition 3.1.2 (ϵ-Wrapper) *An n-slot wrapper $\mathcal{W}$ is an* ϵ-wrapper *iff there exist $x, y \in \mathcal{W}$ and there exists $i \in \{1, \ldots, n\}$ such that*

- $x.i = \epsilon \wedge y.i \neq \epsilon$
- $x.o^{-1}(i) = j \wedge y.o^{-1}(i) = k$
- $x.D = m_0\ x.o(1)\ m_1 \ldots m_{j-1}\, m_j\ \ldots\ m_{n-1}\ x.o(n)\ m_n$ *with* $m_l \in W^+(D)$ *and* $l = 1, \ldots, n$

- $y.D = m'_0\ y.o(1)\ m'_1 \ldots m'_{k-1}\ y.o(k)\ m'_k \ldots m'_{n-1}\ y.o(n)\ m'_n$ with $m'_l \in W^+(D)$ and $l' = 1, \ldots, n$
- $suf(m_{j-1}, m'_{k-1}) \neq \epsilon$
- $pre(m_j, m'_k) \neq \epsilon$

□

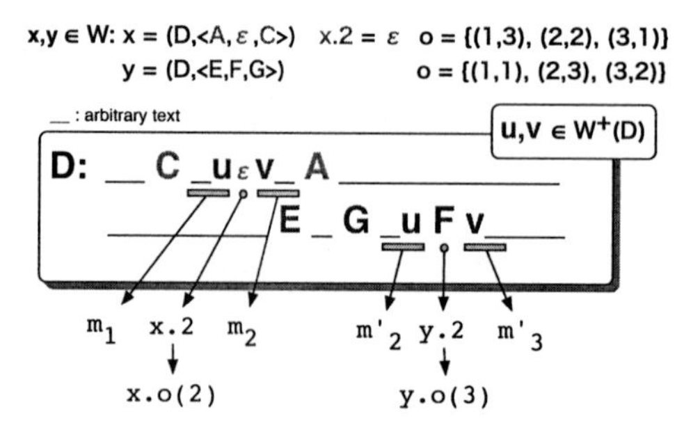

Figure 3.3.: ϵ-wrapper example (non-linear)

Definition 3.1.3 (ϵ-free Wrapper) *An n-slot wrapper $\mathcal{W}$ is an* ϵ-free wrapper *iff it does not contain extraction tuples with the empty word as argument. $\mathcal{W}$ is a* ϵ-free wrapper *iff for all $(D, (w_1, \ldots, w_n)) \in \mathcal{W} : w_i \neq \epsilon$ with $i = \{1, \ldots, n\}$ and ϵ the empty word.* □

3.1.3. Missing Slot Wrappers

More difficult than handling the empty slot fillers is the complete absence of delimiters. For instance, assume a web page containing e-mail addresses, alternative email addresses (like `alt: joe@web.de<br>`) and additional information about customers. In this example not every customer has an alternative e-mail address and therefore it is simply left out which also holds for the delimiters `alt:` and `<br>`. So all information about the alternative e-mail slot is missing.

Definition 3.1.4 (ϵ^+-Wrapper) *An n-slot wrapper $\mathcal{W}$ is an* ϵ^+-wrapper *iff there exists $i \in \{1, \ldots, n\}$ and there exists $x \in \mathcal{W} : x.i = \epsilon$ and for all $y \in \mathcal{W}$ for which $y.i \neq \epsilon$ it holds that*

- $x.o^{-1}(i) = j \wedge y.o^{-1}(i) = k$
- $x.D = m_0\ x.o(1)\ m_1 \ldots m_{j-1} m_j \ldots m_{n-1}\ x.o(n)\ m_n$ *with* $m_l \in W^+(D)$ *and* $l = 1, \ldots, n$
- $y.D = m'_0\ y.o(1)\ m'_1 \ldots m'_{k-1}\ y.o(k)\ m'_k \ldots m'_{n-1}\ y.o(n)\ m'_n$ *with* $m'_l \in W^+(D)$ *and* $l' = 1, \ldots, n$
- *there is no m_{j-1} and m_j such that* $suf(m_{j-1}, m'_{k-1}) \neq \epsilon$ *and* $pre(m_j, m'_k) \neq \epsilon$

□

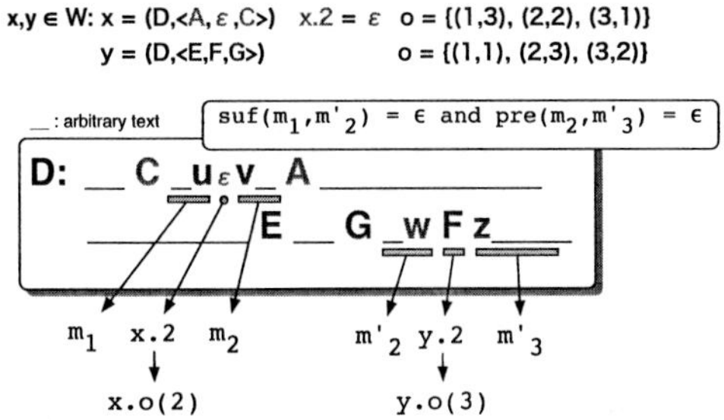

x,y ∈ W: x = (D,<A,ε,C>) x.2 = ε o = {(1,3), (2,2), (3,1)}
y = (D,<E,F,G>) o = {(1,1), (2,3), (3,2)}

__ : arbitrary text
suf(m1,m'2) = ε and pre(m2,m'3) = ε

D: __ C _u ε v_ A ______
______E __ G _w F z____

m1 x.2 m2 m'2 y.2 m'3
x.o(2) y.o(3)

Figure 3.4.: ϵ^+-wrapper example (non-linear)

3.1.4. Multiple Values per Slot Wrappers

Assume a web page listing football teams and the year they managed to win the cup. Thus the intended extraction is a tuple <team,year>. But on the web page the glorious wins are represented as lists like team: year,year
. Obviously an appropriate extraction procedure has to be able to identify such lists containing n elements in the correct way to provide n extractions.

Definition 3.1.5 (MV-Wrapper) *An n-slot wrapper $\mathcal{W}$ is an* MV-wrapper *iff there exists $j \in \mathbb{N} : x_1, \ldots, x_j \in \mathcal{W}$ with $j > 1$ such that*

- $x_1 \neq x_2 \neq \ldots \neq x_j$
- $x_1.D = x_2.D = \ldots = x_j.D$
- *there exists $i \in \{1, \ldots, n\}$ such that:*
  - $x_1.D = m_o\ x_1.i\ m_1\ x_2.i\ m_2 \ldots m_{j-1}\ x_j.i\ m_j$ *with* $m_l \in W^+(D)$ *and* $l = 1, \ldots, j$
  - *there exists no $k \in \{1, \ldots, n\}$ and there exists no $s, r \in \{1, \ldots, j-1\}$ such that $x_s.k$ is an infix of m_r*

□

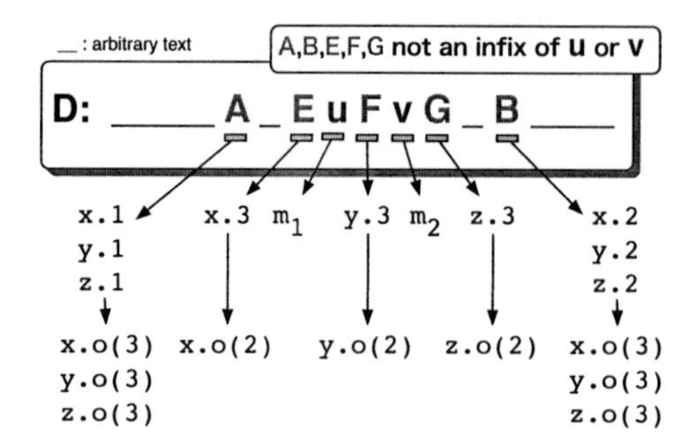

Figure 3.5.: MV-wrapper example (non-linear)

3.1.5. Varying Delimiter Wrappers

The very naive assumption that for a certain slot of a wrapper a unique delimiter can be determined is in most practical cases too optimistic. In practice delimiters for a specific slot vary and in general there does not exist a reasonable minimal version or intersection of them. Hence it is necessary to use a disjunction of delimiters or a more elaborated pattern language describing delimiters.

Definition 3.1.6 (∨-Wrapper) *An n-slot wrapper $\mathcal{W}$ is an* ∨-wrapper *iff there exist $x, y \in \mathcal{W}$ and there exists $i \in \{1, \ldots, n\}$ such that*

- $x.o^{-1}(i) = j \wedge y.o^{-1}(i) = k$
- $x.D = m_0\ x.o(1)\ m_1 \ldots m_{j-1}\ x.o(j)\ m_j \ldots m_{n-1}\ x.o(n)\ m_n$ *with* $m_l \in W^+(D)$ *and* $l = 1, \ldots, n$
- $y.D = m'_0\ y.o(1)\ m'_1 \ldots m'_{k-1}\ y.o(k)\ m'_k \ldots m'_{n-1}\ y.o(n)\ m'_n$ *with* $m'_l \in W^+(D)$ *and* $l' = 1, \ldots, n$
- *suf*$(m_{j-1}, m'_{k-1}) = \epsilon$ *or pre*$(m_j, m'_k) = \epsilon$

□

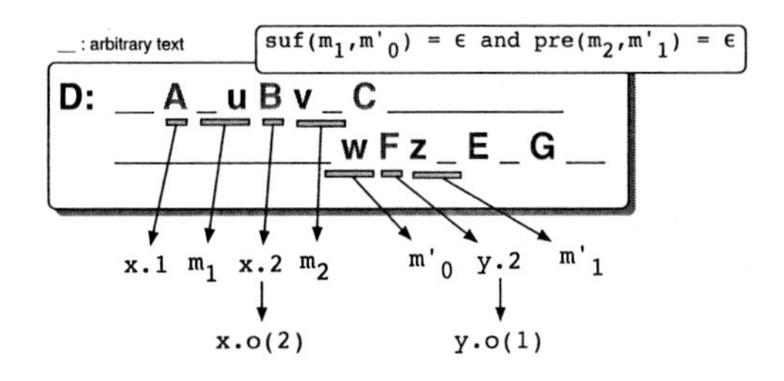

Figure 3.6.: ∨-wrapper example (non-linear)

The dual wrapper class ∧-wrapper is defined as follows:

Definition 3.1.7 (∧-Wrapper) *An n-slot wrapper $\mathcal{W}$ is an ∧-wrapper iff for all $x, y \in \mathcal{W}$ with $x \neq y$ it holds*

- $x.D = m_0\ x.o(1)\ m_1\ x.o(2)\ m_2 \ldots m_{n-1}\ x.o(n)\ m_n$ *with* $m_l \in W^+(D)$ *and* $l = 1, \ldots, n$
- $y.D = m'_0\ y.o(1)\ m'_1\ y.o(2)\ m'_2 \ldots m'_{n-1}\ y.o(n)\ m'_n$ *with* $m'_l \in W^+(D)$ *and* $l' = 1, \ldots, n$
- $suf(m_0, m'_0) \neq \epsilon$
- *for all* $i \in \{1, \ldots, n-1\}$ *it holds that* $pre(m_i, m'_i) \neq \epsilon$ *and* $suf(m_i, m'_i) \neq \epsilon$
- $pre(m_n, m'_n) \neq \epsilon$

□

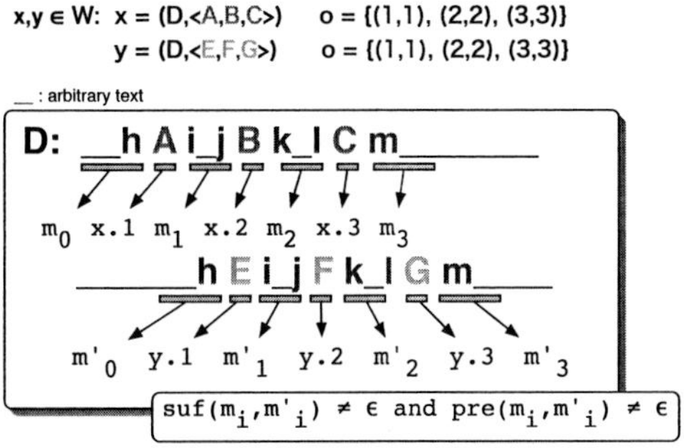

x,y ∈ W: x = (D,<A,B,C>) o = {(1,1), (2,2), (3,3)}
y = (D,<E,F,G>) o = {(1,1), (2,2), (3,3)}
__ : arbitrary text
D: __hAi_jBk_lCm_____
m0 x.1 m1 x.2 m2 x.3 m3
_____hEi_jFk_lGm____
m'0 y.1 m'1 y.2 m'2 y.3 m'3
suf(mi,m'i) ≠ ∈ and pre(mi,m'i) ≠ ∈

Figure 3.7.: ∧-wrapper example

Note, although Definition 3.1.7 does not restrict an $\wedge$-wrapper to be a linear wrapper they are closely related. Roughly speaking an $\wedge$-wrapper possesses only identical delimiters for each slot. Hence it can be modeled by a set of pairs of left and right delimiter text for each slot. In contrast, an $\vee$-wrapper needs a disjunctive pattern (or more than one pair of delimiters) for at least one of its slots. Due to this definition the delimiter text of a non-linear $\wedge$-wrapper with varying slot filler occurrences consists of identical delimiter text for different slots. Obviously this makes it hard for discover the correct mapping of slot fillers to their slot position.

3.1.6. Non Existent Delimiter Wrappers

Sometimes two or more slot fillers occur as one token within a document and therefore no delimiters exist to mark the borders between them. At first glance it seems that this is more an issue of document representation. But even for the weaker case where two slot fillers represented as two tokens are directly adjacent this problem exists. In this case one slot filler takes over the role of a delimiter, but since slot fillers in general vary, this problem is related to the $\vee$-wrapper class.

Definition 3.1.8 (x-Wrapper) *An n-slot wrapper $\mathcal{W}$ is an* x-wrapper *iff there exists $x \in \mathcal{W}$ and there exist $i, j \in \{1, \ldots, n\}$ such that*

- $x.o^{-1}(i) = k$ *and* $x.o^{-1}(j) = l$
- $x.D = u\, x.o(k)\, x.o(l)\, v$ *with* $u, v \in W(D)$

□

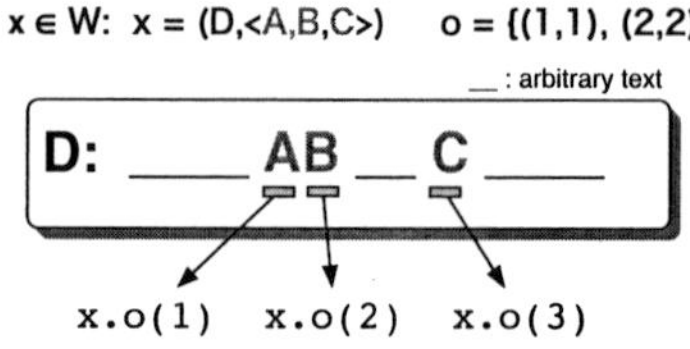

Figure 3.8.: x-wrapper example

3.1.7. Nested Slot Wrappers

Probably the most challenging class is the one of nested hierarchical documents [Kushmerick, 2000; Kushmerick, 1997]. They do not represent relevant information to be extracted in a clean tabular manner, but it is organized in a tree-like structure (Figure 3.9).

Extracting information from this class of documents requires to detect the membership of a slot's filler to its correct slot and the correct grouping of slots. In most cases nested documents vary in their number of slots and order which complicates the task (e.g. `phone`). Like in the *MV-wrapper* class, wrappers of this class also have to handle multiple slot filler values. But in contrast to the MV class multiple values are nested within a structure (e.g. person information) and they have identifiable delimiters (e.g. `phone:`). The most important point about nested hierarchical structures is that the information to be extracted is represented on different levels where each level represents some sort of attribute for which one or more values are listed on the subordinated level. Kushmerick proposes to describe example extractions of such nested wrappers also as a tree-like or nested structure as shown in Figure 3.9.

```
name: John
  address: 9 Maple Lane
    phone:  666-777
  address:  12 Main St
    phone: 123-4567
    phone: 444-555
name: Sally
name: Jane
   phone: 453-7383
```

$$\left[\mathit{John} \left[\begin{array}{l} 9\ \mathit{Maple}\ldots \left[666-777 \right] \\ 12\ \mathit{Main}\ldots \left[\begin{array}{c} 123-4567 \\ 444-555 \end{array} \right] \end{array} \right] \right]$$

Figure 3.9.: hierarchical structure and nested example extraction

In fact he uses some sort of nested lists of delimiters to represent nested hierarchical wrappers. To keep a uniform representation for wrappers and example extractions as tuples of words from a document we chose a *flat* representation for nested wrappers. This can be easily achieved by simply *unfolding* a hierarchical structure into a tabular like relational representation, e.g. : <John, 9 Maple ..., 666-777>, <John, 12 Main ..., 123-4567>, <John, 12 Main ..., 444-555>.

Definition 3.1.9 (nested-Wrapper) *An n-slot wrapper $\mathcal{W}$ is a* nested-wrapper *iff there exist $x_1, x_2 \in \mathcal{W}$ with $x_1 \neq x_2$ such that*

- $x_1.D = x_2.D$
- *there exists $s, i \in \{1, \ldots, n\}$ such that:*
  - $x_1.s = x_2.s$ *with* $x_1.o^{-1}(s) = k$
  - $x_1.s = x_2.s$ *with* $x_1.o^{-1}(s) = k$
  - $x_1.i \neq x_2.i$
  - $x_1.D = u\ x_1.o(k)\ m\ x_2.i\ v$
  - $m = z_0\ x_1.o(k+1)\ z_{k+1}\ x_1.o(k+2)\ z_{k+2} \ldots z_{n-1}\ x_1.o(n)\ z_n$

□

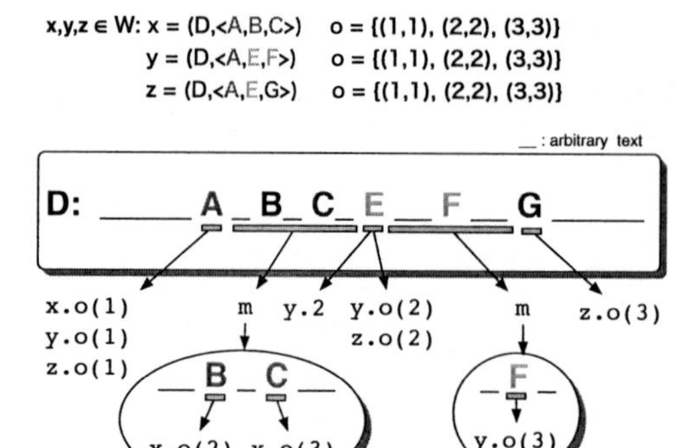

Figure 3.10.: nested wrapper example

3.1.8. Slot Inclusion Wrappers

Some extraction tasks require the use of extraction tuples with inclusion conditions among slot fillers. For instance, a document is given that contains several paragraphs where within each paragraph a telephone number appears. Assuming that the extraction task consists of providing tuples like `<paragraph text, telephone>` then a wrapper procedure must be capable of detecting such inclusions among slot fillers. Consequently this will involve a two or more level pass of the text or a more elaborated technique to represent documents, delimiters or wrappers in general.

Definition 3.1.10 (inc-Wrapper) *An n-slot wrapper $\mathcal{W}$ is an* inc-wrapper *iff there exists $x \in \mathcal{W}$ and there exist $i, j \in \{1, \ldots, n\}$ with $i \neq j$ such that $x.D = u\, x.i\, v$ with $u, v \in W(D)$ and $x.i = s\, x.j\, r$* □

x ∈ W: x = (D,<ABC,B>) o is undef.

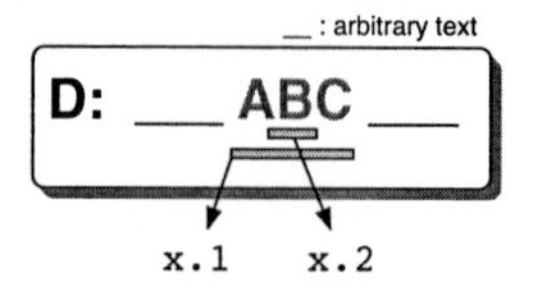

Figure 3.11.: inc-wrapper example

3.1.9. Single Slot Wrappers

Completing the set of wrapper classes presented in this section the *single slot* class contains wrappers consisting only of one slot.

Definition 3.1.11 (Single Slot Wrapper) *An n-slot wrapper $\mathcal{W}$ is* a single slot wrapper *iff* $n = 1$. *Alternatively we say* a 1-slot wrapper *is* a single slot wrapper. □

3.1.10. Notes on Wrapper Classes

Figure 3.12 displays the relationship between the various wrapper classes presented in this chapter. Some of them are briefly discussed in the following. From Definition 3.1.7 it follows that no wrapper $\mathcal{W}$ can be an $\wedge$-wrapper and $\vee$-wrapper or x-wrapper, because this would violate the $\wedge$-wrapper condition that every i-th delimiter for any $e \in \mathcal{W}$ has at least a common suffix and prefix respectively with the i-th delimiter of every other $e' \in \mathcal{W}$. For the same reason no $\wedge$-wrapper can be an ϵ^+-wrapper. To show that a wrapper belongs to the $\wedge$-wrapper class and to the *nested*-wrapper or MV-wrapper class or to the intersection of these two, we simply can construct a document and a wrapper fulfilling the discussed conditions of identical slot delimiters. This also holds for the relationship of ϵ-wrappers and $\wedge$-wrappers, for which we can construct a trivial 1-slot wrapper containing two example extractions having the same delimiters but one slot filler with the empty word. From Definition 3.1.8 it follows that a x-wrapper can not be a *single-slot* wrapper, because a x-wrapper contains extraction examples with at least two slot fillers. This also holds for the *inc*-wrapper. Hence a *single-slot* wrapper does not belong to the class of *inc*-wrapper. From the n-slot wrapper Definiton 2.1.2 and Definition 3.1.10 it follows that there is no *inc*-wrapper for which the slot filler occurrence order is defined. Because all other wrapper classes except the MV-wrapper class require a slot filler occurrence order this is the only class an *inc*-wrapper can additionally belong to.

3.2. Wrapper Properties

For several reasons it is necessary and desirable to characterize automatically constructed wrappers concerning the quality and quantity of their provided extractions. If wrappers are to be used as integrative part in a larger information system it is of great importance to be able to predict how good the extraction quality will be at least. It is of fundamental importance how trustworthy the provided results are and which post processing steps are needed to incorporate the extracted information in the overall workflow.

The previous wrapper definitions are exclusively given in a declarative manner, defining a wrapper to consist of a possibly infinite set of tupels $(D, (w_1, \ldots, w_n))$. Though the practical motivation of information extraction understands wrappers as extraction procedures this perception has not be considered so far. In practice a wrapper is a function applied to a document yielding extraction tuples. So learning a wrapper is in fact the automatic construction of some sort of extraction procedure. For now it is sufficient to think of an arbitrary procedure to be learned. More important is the observation that the wrapper definition given so far describes *ideal* wrappers in the sense that they are based on an *intended semantics*, e.g. an ideal address extraction wrapper. This is analogue to the idea of a target concept presented in Section 2.2. Thus we differentiate for the rest of this thesis between a *target wrapper* $\mathcal{W}$ and a learned or

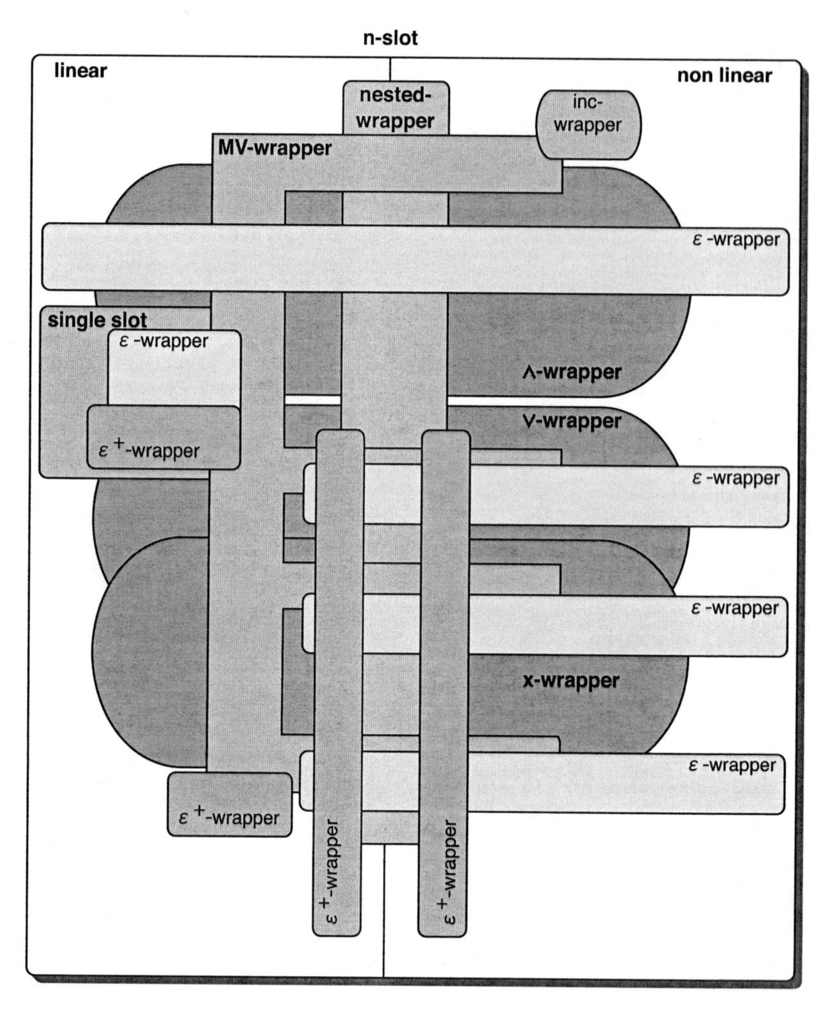

Figure 3.12.: wrapper classes

approximated wrapper W for $\mathcal{W}$. As presented in Section 2.2 the automatic construction or learning of W is equivalent to finding a hypothesis that best fits a target concept (here the target wrapper $\mathcal{W}$).

From a practical point of view this may cause a dilemma. One can imagine that only a subset of the *target wrapper* $\mathcal{W}$ is present. This leads to the problem how to determine if extractions provided by W are correct, if they are not enclosed in the known subset of $\mathcal{W}$. Obviously it makes sense to restrict the evaluation to a known subset of $\mathcal{W}$. In practice the set of positive (E^+) and negative examples (E^-) as defined in Definition 2.2.1 and 2.2.2 are used to evaluate a learned wrapper W. In general this is known as the inductive learning hypothesis, saying that any found hypothesis which approximates the target function well (having seen a sufficiently large training set) will also approximate the target function well for future examples.

3.2.1. Completeness, Soundness and Consistency

The terms *soundness* and *completeness* are commonly used in the field of *Automated Reasoning* [Bibel and Schmitt, 1998] to state if a logical calculus derives only correct logical consequences and if it is capable to enumerate all possible consequences. These properties can be formulated for wrappers as well and refer to procedural properties.

Definition 3.2.1 (Soundness) *Given a target wrapper $\mathcal{W}$, a set of positive examples E^+ for $\mathcal{W}$ and a wrapper W. W is* totally sound *wrt. the target wrapper $\mathcal{W}$ iff $E^+ \subseteq W \subseteq \mathcal{W}$. W is* sound *wrt. a set of positive examples iff $W \subseteq E^+$* □

Definition 3.2.2 (Completeness) *Given a target wrapper $\mathcal{W}$, a set of positive examples E^+ for $\mathcal{W}$ and a wrapper W. W is* totally complete *wrt. the target wrapper $\mathcal{W}$ iff $\mathcal{W} \subseteq W$. W is* complete *wrt. a set of positive examples iff $E^+ \subseteq W$* □

Wrappers are learned from a set of positive examples and depending on the learning technique also from an explicitly provided set of negative examples. Since in general these sets of example extractions are the only available information for learning and evaluation, they are commonly divided into training and testing sets. Thus for the determination of soundness and completeness properties of a wrapper the training and testing set can be used. In general the quality evaluation is done regarding the testing set, since one is interested in how good the wrapper performs on new documents. Quality measures regarding the training set are in general used to determine stopping criteria of the learning phase. In almost all cases no total completeness or soundness wrt. to the properties are examined, since most documents and extraction tasks in practice do not allow us to enumerate all possible documents nor all possible extractions regarding a wrapper.

In the context of Machine Learning (Section 2.2) we already stated the notion of a *consistent hypothesis* (see Definition 2.2.4). Saying that a learned hypothesis is consistent if it shows the same classification results on all presented examples (positive and negative) as the target function. In the context of wrapper learning we can define consistency solely on positive examples and in terms of soundness and completeness such that a wrapper W is consistent if it is sound and complete.

Definition 3.2.3 (Consistency) *Given a target wrapper $\mathcal{W}$, a set of positive examples E^+ for $\mathcal{W}$ and a wrapper W. W is* consistent *iff it is* sound *and* complete *wrt. E^+. We say W is* perfect *iff it is* totally sound *and* totally complete *wrt. $\mathcal{W}$.* □

Because complete $\mathcal{W}$ is in general not known the notion of *perfect* wrapper is in practice used if a wrapper W is able to extract all relevant data correctly without any false positive extractions from a testing set.

3.2.2. Precision and Recall

The terms *precision* and *recall* are used in the *Information Retrieval* area [Baeza-Yates and Ribiero-Neto, 1999] to define quantitative measurements for the quality of retrieval processes. The notion of *precision* provides information about the number of correct retrievals of all performed retrievals and *recall* provides information about the ratio of correct retrievals to the total number of possible correct retrievals. For estimating the quality of a wrapper regarding these two measurements we define precision and recall as stated in Definition 3.2.4 and 3.2.5.

Definition 3.2.4 (Precision) *Given a wrapper W and an exhaustive set of examples E^+ for a target wrapper $\mathcal{W}$. Let $N = \{(D,x)|(D,x) \in W$ and $(D,x) \notin E^+\}$. Let $P = W \cap E^+$. The* precision *of W wrt. E^+ is defined as $pre = \frac{|P|}{|N|+|P|}$.* □

Definition 3.2.5 (Recall) *Given a wrapper W and an exhaustive set of examples E^+ for a target wrapper $\mathcal{W}$. Let $P = W \cap E^+$. The* recall *of W wrt. E^+ is defined as $rec = \frac{|P|}{|E^+|}$.* □

In some cases it is of interest not only to know how precise and enclosing a wrapper is but also to know how many documents have been processed successfully. Given that kind of information it might be possible to figure out classes of documents which are problematic for a certain wrapper class.

Definition 3.2.6 (Coverage) *Given a wrapper W and an exhaustive set of examples E^+ for a target wrapper $\mathcal{W}$. Let $P = W \cap E^+$, N_{E^+} the number of different documents in E^+ and N_P the number of different documents in P. The* coverage *of W wrt. E^+ is defined as: $cov = \frac{N_P}{N_{E^+}}$* □

As a fourth metric to determine the quality of an information retrieval process the notion of *accuracy* can be used.

Definition 3.2.7 (Accuracy) *Let tp be the true positive retrieved information, tn the true negative information, fp the false positive and fn the false negative retrieved information.* Accuracy *is defined as $acc = \frac{tp+tn}{tp+tn+fp+fn}$.* □

Obviously, in the here discussed wrapper context where no information about *negatives* are given (i.e. wrapper provides only *true* or *false positives*) the notion of *accuracy* is identical to that of *precision*.

Considering precision and recall values independently from each other can be missleading regarding the overall quality of a wrapper. For example, if a test set E^+ is given with n examples and the learned wrapper only extracts one text tuple, but this extraction is enclosed by E^+, then the wrapper is 100 % precise.

Obviously solely considering the precision rate is not very helpful to evaluate the wrapper's quality. Vice versa a 100 % recall score gives unsufficient information about the precision of the wrapper. Consequently, these two measurements are used to define a more significant measurement regarding the overall quality of a wrapper. Therefore the harmonic mean of precision and recall is defined as the F_1 value in Definition 3.2.8. This is also known as the *F-measure* [Chinchor, 1992] which is defined as $F_\beta = \frac{(\beta^2+1)\times P\times R}{\beta^2\times P+R}$ where β is the relative importance of recall to precision. Since in general wrappers are desirable that perform equally good on precision and recall equal weights given $\beta = 1$ are chosen.

Definition 3.2.8 (F_1) *Given a wrapper W, an exhaustive set of examples E^+ and precision $prec$ and recall rec values of W regarding E^+. The F_1 value of W regarding E^+ is defined as:* $F_1 = \frac{2}{(1/prec)+(1/rec)}$ □

3.2.3. Error Rate

This short subsection is intended to clarify the relation of a wrapper W and the notions of wrong extractions. Reconsidering Definition 3.2.4 (precision of wrappers), we defined the set of *false positive* extractions to consist of those extractions (D, x) from documents D for which extractions are given in the sample data set but the extraction x is not contained. In general this is known as the *sample error*. Since in practice only a limited set of examples for the target wrapper is available the conclusions about the error of W is solely based on the available sample data.

Definition 3.2.9 (Sample Error Rate) *Given a wrapper W and set of examples E^+ for a target wrapper $\mathcal{W}$. Let $P = W \cap E^+$. The* sample error rate *of W wrt E^+ is defined as follows:* $error_s = \frac{|E^+|-|P|}{|E^+|}$. □

Actually the information we are really interested in is the *true error* which should state how good the wrapper will be for future documents. Unfortunately this can only be estimated given the probability for W how likely it performs a correct extraction from an arbitrarily chosen future document. Thus we cannot really measure the quality of a wrapper for future examples. More detailed investigations on these topics lead to the machine learning area of *evaluating hypothesis*. For an overview the interested reader is referred to [Mitchell, 1997].

3.3. Wrapper Models

This section introduces three basic wrapper models, for which ILP based learning algorithms in Part II are presented.

So far several formal concepts have been introduced concerning wrappers. The interrelation between those are clarified in the following. The wrapper concept as defined in Definition 2.1.2 is a declarative definition stating that a wrapper may be any arbitrary set of tuples consisting of a document and an arbitrary number of words drawn from this document. Based on this definition we introduced the concept of a *target wrapper* (Section 2.2 and Section 3.2) which is in fact a universal wrapper capturing all documents and all extractions regarding an intended concept. From a machine learning perspective a third kind of wrapper is defined, a *learned wrapper* (hypothesis) or *partial wrapper* (Definition 3.3.5) with respect to a *target wrapper*. By observations taken from practical extraction problems the basic wrapper definition of an n-slot wrapper was separated into several wrapper classes (Section 3.1).

Until now, no procedural or functional interpretation of wrappers have been given, saying how extractions can be obtained or computed. This is in fact what the practitioner is interested in. He wants to know how to *model* a wrapper belonging to a certain wrapper class with specific properties. Thus a *wrapper model* defines a procedural concept how information can be extracted. This does not predefine any implementation technique according to a chosen wrapper model. It defines a structural or conceptual description of a particular wrapper.

In practice the automatic modeling process uses a split set of all available examples, namely the training examples. The wrapper model is constructed regarding this training set. Consequently from a practical point of view the so far discussed wrapper W denotes the union of *training* and *testing* data. Defining a wrapper model M to include all extractions of W also implies a wrapper model to be *complete* with respect to the *training* and *testing set.* Depending on W and in particular on the exhaustive (see Definition 2.2.3) or non-exhaustive enumeration of extractions in W the model M can be determined to be *constistent* with respect to the *training* and *testing set.*

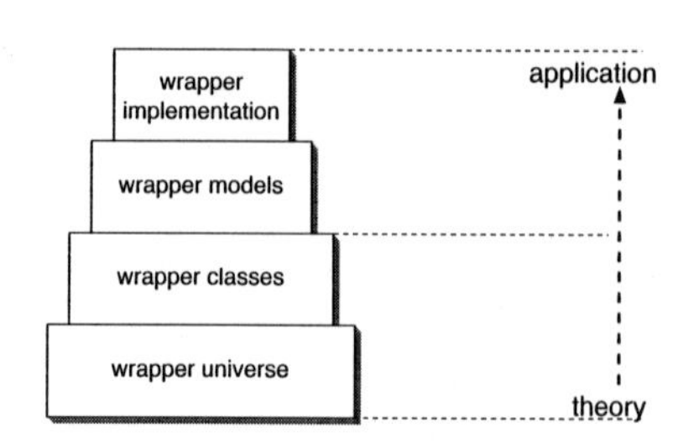

Figure 3.13.: wrapper world

Figure 3.13 shows the *wrapper world* view used in this thesis. The wrapper universe and wrapper classes belong to the theoretical foundation (introduced in Section 2.1.1 and 3.1), the wrapper models provide an abstract description for the implementation of wrappers (introduced in Section 3.3) and the top of the world is the practical realization of the wrapper models (briefly discussed in Section 3.5).

3.3.1. Attribute-Value Delimiter Wrapper

The Attribute-Value representation (*AV-rep*) introduced in Section 2.3.2 interprets every document as a sequence of tokens. Based on this document representation an *AV-Delimiter wrapper* follows three simple assumptions:

1. preceding (left) and succeeding (right) tokens regarding the text to be extracted determine relevant text parts (delimiters) for extraction

2. tokens offer a rich enough *AV*-representation for the description of differing texts

3. significant AV-pairs can be figured out for building patterns and rules to determine delimiters.

In a more formal way an *AV-Delimiter wrapper* is defined as an n-tuple of triples with each triple corresponding to one slot of the wrapper. Each triple consists of a left and right delimiter rule and a pattern for the slot filler. To simplify it, the key idea to extract information is to search for a token sequence that matches a left delimiter pattern or is described by a left delimiter rule. Once we found such a sequence an immediately following token sequence must be matched by the slot filler pattern and a directly succeeding sequence must be matched by the right delimiter pattern. Then the resulting matches for the slot filler patterns are yielding the wanted extractions.

In general, every *AV-Delimiter wrapper* is characterized by a *maximal delimiter length* restricting the number of tokens a left and right delimiter is allowed to contain. Then the question arises how large intervals are allowed to be between right and left delimiters? One basic idea is to assume that only a certain length of left and right delimiter tokens are relevant for determining the correct slot fillers. Token sequences between right and left delimiter sequences are considered to be gap tokens, which in general vary too much (as that it makes sense) to describe them by a pattern. Hence a gap might be defined by an average estimated length observable from the examples. This shows that several additional properties of *AV-Delimiter wrappers* have to be set up. This leads to the question if this has any impact on the learning task? Does this have any consequences on the wrapper's applicability? In Part III some of these issues are discussed based on empirical observations. For now it is sufficient to know that there are several properties of *AV-Delimiter wrappers* necessary to be investigated regarding learning and applicability.

Yet it might not be clear why this wrapper model is called *attribute-value delimiter.* The basic idea behind an *AV-delimiter wrapper* is to describe its delimiters by means of attribute-value pairs. In more detail, delimiters, which are in fact sequences of tokens, can be expressed by patterns or rules over attribute-value pairs. Consequently a wrapper model is defined over words of a wrapper language. This requires some language for the description of *AV-patterns.* In the case of AV-delimiter wrappers a wrapper is defined by an AV-pattern language introduced in Section 3.4.1. This is only one possible wrapper language for modeling *AV-Delimiter wrappers.* Hence for the definition of a general *AV-Delimiter wrapper* we assume an arbitrary wrapper language $\mathcal{L}$ consisting of expressions or words $p \in \mathcal{L}$ to describe patterns. In the following we refer to an *AV-Delimiter wrapper* as an *AV-wrapper.*

Definition 3.3.1 (AV-Delimiter Wrapper Model) *Given an n-slot wrapper W with $n \in \mathbb{N}$ for documents D from Σ^*. Further let $\mathcal{L}$ be an arbitrary language for the description of sequences from $Tok(D)$ with $p \in \mathcal{L}$ s.t. $Sem(D,p) \subseteq \{s|Tok(D) = XsY\}$. $Sem(D,p)$ denotes the application or semantics of p on D. X and Y are arbitrary possibly empty sequences. An* AV-Delimiter Wrapper model *$AVW(W)$ for W with* maximal delimiter length *($mdl \in \mathbb{N}$) is defined as:*

$$AVW(W) = \{R_1, \ldots, R_m\} \text{ with } m \leq |W| \text{ and for every } j = 1, \ldots, m:$$
$$R_j = ((l_1, s_1, r_1), g_1, (l_2, s_2, r_2), g_2, \ldots, g_{n-1}, (l_n, s_n, r_n))$$

and for every $e \in W$ exists a $R \in AVW(W)$ with i=1,...,n such that
$Tok(e.D) = u\ d_1^l\ f_1\ d_1^r\ m_1\ d_2^l\ f_2\ d_2^r\ m_2\ \ldots\ m_{n-1}\ d_n^l\ f_n\ d_n^r\ v$ *with*

- $f_i \in Sem(e.D, s_i) : Tok(e.o(i)) = f_i$
- $m_i \in Sem(e.D, g_i)$
- $d_i^l \in Sem(e.D, l_i), d_i^r \in Sem(e.D, r_i)$
- $|d_i^l| \leq mdl$ and $|d_i^r| \leq mdl$

□

Note, by requiring $Sem(e.D, s_i)$ to represent the slot filler that occurs as i-th filler text in D ($e.o(i)$) an *AV-wrapper* is able to represent *non-linear* wrappers. In the worst case for every possible filler occurrence order a different rule R_j has to be defined. By modifying the definition such that $\exists f_i \in Sem(e.D, s_i) : Tok(e.i) = f_i$ we obtain a wrapper model that is only capable to represent *linear* wrappers, because in this case the order of slot fillers $e.i$ given by e is assumed to represent the occurrence order of the filler text in D.

Example 3.3.1 (Example $AVW(W)$) *Assume the following part of a document D is given and its tokenized version $T(D)$:*

```
D:      ...<tr><td>16.8.</td><td>Live Music</td></tr>...
        ...<tr><td>17.8.</td><td>DJ Tom</td></tr>...

Tok(D): ...((type,html),(tag,tr)), ((type,html),(tag,td)),
         ((type,int),(val,16)), ((type,punct),(txt,'.')),
         ((type,int),(val,8)), ((type,punct),(txt,'.')),
         ((type,html_end),(tag,td)), ((type,html),(tag,td)),
         ((type,word),(txt,'Live')), ((type,word),(txt,'Music')),
         ((type,html_end),(tag,td)), ((type,html_end),(tag,tr))...
         ((type,html),(tag,tr)), ((type,html),(tag,td)),
         ((type,int),(val,17)), ((type,punct),(txt,'.')),
         ((type,int),(val,8)), ((type,punct),(txt,'.')),
         ((type,html_end),(tag,td)), ((type,html),(tag,td)),
         ((type,word),(txt,'DJ')), ((type,word),(txt,'Tom')),
         ((type,html_end),(tag,td)), ((type,html_end),(tag,tr))...
```

Let us assume that words (patterns) of the description language $\mathcal{L}$ consist of sequences of tokens and $Sem(D,p)$ with $p \in \mathcal{L}$ denotes the set of token subsequences from $Tok(D)$ such that the set of attribute value pairs of every token $p.i$ is a subset of the set of attribute value pairs of every i-th token $p'.i$ with $p' \in Sem(D,p)$. For instance,

```
        p =  <((type,int)), ((type,punct))>
Sem(D,p) =  { <((type,int),(val,16)), ((type,punct),(txt,'.'))>,
              <((type,int),(val,8)), ((type,punct),(txt,'.'))>,...}
```

Let $W = \{$`(D,<'16.8.','Live Music'>)`,`(D,<'17.8.','DJ Tom'>)`$\}$ *be a wrapper. A possible AV-delimiter wrapper model* $AVW(W)$ *for* W *is given by* $AVW(W) = \{R\}$ *with* $R = ((l_1, s_1, r_1), g1, (l_2, s_2, r_2))$ *and*

```
l1 =  <((tag,tr)), ((tag,td))>
s1 =  <((type,int)), ((type,punct)), ((type,int)), ((type,punct))>
r1 =  <((tag,td))>
g1 =  <>
l2 =  <((tag,td))>
s2 =  <((type,word)), ((type,word))>
r2 =  <((tag,td)), ((tag,tr))>
```

┘

3.3.2. Constraint Delimiter Wrapper

The basic scheme of *AV-Delimiter wrappers* have a serious shortcoming which leads to false extractions in many cases. Since none of the delimiter and slot filler rules are constrained it is possible that for instance a slot filler rule may cover a subset of sequences of a right delimiter rule. Such an inclusion relationship results in a wrapper *ignoring* the right delimiter of the slot filler. Depending on the text succeeding the right delimiter the false extractions might contain very large document parts or the overall extraction process fails if the right delimiter is not matched on the remaining document. Thus a reasonable way to increase the precision of a wrapper is to enhance the basic concept of a *AV-Delimiter wrapper* with constraints regarding its delimiter and slot filler rules.

One way to do this is to define a constraint rule stating which token sequences are not allowed to be contained in the match of a certain delimiter rule. For example, assume the most general slot filler rule describing an arbitrary sequence of tokens. Further assume that a right delimiter rule for this slot is given, stating that the slot filler succeeding text must be an *HTML* tag. Logically the slot will be instantiated with all possible token sequences succeeded by an *HTML* tag. Unfortunately the slot instantiations will also contain *HTML* tags under the assumption that there are more than one contained in the document. But these tokens were never meant to be part of a slot filler, since the basic idea was that they mark the end of slot fillers.

Having in mind that the construction of delimiter and slot rules is done automatically a reasonable first step is to add to each rule of an *AV-Delimiter* wrapper an additional *constraint rule*. Definition 3.3.2 introduces this concept of a *Constraint AV-Delimiter* wrapper model which is called in the following *CD-wrapper*.

Definition 3.3.2 (Constraint Delimiter Wrapper Model) *Given an n-slot wrapper W with $n \in \mathbb{N}$ for documents D from Σ^*. Further let $\mathcal{L}$ be an arbitrary language for the description of sequences from $Tok(D)$ with $p \in \mathcal{L}$ s.t. $Sem(D,p) \subseteq \{s|Tok(D) = XsY\}$. $Sem(D,p)$ denotes the application or semantics of p on D and X, Y are arbitrary possibly empty sequences. A* CD-wrapper model *$CDW(W)$ for W with* maximal delimiter length *($mdl \in \mathbb{N}$) is defined as:*

$$CDW(W) = \{R_1, \ldots, R_m\} \text{ with } m \leq |W| \text{ and for every } j = 1, \ldots, m:$$
$$R_j = ((l_1, c_1^l), (s_1, c_1^s), (r_1, c_1^r)), (g_1, c_1^g), \ldots, (g_{n-1}, c_{n-1}^g), ((l_n, c_n^l), (s_n, c_n^s), (r_n, c_n^r))$$

and for every $e \in W$ exists a $R \in CDW(W)$ with $i = 1, \ldots, n$ such that
$$Tok(e.D) = u\ d_1^l\ f_1\ d_1^r\ m_1\ d_2^l\ f_2\ d_2^r\ m_2\ \ldots\ m_{n-1}\ d_n^l\ f_n\ d_n^r\ v \text{ with}$$

- $f_i \in Sem(e.D, s_i) : Tok(e.o(i)) = f_i$
- $d_i^l \in Sem(e.D, l_i), d_i^r \in Sem(e.D, r_i)$
- $|d_i^l| \leq mdl$ and $|d_i^r| \leq mdl$
- $m_i \in Sem(e.D, g_i)$

and for every constraint $c_i^l, c_i^s, c_i^r, c_i^g$ *with* $i = 1, \ldots, n$ *it holds that:*

- $\neg\exists s \in Sem(e.D, c_i^s) : Tok(e.o(i)) = wsz$
- $\neg\exists s \in Sem(e.D, c_i^l) : d_i^l = wsz$
- $\neg\exists s \in Sem(e.D, c_i^r) : d_i^r = wsz$
- $\neg\exists s \in Sem(e.D, c_i^g) : m_i = wsz$

□

The *CD-wrapper model* provides a wide window of opportunities to define several submodels depending on the degree of complexity chosen for each constraint rule (i.e. constraints based on features of preceding and succeeding delimiter patterns with specific variable dependencies). But separating a delimiter into a *describing rule* and *constraining rule* offers another advantage, it simplifies the automatic construction of wrappers. A learning algorithm may construct *describing rules* during a generalization process and *constraining rules* during its specification process. In fact the *AV-wrapper model* and *CD-wrapper model* based on the principle of the *maximal delimiter length* was introduced under the name *island-wrapper* in [Thomas, 1999a]. General learnability issues on this wrapper class are discussed in [Grieser *et al.*, 2000].

Example 3.3.2 (Example $CDW(W)$**)** *Assume the following part of a document* D *is given and its tokenized version* $T(D)$*:*

D:

```
...<p><b>17.8.</b>:  Stage 1 <b>DJ Tom</b><br>...
...<br><b>18.8.</b><h1>only concert</h1>Main Hall<b>XYZ</b>...
```

$Tok(D)$:

```
...((type,html),(tag,p)), ((type,html),(tag,b)),
((type,int),(val,17)), ((type,punct),(txt,'.')),
((type,int),(val,8)), ((type,punct),(txt,'.')),
((type,html_end),(tag,b)), ((type,punct),(txt,':')),
((type,word),(txt,'Stage')), ((type,int),(val,1)),
((type,html),(tag,b)), ((type,word),(txt,'DJ')),
((type,word),(txt,'Tom')), ((type,html_end),(tag,b)),
((type,html),(tag,br))...
```

Let us reconsider the description language given in Example 3.3.1. Now assume that this language is extended by an iterator operator `*` *similar to those used in regular expressions matching arbitrary many tokens. Additionally the term* `any` *is introduced that matches any token.*

```
        p = <((tag,b)), * ((type,word)), ((tag,b))>
Sem(D,p) = { <((type,html),(tag,b)), ((type,word),(txt,'DJ')),
               ((type,word),(txt,'Tom')), ((type,html_end),(tag,b))>,
             <((type,html),(tag,b)), ((type,word),(txt,'XYZ')),
               ((type,html_end),(tag,b))>}
```

Further we denote by $c = \times$ *that a constraint* c *is not defined and* $Sem(D, c) = \emptyset$*. Note that this is different from* $c = <>$ *(denoting the empty sequence) for which usually* $Sem(D, c) = \{<>\}$*.*

Let W = {`(D,<'17.8.','DJ Tom'>),(D,<'18.8.','XYZ'>)`} *be a wrapper. A possible* CD-delimiter wrapper model $CDW(W)$ *for* W *is given by* $CDW(W) = \{R\}$ *with* $R =$

$((l_1, c_1^l), (s_1, c_1^s), (r_1, c_1^r), (g_1, c_1^g), (l_2, c_2^l), (s_2, c_2^s), (r_2, c_2^r)$ *and*

$$
\begin{aligned}
l_1 &= \texttt{<((tag,b))>} \quad c_1^l = \times \\
s_1 &= \texttt{<((type,int)), ((type,punct)), ((type,int)), ((type,punct))>} \\
c_1^s &= \times \\
r_1 &= \texttt{<((tag,b))>} \quad c_1^r = \times \\
g_1 &= \texttt{<*any>} \quad c_1^g = \texttt{<((tag,b))>} \\
l_2 &= \texttt{<((tag,b))>} \quad c_2^l = \times \\
s_2 &= \texttt{<*(type,word))>} \quad c_2^s = \texttt{<>} \\
r_2 &= \texttt{<((tag,b))>} \quad c_2^r = \times
\end{aligned}
$$

The constraint c_1^g allows the gap pattern g_1 to match any token upto the first occurrence of a `b` *tag token. The constraint c_2^s makes sure that only not empty slot fillers are matched.* ⌟

3.3.3. Relational TDOM Wrapper

From Section 2.3.3 and Definition 2.3.6 we already know that text occurring in a document can be determined or referred to by a minimal span within a *TDOM-tree* regarding the document. In general a minimal span is determined by a node identifier and left and right child node boundaries for a given text occurring in the document. So far it seemed to be a nice property that we can always find a minimal span for a given text. Unfortunately for some text selections an according minimal span includes more than only the selected text. For instance, if the selected text overlaps several annotations in an *HTML* document as shown in Figure 3.14. The minimal span for such a text includes text parts not intended to be extracted. Consequently it is not sufficient to use only minimal spans for detecting relevant text parts within a *TDOM-tree* for extraction. Much more reasonable is it to combine the basic ideas of *AV-Delimiter wrappers* to investigate the left and right surrounding text (nodes) with the concept to examine certain properties of a minimal span in relation to other nodes occurring in the document tree. Such examinations can be formalized as a set of descriptions for each example text. In return slot descriptions can be used to determine minimal spans plus the correct left and right text boundaries within a *TDOM-tree*. So, the essential idea of *Relational TDOM wrappers* is to use a set of rules to determine minimal spans, which include relevant text to be extracted. These rules that describe relational properties between nodes in a *TDOM* and the relevant minimal span are constructed from observable properties of the sample text fragments. This approach combines the observable information of surrounding text parts similar to *AV-Delimiter wrappers* with structural information observable from the document's structure and layout.

Bearing in mind the overall goal to learn wrappers and that the learning result shall be an automatically constructed wrapper model it becomes clear that these descriptions or sets of rules provide much more flexibility and possibilities then restricting this wrapper model to consist solely of minimal spans. The three essential ideas of a *RTD-wrapper* are summarized as follows:

1. minimal spans, which include relevant text to be extracted (slot fillers), are determined by relational descriptions (rules) between nodes, paths, and spans.

2. though based on the *TDOM* representation, the basic concept of an *RTD-wrapper* is not strictly bound to this representation

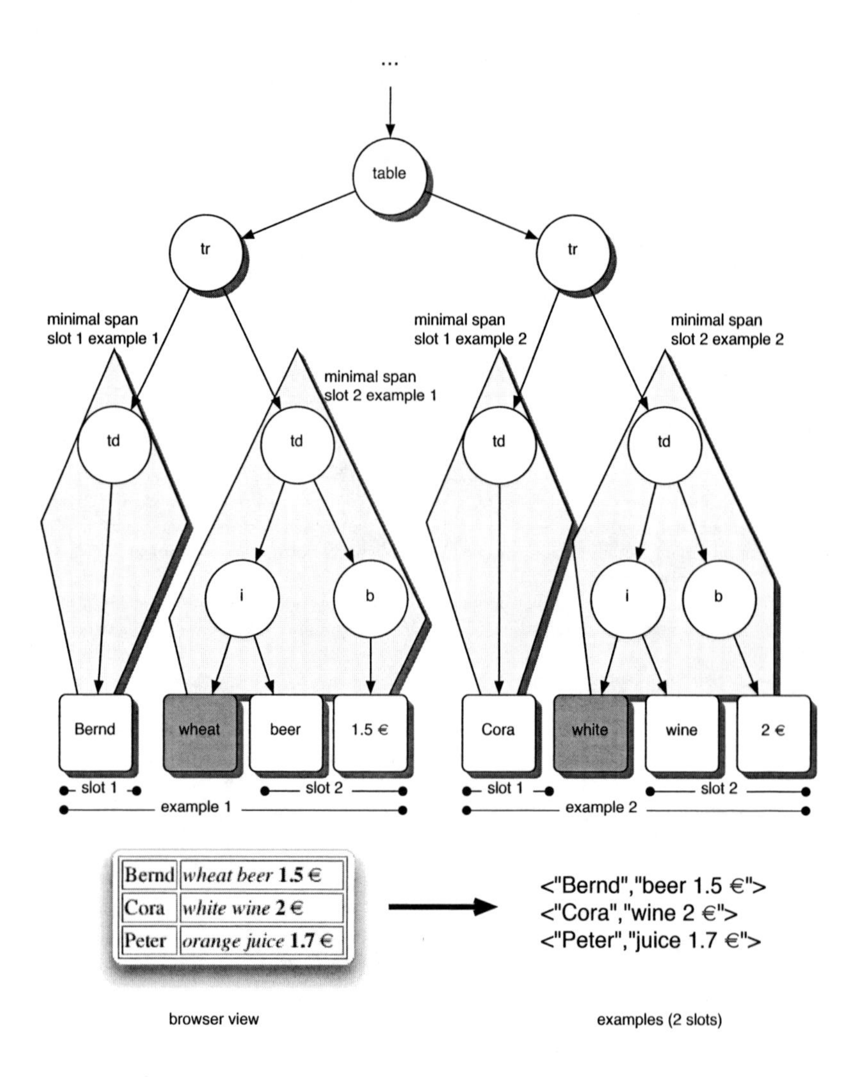

Figure 3.14.: minimal span covering too much text

3. the language for relational descriptions of node properties, subtree relationships, paths or text properties can be freely chosen and extended.

Definition 3.3.3 formally defines the idea of a relational description in terms of the logic programming paradigm, which is then used for the definition of a *Relational TDOM Wrapper model* in Definition 3.3.4.

Definition 3.3.3 (TDOM Predicates) *Let P_H be a normal program with some predicate definitions H such that each $p \in H$ describes relationships between text, nodes or spans of a TDOM representation T for an arbitrary document D. We call H a set of TDOM predicates if for some $p \in H$ and an arbitrary document D with TDOM-representation T*

- *at least one correct answer θ for $P_H \cup T \vdash p\theta$ with query p not ground can be computed*
$$(3.1)$$
- *p is based on unit clauses from T. This means that at least one $t \in T$ is used to compute $P_H \cup T \vdash p\theta$.*
$$(3.2)$$

□

Definition 3.3.3 simply states that in general an arbitrary set of predicates describing properties of a *TDOM-representation* can be chosen. The second proposition demands that these predicates are sensibly chosen, or more precisely that at least some properties can be observed regarding any given document. The relevant point here is as follows: if P_H would not fullfil constraint 3.1 then P_H would simply be too weak to describe any example extraction regarding an arbitrary document and consequently P_H would be useless for building relational *TDOM* wrappers. On the other hand, if P_H would allow to derive answers for $p\theta$ independent of T (not satisfying constraint 3.2) then H would clearly be too general or roughly speaking also not useful for the description of relational properties of text examples.

Definition 3.3.4 (Relational TDOM Wrapper Model) *Given an n-slot wrapper W with $n \in \mathbb{N}$ for documents D from Σ^* and a normal program P_H with TDOM predicates H. Let R be a predicate definition of the form $extract(D_{ID}, \vec{W}, \vec{S}) \leftarrow r_1, \ldots, r_m$ with $\vec{S} = [S_1, \ldots, S_n]$, $\vec{W} = [W_1, \ldots, W_n]$ and for every $i \in \{1, \ldots, m\}$ and $j \in \{1, \ldots, n\}$: $r_i \in H$ and S_j is the corresponding minimal span for the text W_j in the TDOM-representation of a document referred to by D_{ID}.*

We call R a set of extraction rules *and by $[A_1, \ldots, A_n]$ we denote a list of terms A_1 to A_n which corresponds to the common* Prolog *notation for lists. A normal program P_H with TDOM predicates H and a set of extraction rules R is an* relational TDOM wrapper model *for wrapper W written $RTD(W) = (P_H, R)$ iff for every $(D, (w_1, \ldots, w_n)) \in W$ it holds that*

- *D_{ID} is a document identifier for D*
- *$TD(D) = T$*
- *there exists a correct answer θ for $T \cup P_H \cup R \vdash extract(D_{ID}, \vec{W}, \vec{S})\theta$ such that $\vec{W}\theta = [W_1, \ldots, W_n]\theta = (W_1\theta, \ldots, W_n\theta) = (w_1, \ldots, w_n)$.*

□

In Section 3.4.4 the construction of an example *RTD-wrapper* is illustrated in detail.

3.3.4. Short Discussion on Wrapper Models

Following the definitions of wrapper models from previous sections a wrapper model is defined as a superset regarding the given wrapper W, because it is only stated that its rules or patterns extract a set W' such that $W \subseteq W'$. Hence the wrapper model might provide further extractions such that $|W'| > |W|$. This fact becomes more obvious considering that token patterns contain iteration operators (`?,*,+,maxlength`) or variables and extraction rules that do not contain ground body literals and node identifier patterns (see Section 2.3.3). Wrapper models based on such patterns and rules will probably be able to yield extractions from documents not present in W. Or in other words probably the wrapper model M constructed for one wrapper W hopefully will provide feasible extraction results if applied to future documents.

This captures the overall inductive learning idea, that a wrapper W will hopefully show a good performance on future sample data. Having a closer look on the wrapper model definitions it is important to note, that if a model covers more extractions for a specific document than defined by W, it is not clarified so far if these extractions have to be counted as false positives.

Assume the following: a wrapper W is exhaustively determined as described in Definition 2.2.3 and a wrapper model M covers partially different extractions from a document D than W. Then it immediately follows that M is not *sound* or in a quantitative saying is not 100% precise. Hence exhaustive testing sets are a necessity for determining the *consistency* (Definition 3.2.3) of a wrapper model.

Sometimes it might be desirable to construct a wrapper model for a wrapper W that exclusively covers extractions defined by W. We call this wrapper model a *minimal wrapper model* if it does not provide extractions from documents other than those contained in W. Obviously learning a wrapper model with this intention can be easily achieved by simply storing all extraction examples. But as soon as W is split into a training and testing set for construction (learning) and evaluation with the aim to learn a *minimal wrapper* this simple method will not suffice anymore. Thus learning a *minimal wrapper model* can be characterized as learning a model from a subset of W with the aim to cover only extractions determined by W. In contrast a non-minimal consistent wrapper model has the ability to provide extractions from all documents determined by W plus previously unseen documents, which exactly refers to the notion of a target wrapper $\mathcal{W}$ introduced in Section 3.2 if the provided extractions are intended ones. In general there is a simple method to test if a learned wrapper model M is minimal or not. Therefore we choose a document D from W and modify it. If M still provides extractions from the modified version of D, than it follows that M is not minimal. The crucial point of this test is to decide to which degree and how a document has to be changed.

Another interesting point about wrapper models is the observation mainly made from learning wrappers. Assuming an arbitrary wrapper model M is given, that is not consistent with a given wrapper W. Furthermore assume M covers some examples in W and either is not complete meaning M covers only a subset of W or M is not sound meaning M yields extractions not in W. Such a wrapper model M is called a *partial model* (Definition 3.3.5) of W since it is either not sound or not complete or both. In this case M can also be considered to be a hypothesis for W.

Definition 3.3.5 (Partial Wrapper Model) *Given a wrapper W and a wrapper model M. Let $Cov(D, M)$ be the set of extractions (D, x) obtained if M is applied on a document D and $\mathcal{X} = \bigcup_{(D,x) \in W} Cov(D, M)$. A wrapper model M is a* partial wrapper model *iff $W \supset \mathcal{X}$ or $\mathcal{X} \setminus W \neq \emptyset$.* □

I want to conclude that learning a wrapper model regarding a given wrapper W will in almost all cases result in learning a *partial wrapper model* M for W, unless a learning algorithm is discovered that can construct sound and complete wrapper models for arbitrary wrappers W. Unfortunately this thesis won't bring home the bacon.

3.4. Wrapper Languages

The introduced wrapper models can be described by arbitrary languages where the choice for the *right* language is solely restricted by its expressiveness to capture all requested features of the structure of the selected wrapper model. This section discusses mainly two languages for the description of wrapper models: an *attribute-value pattern* language [Thomas, 2000], which is strongly related to the token concept and the *attribute-value representation* of documents as introduced in Section 2.3.2. Secondly, a first-order predicate logic based language [Thomas, 2003], consisting of a set of literals for the description of relational features observable from *TDOM-representations* of documents (Section 2.3.3).

3.4.1. Attribute-Value Patterns

Depending on the choice of attributes and values a ground token can either represent exactly one word or a set of words matching certain features described by its attributes and values. In general it is reasonable to choose attributes such that a token uniquely determines the text in a document it was constructed from. But for matching or detecting texts that are similar according to certain attributes a token can also deal as pattern for finding such texts in a document. Assuming a tokenization function is mapping texts from a document onto 4-tuples consisting of the attributes *text type, original text, start position, end position.* The position attributes are a simple way to uniquely determine the original text represented by such a token. As soon as these two attributes are omitted a token consisting of the remaining attributes refers to all text fragments identical to the examples occurring within the document.

Instead of omitting certain attributes it is also reasonable to keep the attributes and to use variables. For example, a token like `token([type=punctuation,text=X,spos=S,endpos=E])` forms a pattern to describe all possible punctuation marks in a document. Using variables for values instead of dropping attributes has another advantage, we can demand that a matching text part offers a certain attribute. In the case of omitting attributes we loose this property. This of course makes only sense if tokens vary in their number and type of attributes, which leads to the definition of an *atomic token pattern.*

Definition 3.4.1 (Atomic Token Pattern) *Given a set of attribute names A, a corresponding set of attribute values V and a set of variables $\mathcal{X}$. A tuple $((a_1, v_1), \ldots, (a_n, v_n))$ with $a_i \in A$ and $v_i \in (V \cup \mathcal{X})$ is called an* atomic token-pattern. □

Reconsidering the attribute-value representation of documents $AV(D)$ a token pattern can be used to identify certain texts in a document by simple unification techniques in the following way. Since $AV(D)$ is a set of terms of the form $token(D_{ID}, i, t)$ where t is a token, that can be represented as a *Prolog* term, it is straightforward to compute all matches for a simple token pattern p from an $AV(D)$ representation of D. All that has to be done is to test for every u contained in $AV(D)$ if its token t is *token-unifiable* with the token pattern p[1]. If such

[1]or by computing the answers to the query $AV(D) \vdash token(D, I, T)\theta$

elements in $AV(D)$ exist with $p\theta = t$ where θ is the most general unifier, this is a convenient way to obtain information requested by the pattern. The information is provided by the calculated substitutions, the mgu θ. The following Definition 3.4.2 introduces the notion of *token unification* based on the definition of an *atomic token pattern* (see Definition 3.4.1).

Definition 3.4.2 (Token Unification) *The reduced term notation of a token t wrt. a attribute set $A = \{a_1, \dots, a_n\}$ is defined as: $token(a_1(v_1), \dots, a_n(v_n))$ with (a_i, v_i) occurs in t and $a_i \in A$ with $i = 1, \dots, n$. Given an* atomic token pattern *t_1 and an arbitrary token t_2. Let A be the set of attributes of t_1. Further let t'_2 be the reduced term notation of t_2 wrt. A. Then t_1 and t_2 are* token unifiable *iff t_1 notated in reduced term notation as $t_1 = token(a_1(v_1), \dots, a_n(v_n))$ and $t'_2 = token(a_1(v'_1), \dots, a_n(v'_n))$ are unifiable with mgu θ. We write $t_1\theta \overset{u}{=} t_2\theta$ if t_1 and t_2 are* token-unifiable *with mgu θ.*

If t_2 is a sequence of tokens, we say $t_1\theta \overset{u}{=} t_2$ iff there exsists a mgu θ such that for all tokens t_i from t_2 it holds that $t_1\theta \overset{u}{=} t_i\theta$. □

A few more things have to be noted about *token unification.* At first, unifying tokens is more related to the idea of matching than of unification. Because $t_1 \overset{u}{=} t_2$ is a directed operation in the sense that t_1 is *applied* to t_2. So, this is not a reflexive operator as unification is, where it does not matter if t_1 is unified with t_2 or vice versa. Thus the intention of *token unification* is to interpret the term t_1 to be a pattern and t_2 to be an arbitrary token, which, if reduced according to t_1's attribute set, can be unified with t_1. In contrary to *feature unification* [Carpenter, 1991; Aït-Kaci *et al.*, 1997; Carpenter, 1992] the defined token unification is more restrictive, since the unification fails if an attribute occurring in the *token pattern* (e.g. t_1) is not present in the other token.

So far an *atomic token pattern* does not differ too much from an ordinary token, the only difference is that it possibly contains variables and might have less attributes than tokens obtained by a given tokenization function regarding the same text. Simple token patterns like those discussed until now are used to build more complex token patterns consisting of iteration operators, negation, conjunctive and disjunctive junctors. For instance, to match a sequence of upper case written words of arbitrary length in a document a token pattern like `* token([type=word,upper=true])` similar to the syntax of regular expressions can be defined. Often it is necessary to collect the matched tokens thus the presented token pattern language is extended with so called *extraction variables.* Extraction variables are instantiated with token sequences matched by token patterns. Assuming we want to extract all words from an *HTML* document nested in `<b>` tags. A reasonable token pattern looks like: `token([type=html, tag=b]), X=*token([type = word]), token([type=html_end, tag=b])`, where the variable `X` is instantiated with a token sequence covered by the token pattern `*token([type=word])`. Computing the set of all possible matches and instantiations for `X` then provides the intended extractions from a document. In fact this example token pattern is an *AV-delimiter model* for a wrapper W with $AVW(W) = (l_1, r_1, s_1)$ and l_1=`token([type=html, tag=b])`, r_1=`token([type=html_end, tag=b])` and s_1=`*token( [type=word])`.

Keeping in mind that the token pattern language forms the basis for the delimiter and slot filler patterns to be learned, it is reasonable to define it regarding to the intended learning techniques presented in the subsequent chapters of this thesis. In Definition 3.4.3 and 3.4.4 the formal definition for the token pattern language is given. The language of token patterns introduced in this thesis is a subset and slightly modified version of the language first published in [Thomas, 2000; Thomas, 1999a; Thomas, 1999b]. The general concept is strongly based on regular expressions but also incorporates concepts taken from the logic programming area like the use of non disjunctive variables and variable binding constraints to restrict the matching in combination with unification techniques. A suitable ILP representation of a subset of this language for learning *AV* and *CD-wrappers* is presented in Part II, Section 5.4.

Definition 3.4.3 (Token Pattern Language: Syntax) *Given an arbitrary set of attribute names A, a corresponding set of attribute values V and a set of variables $\mathcal{X}$. The token pattern language $\mathcal{L}_T$ is defined as follows:*

- *if p is an atomic token pattern wrt. A, V and $\mathcal{X}$ then $p \in \mathcal{L}_T$*
- *if $p \in \mathcal{L}_T$ then $(\Box p) \in \mathcal{L}_T$ with $\Box \in \{?, *, +\}$*
- *if $p \in \mathcal{L}_T$ then $times(n, p) \in \mathcal{L}_T$ and $upto(n, p) \in \mathcal{L}_T$ with $n \in \mathbb{N}$*
- *if $p_1, p_2 \in \mathcal{L}_T$ then $(p_1, p_2) \in \mathcal{L}_T$ and $(p_1; p_2) \in \mathcal{L}_T$*
- *if $p \in \mathcal{L}_T$ then $(X = p) \in \mathcal{L}_T$ with $X \in \mathcal{X}$*
- *if $p \in \mathcal{L}_T$ then $not(p) \in \mathcal{L}_T$*
- *if $p \in \mathcal{L}_T$ then $maxlength(n, p) \in \mathcal{L}_T$*

$p \in \mathcal{L}_T$ is called token-pattern. *Atomic token patterns are alternatively denoted as* Prolog *term: $token([a_1 = v_1, \ldots, a_n = v_n])$* □

Definition 3.4.4 (Token Pattern Language: Semantics) *Given a document D, the corresponding tokenized document $Tok(D)$ and a token pattern p. The semantics of p wrt. $Tok(D)$ and substitution θ written $Sem(D, p)\theta$ is defined as follows.*

- *if p is an atomic token pattern then $Sem(D, p)\theta = \{< t > | t = Tok(D).i$ and $p\theta \stackrel{u}{=} t$ and $i \in \mathbb{N}\}$*
- *if p is a token pattern of the form (p_1, p_2) with*
  - *p_1 and p_2 are atomic patterns then $Sem(D, p)\theta = \{< t_1, t_2 > | Tok(D).i.(i+1) = t_1 \circ t_2$ and $p_1\theta \stackrel{u}{=} t_1$ and $p_2\theta \stackrel{u}{=} t_2$ and $i \in \mathbb{N}\}$*
  - *p_1 is an atomic pattern then $Sem(D, p)\theta = \{< t_1 > \circ S | t_1 = Tok(D).i$ and $p_1\theta \stackrel{u}{=} t_1$ and $S \in Sem(D, p_2)\theta$ and $Tok(D).i.j = t_1 \circ S$ and $i, j \in \mathbb{N}$ and $i \leq j\}$*
  - *p_2 is an atomic pattern then $Sem(D, p)\theta = \{S \circ < t_2 > | t_2 = Tok(D).j$ and $p_2\theta \stackrel{u}{=} t_2$ and $S \in Sem(D, p_1)\theta$ and $Tok(D).i.j = S \circ t_2$ and $i, j \in \mathbb{N}$ and $i \leq j\}$*
  - *else $Sem(D, p)\theta = \{S_1 \circ S_2 | S_1 \in Sem(D, p_1)\theta$ and $S_2 \in Sem(D, p_2)\theta$ and $Tok(D).i.j = S_1 \circ S_2$ and $i, j \in \mathbb{N}$ and $i \leq j\}$*
- *if p is a token pattern of the form $(p_1; p_2)$ then $Sem(D, p)\theta = Sem(D, p_1)\theta$ or $Sem(D, p)\theta = Sem(D, p_2)\theta$*
- *if p is a token pattern of the form $?p'$ then $Sem(D, p)\theta = Sem(D, p')\theta \cup \{<>\}$*

- *if* p *is a token pattern of the form* $+p'$ *then* $Sem(D,p)\theta = \{S | S = (S_1 S_2 \ldots S_n)\theta$ *with* $S_i \in Sem(D,p')\theta$ *and* $Tok(D).j.k = S$ *and* $j,k,n \in \mathbb{N}$ *and* $j \leq k$ *and* $i = \{1,\ldots,n\}\}$
- *if* p *is a token pattern of the form* $*p'$ *then* $Sem(D,p)\theta = Sem(D,+p')\theta \cup \{<>\}$
- *if* p *is a token pattern of the form* $times(n,p')$ *then* $Sem(D,p)\theta = \{S | S = (S_1 S_2 \ldots S_n)\theta$ *with* $S_i \in Sem(D,p')\theta$ *and* $Tok(D).j.k = S$ *and* $j,k \in \mathbb{N}$ *and* $j \leq k$ *and* $i = \{1,\ldots,n\}\}$
- *if* p *is a token pattern of the form* $upto(n,p')$ *then* $Sem(D,p)\theta = \bigcup_{i=1}^{n} Sem(D, times(i,p'))\theta_i$
- *if* p *is a token pattern of the form* $not(p')$ *then* $Sem(D,p)\theta = \{S | S = Tok(D).i.j$ *and* $S \notin Sem(D,p')\theta'$ *and* $i,j \in \mathbb{N}$ *and* $i \leq j\}$ *and* $\theta = \{\}$
- *if* p *is a token pattern of the form* $maxlength(n,p')$ *then* $Sem(D,p)\theta = \{S | S = Sem(D,p')\theta$ *and* $|S| \leq n\}$

□

According to Definition 3.4.4 applying a token pattern p to a document D results in a set of token sequences. Each of these token sequences (the matching set) are subsequences of $Tok(D)$. Strictly following this semantics definition of token patterns, there is no order given among the matched sequences. For practical reasons it is often useful to request that matches are enumerated according to their length (e.g. the matching operators work in a non greedy way). For instance, if the $*$ operator is greedy the first match of pattern $*p$ consists of the longest possible token sequence unifiable with p. But if it is non greedy the first match is the empty sequence. Without giving a formal refinement of Definition 3.4.4 we assume all operators to be non-greedy. Differentiating between greedy and non greedy operators have no influence on the computed matching set. But regarding performance issues for the later matching application it is reasonable to use a fixed enumeration order.

3.4.2. Relational Descriptions

So far a *RTD-wrapper* has been introduced to consist of a logic program plus a set of extraction rules based on predicates describing certain text parts in relation to other texts, nodes or spans regarding a given *TDOM-representation.* In this section we introduce the underlying set of predicates used for describing certain observable relationships in more detail.

For instance, assume it can be observed that a root node of a minimal span (see Definition 2.3.6) of a first argument of an example tuple has two left brother nodes of the same type and one right brother node which is an html tag `<b>` as depicted in Figure 3.15. Assuming further that many of the provided learning examples have similar structural properties then it seems reasonable to define a set of predicates describing certain relationships among entities of a *TDOM-tree* covering these observations.

Obviously there are many possible relationships one can start to investigate, depending on the used markup language; different emphasis set on the relation among example arguments; the relation between an argument and other nodes; or the ancestor and descendant nodes of a minimal span. These are only a few possible relationships that can be chosen for describing text examples by means of relations among entities of an *TDOM-tree* (representation). In general, we distinguish four levels for the description of examples in the context of *TDOM-trees*: delimiter, textual or content, structural and relational level. With $s.n$, $s.l$ and $s.r$ we

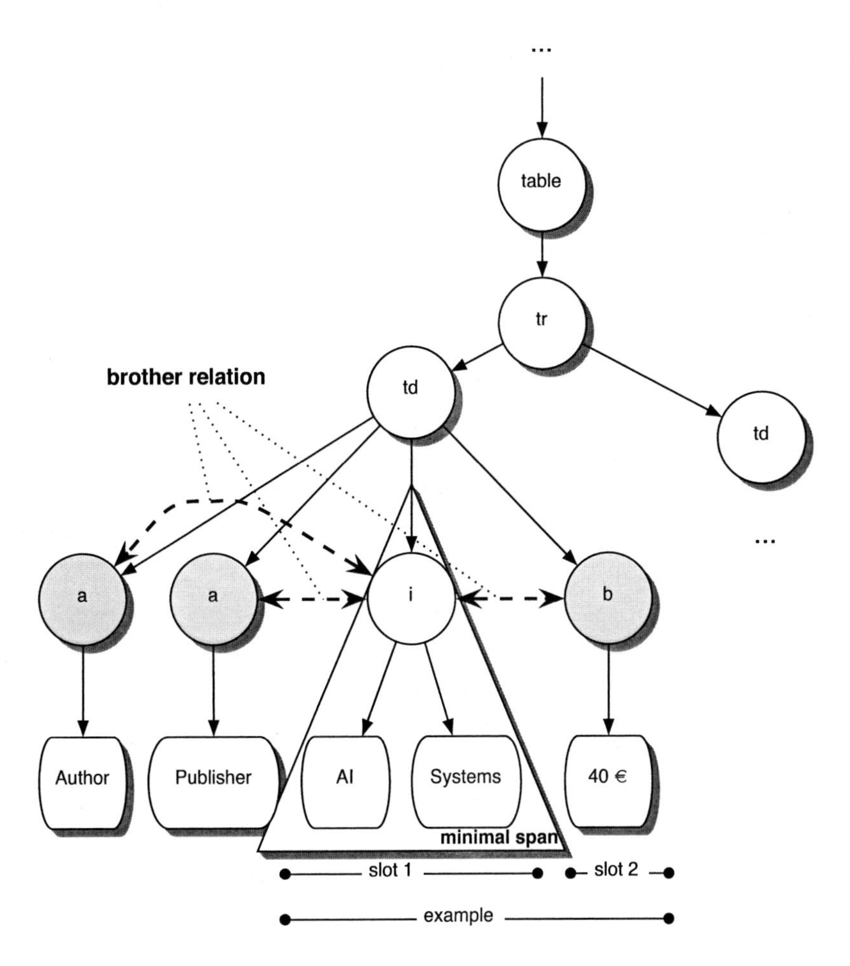

Figure 3.15.: brother node relation

denote the components of a span $s = (n, l, r)$ with w_i we refer to the i-th argument of a text example $w = (w_1, \ldots, w_n)$ $(D, w) \in W$ and D_{ID} the regarding document identifier of D (see Definition 2.3.4). When talking about a node, in general its uniquely determined *node identifier* is meant.

The Delimiter Level

The *delimiter level* is based on the idea of *AV-delimiter wrappers* where the preceding and succeeding nodes of example texts are described or determined by patterns for later detection of relevant text parts. But also variants more adapted to the tree structure are considered, for instance the ancestor and descendant nodes are taken into account. The difference of preceding and ancestor nodes is the following: nodes having been visited in an depth first traversal of the tree before reaching the corresponding node of an argument example are called preceding whereas nodes visited on the direct path from the root node to the argument example node are called ancestor nodes. Thus predecessor and successor nodes are more related to the visualized text version of an *HTML* page. For example, a word occurring before the example text in the visualized *HTML* document is the direct predecessor node. Whereas the ancestor node of the example text can be quite far away in terms of text distance from the example text. Figure 3.16 illustrates the difference between predecessor and ancestor nodes.

So, there are in fact two different levels of delimiter related predicates, those which are based on the idea of surrounding text elements observable from the document surface (i.e. in the case of *HTML* the rendered web page). The second level of delimiter predicates are those predicates based on the structural or hierarchical elements of a document regarding the given annotation language. In semi-structured languages like *HTML* or *LaTeX* you can find many nesting environments providing a lot of information about the relation of text parts to each other. Those structures are difficult to detect if you strictly follow the idea of surrounding texts with a given length solely understanding the document as a stream of words. Table 3.1 summarizes the set of delimiter level predicates used in this thesis.

$\mathbf{start\_end\_nodes(D_{ID}, s, n_l, n_r)}$ holds if n_l is the start node and n_r the end node of some text w in the span s of $TD(D)$ referred to by D_{ID}.
$\mathbf{xpredecessor(D_{ID}, n, n_i, tl)}$ holds if the token list tl contains the tokens associated with the first n nodes we meet going backwards from node n_i in a depth first tree traversal of the tree regarding D_{ID}.
$\mathbf{xsuccessor(D_{ID}, n, n_i, tl)}$ holds if the token list tl contains all n successor tokens met by a depth first traversal starting at node n_i.
$\mathbf{xancestors(D_{ID}, n, n_i, tl)}$ holds if the token list tl contains the first n tokens associated with the nodes met following the path from n_i to the root node of D_{ID}.
$\mathbf{xdescendants(D_{ID}, n, n_i, tl)}$ holds if the node n_i has at least n child nodes and tl contains the first n tokens associated with the nodes met by a depth first traversal starting at node n_i.
for all predicates n is called the *context length*

Table 3.1.: delimiter level description predicates

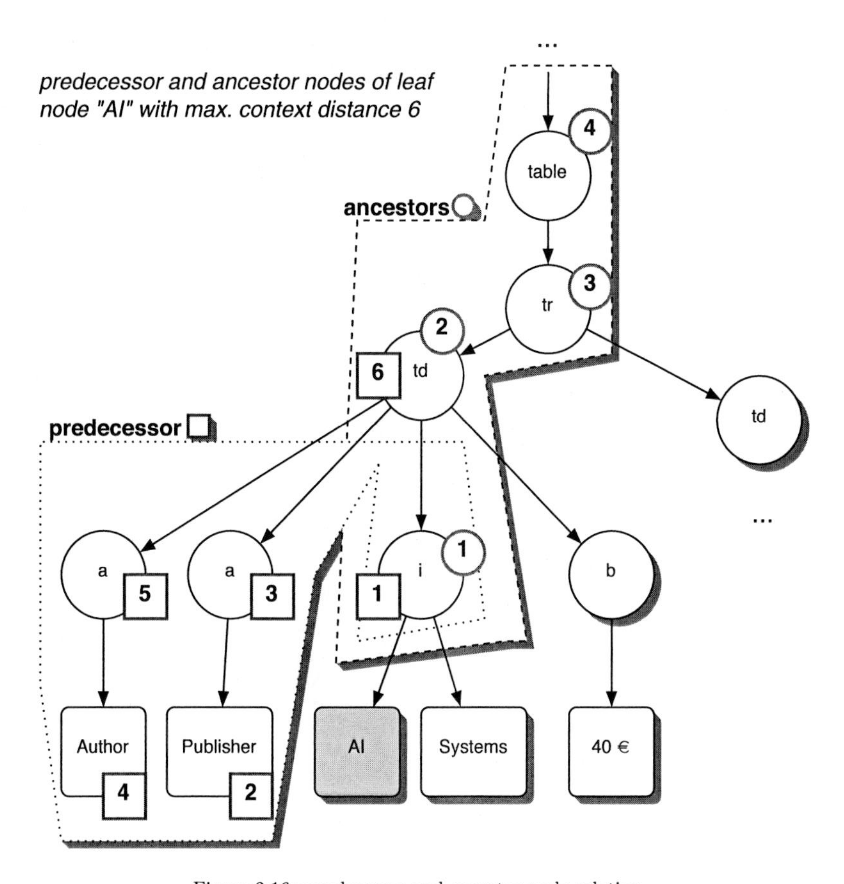

Figure 3.16.: predecessor and ancestor node relation

Textual or Content Level

On the *textual or content level* relations between the example text, its tokens associated with the leaf nodes and its span are described. In addition to these structural features more sophisticated observations are formalized based on word semantic background knowledge. In this thesis we used a quite simple predicate relating the text of a slot to the minimal span it is located in, namely: $span\_text\_and\_tokens(D_{ID}, s, w, tl)$ which holds if tl is a list of tokens associated with all leaf nodes of span s for text w. The list tl differs from the complete minimal span in that it only consists of nodes (leafs) appearing on the text surface of the document.

In fact the content level offers many more possibilities to incorporate other predicates into extractions rules. Especially the use of semantic content descriptions seems to be very reasonable. For instance, the use of word sense analysis, ontologies or a simple thesaurus can be used to classify, determine or describe a word or example text in a more precise way.

A predicate like $subclass(D_{ID}, s, w, C)$ could be used to determine if w is contained in a subclass of C based on a given ontology. For example, assuming that for several example texts like `robbery, rape, murder` this predicate holds with classes C_1, C_2, C_3 and from the ontology it is derivable that all C_i of the examples are subclasses of a class C which is called `punishable act`, this might be a strong semantic based evidence that the user is interested in extracting information about crimes. Thinking about learning wrappers, which means in the context of *RTD-wrappers* to learn extraction rules, this generalization process based on computing subsumption classes from an ontology related to the given example texts, obviously can assist the learner in constructing better extraction rules. In fact this is not a new approach. A similar idea has been used in the IE system called RAPIER [Califf, 1998].

The Structural Level

Shifting the focus of investigations solely on nodes (span nodes), neighbor relationships of them and the reachability in relation to the root node, leads to descriptions on a *structural level*. In contrast to *AV-wrappers* purely based on token patterns *structural level predicates* yield the possibility to estimate spans and therefore text examples in relation to elements of the documents structure (e.g. layout). For instance, if a set of examples have the same surrounding texts and thus can be described by the same token delimiter patterns it may not necessarily be the case that these patterns guarantee to match always the intended texts. There may be many documents where only those texts identifiable by such delimiter patterns shall be extracted which are nested in a certain environment. If those environment tags (annotations) are not within the surrounding text parts of the examples an *AV-wrapper* will not be able to identify the constraint to extract only texts occurring in this specific environments. Instead an *RTD-wrapper* using structural level predicates is able to determine such nestings and thus can constrain the set of possible extractions based on structural information observable from the text examples. An overview of the used predicates is given in Table 3.2.

The Relational Span Level

On the *relational span level* relations between spans are described. While predicates on the structural level state relations among the root nodes of a span and other nodes in a tree, this description level investigates properties of complete spans and their position to each other. This has a strong importance if the text to be extracted or more precisely the slot fillers are

$\mathbf{xpath(D_{ID}, n, tl)}$ holds if n is a node in document D_{ID} and tl is the list of tokens associated with each node following the path from the root node to the node n.
$\mathbf{xspan(D_{ID}, s, tl)}$ holds if tl is the associated list of tokens of all nodes of span s in D_{ID}.
$\mathbf{xcomplete\_span(D_{ID}, s, tl)}$ holds if tl is the associated list of tokens of all nodes of span s in D_{ID} and s is the maximum span of $s.n$. That means $s = (n, 0, m)$ and there is no child node $m' > m$ of n.
$\mathbf{xright\_brother(D_{ID}, n, n_b, t_r)}$ holds if n is a node identifier and t_r is the associated token of the right brother node n_b of n. Analogously $xleft\_brother$ is defined.
$\mathbf{xnright\_brother(D_{ID}, i, n, n_{br}, t_r)}$ holds if n is a node identifier and t_r is the list of associated tokens of the first i right brother nodes of n in the list n_{br}. Analogously $xnright\_brother$ is defined.
$\mathbf{xfather(D_{ID}, n, v)}$ holds if node v (node identifier) is the father node of n (node identifier).
$\mathbf{xchild(D_{ID}, n, i, c)}$ holds if node identifier c is the i-th child node of node n (node identifier) in left to right order.
$\mathbf{xmax\_child(D_{ID}, n, m, c)}$ holds if node identifier c is the $m - th$ child of n and there is no child m' such $m' > m$.

Table 3.2.: structural level description predicates

nested in *sub-related*, *sub-spans*, or in *overlapping spans* to each other. Consider Figure 3.17 where the slot filler for slot two is subordinated to the environment of slot filler one.

As can be observed from this figure each event is associated with a specific week-day and the extractions should contain the week-day as first slot filler. By means of a relational span level predicate like $sub\_related\_span$ (see Table 3.3) it can be determined that the slot fillers two to five are all sub related to the span containing slot filler one. Many other span relationships can be identified, as for example, the case that the intended extractions consists of three slots where the first slot contains the concatenation of the fillers for slot two and three. Irrespectively of the fact if this makes sense or is a redundant representation such cases will probably appear in practice. For instance, if someone is interested in extracting short news articles from an online newsmagazine as whole including additional links and so on. But additionally he is interested in some partial extractions from the extracted article. By identification of *span-in-span* relations between the slot fillers it is possible to process the document in one step. Figure 3.18 exemplifies this type of relation.

A special remark has to be given concerning the predicate $xsame\_span\_node$ from Table 3.3. This predicate defines the root nodes of two spans to be the same if they are equal under a modified version of unification. For instance, if $n_i = [1, X, 0, Y]$ and $n_j = [1, 2, 0, 1]$ are the root nodes of spans s_i and s_j. It is reasonable to say that the span s_i has the same root node as s_j, because n_j is an element in the set of possible span root nodes represented by n_i. Consequently this predicate can be used to identify spans belonging to a set of spans determined by non ground node identifiers.

Having a closer look at the definition of the $xnode\_less$ predicate shows that this predicate

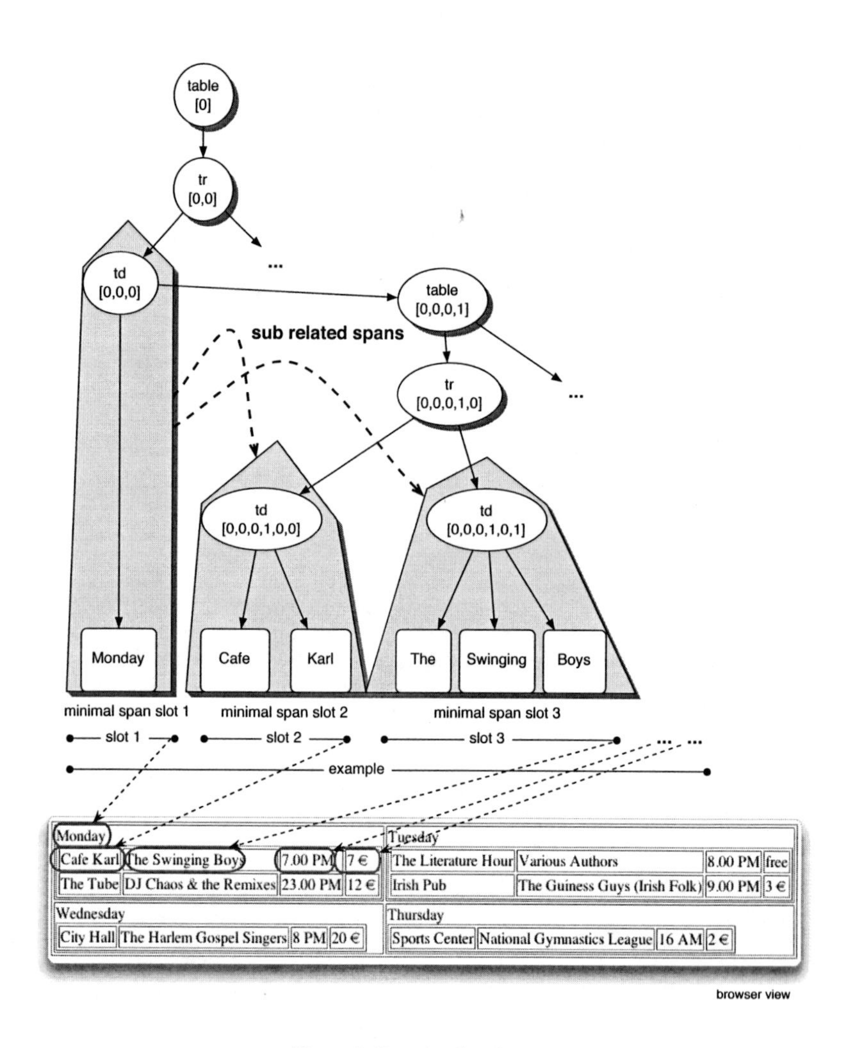

Figure 3.17.: sub related spans

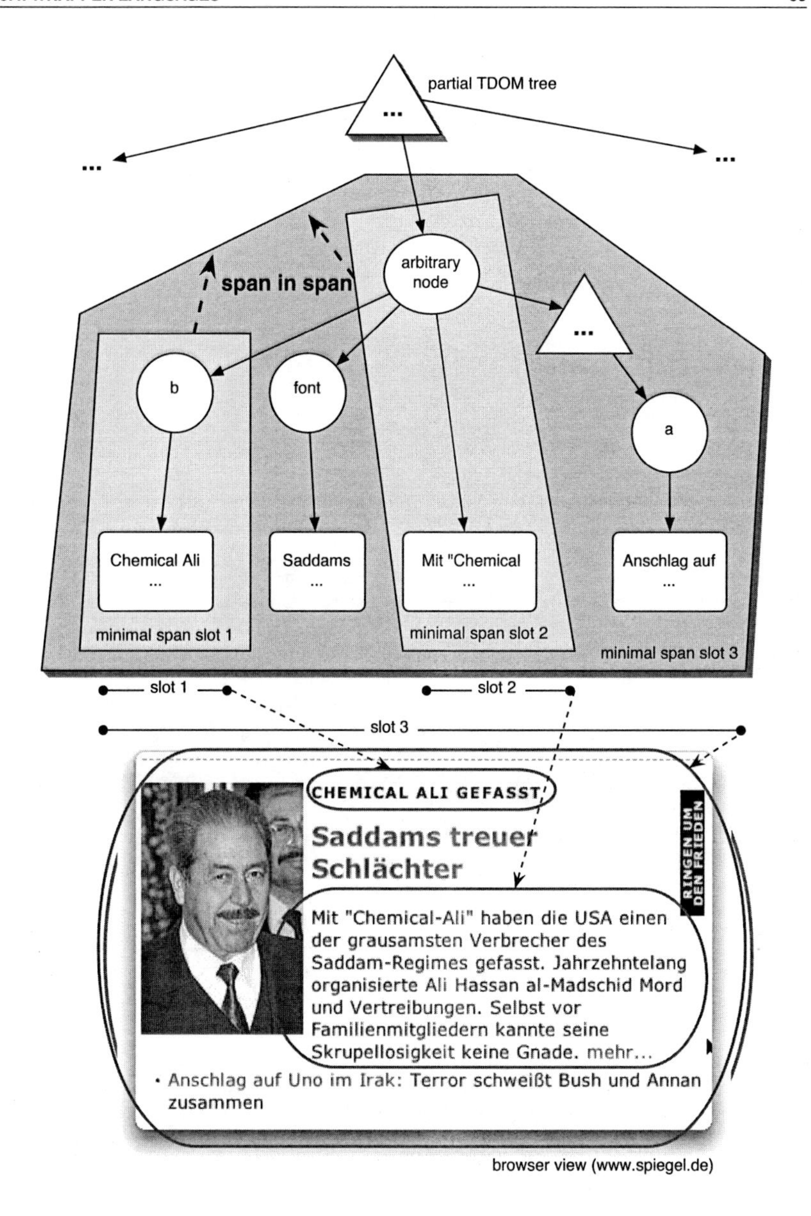

Figure 3.18.: span-in-span relation

can also be used to describe relations between sets of spans. For instance, the list of differences of two node identifiers is simply the calculated difference of each component of the two node identifiers. Given $n_i = [1, 4, 0]$ and $n_j = [2, 3, 0, 2]$ then $n_i < n_j$. The difference list is given by $[2-1, 4-3, 0-0] = [1, -1, 0]$. So the difference list has as many elements as the shortest node identifier has. As with the *xsame_span_node* this predicate can also describe sets of node identifiers related to each other, for example $xnode\_less(D_{ID}, [X_1, X_2, X_3], [Y_1, X_2, Y_3], [2, 0, -3])$ is the set of all possible pairs of node identifiers with the same second node identifier component and where the difference between the first components is 2 and between the third components is -3. So, these two predicates allow us to state relative orderings among nodes with respect to the structure of a *TDOM-tree*.

xsame_span_node($\mathbf{D_{ID}, s_i, s_j}$) holds if n_i and n_j of spans $s_i = (n_i, l_i, r_i)$ and $s_j = (n_j, l_j, r_j)$ are unifiable.
xnode_less($\mathbf{D_{ID}, n_i, n_j, dist}$) holds if $n_i < n_j$. *dist* is a list of differences between the components of n_j and n_i (e.g. $xnode\_less(0, [1, 4, 0], [2, 3, 0, 2], [1, -1, 0])$). Analogously we define *xnode_greater*.
overlapping_span($\mathbf{D_{ID}, s_i, tl_i, s_j, tl_j}$) holds if $(s_i.l < s_j.l) \wedge (s_i.r \geq s_j.l) \wedge (s_i.r \leq s_j.r)$ where tl_i and tl_j are the corresponding token lists of s_i and s_j.
span_in_span($\mathbf{D_{ID}, s_i, s_j}$) holds if span s_i is a subtree of span s_j.
xdirect_neighbor_spans($\mathbf{D_{ID}, s_i, s_j}$) holds if s_i and s_j are spans and following conditions are satisfied: both node identifiers $s_i.n$ and $s_j.n$ have the same length and are left resp. right brothers or their father nodes differs ± 1 and if $s_i.n < s_j.n$ then $s_i.n$ is a max. child and $s_j.n$ a first child or if $s_i.n > s_j.n$ then $s_i.n$ is a first child and $s_j.n$ a max child.
xneighbor_spans($\mathbf{D_{ID}, s_i, s_j}$) holds if s_i and s_j are spans and the node identifiers $s_i.n$ and $s_j.n$ have the same length (span root nodes are on the same level in the tree).
xsub_related_span($\mathbf{D_{ID}, s_i, s_j}$) holds if $s_j.n$ is a prefix of $s_i.n$ (e.g. $[1, 2]$ is a prefix of $[1, 2, 3]$).
xsmallest_cspan($\mathbf{D_{ID}, [s_1, \ldots, s_n], s_x, tk_x}$) holds if s_x is the smallest common span (wrt. to its number of nodes) in D_{ID} such that each span s_i with $i = 1, n$ is a subtree of s_x and tk_x is the token associated with $s_x.n$.

Table 3.3.: relational span level description predicates

Note, though *TDOM-trees* until now are solely used in the context of *HTML* they can also be used for any other tag-based or semi-structured language. Whenever a document contains environments similar to tag based annotations as for example LaTeX bibliography files or LaTeX in general, they can be represented as *TDOM-trees* and the proposed predicates can be applied. All that has to be done for adapting the *TDOM* approach to other types of documents than *HTML* is to modify the tokenization function and the definition of *start* and *end tags* as informally introduced in Section 2.3.3. Throughout this thesis we will solely use the previously introduced *TDOM predicates* to describe properties of text examples by means of logic programs. In fact these literals define the hypothesis language which is used by the learning algorithms introduced in Chapter 6, 7 and 8.

3.4.3. Example AV-wrapper

In the following a simple example illustrates the construction of a handmade AV-wrapper for a highly structured web document. Figure 3.19 shows a web page listing machine learning related books offered by amazon.com. Assuming someone wants to connect amazon's product offers to an online price comparison system (so called shopbots [Doorenbos *et al.*, 1997; Greenwald and Kephart, 1999]) some specific information regarding each book has to be extracted. For this example the shopbot is interested in the title, author, book type and the price offered by amazon.

According to Defintion 3.3.1 of *AV-Delimiter wrappers* the essential task in building a wrapper is to estimate left and right delimiter sequences that mark the start and end of slot fillers. So the first thing in handcrafting a wrapper is to have a closer look at the document source and to determine the slot fillers we are interested in. Figure 3.19 depicts determined slot fillers in the *HTML* source code of the example web page. The example extraction <Machine Learning, Tom M. Mitchell, Hardcover, 129.95> will serve as exemplification. Starting from the slot filler positions the relevant patterns (l_i, s_i, r_i) for the left, slot and right delimiter of each slot i have to be constructed. In general the human wrapper constructor will compare several example slot fillers and its surrounding text parts to build general patterns for (l_i, s_i, r_i). In this demonstration we will focus on this one example and the human *generalization step of patterns* is consciously postponed to be discussed in Part II where appropriate learning algorithms are presented.

Setting the maximum delimiter length to $mdl = 2$ following text surroundings for each slot filler are determined:

l_1 :<b><a href=/exec/obidos/tg/detail/-/0070428077/qid=1067354178/sr=5-2/
ref=cm_lm_asin/103-7354174-6368618?v=glance>
r_1 :</a></b>
l_2 :<font face=verdana,arial,helvetica size=-1>by
r_2 :</font>(
l_3 :</font>(
r_3 :)
l_4 :<b class=price>$
r_4 :</b>

As *AV-wrapper models* are basd on token pattern languages and tokenized documents the text surroundings are to be represented as tokens. For better readability and reasons of illustration only a few attributes in the token representation are used.

l_1 :token([ttype=html,tag=b]), token([ttype=html, tag=a])
r_1 :token([ttype=html_end,tag=a]), token([ttype=html_end,tag=b])
l_2 :token([ttype=html,tag=font]), token([ttype=word,value=by])
r_2 :token([ttype=html_end,tag=font]),token([ttype=sym,value='('])
l_3 :token([ttype=html_end,tag=font]),token([ttype=sym,value='('])
r_3 :token([ttype=special, value=')']), token([ttype=html,tag=br])

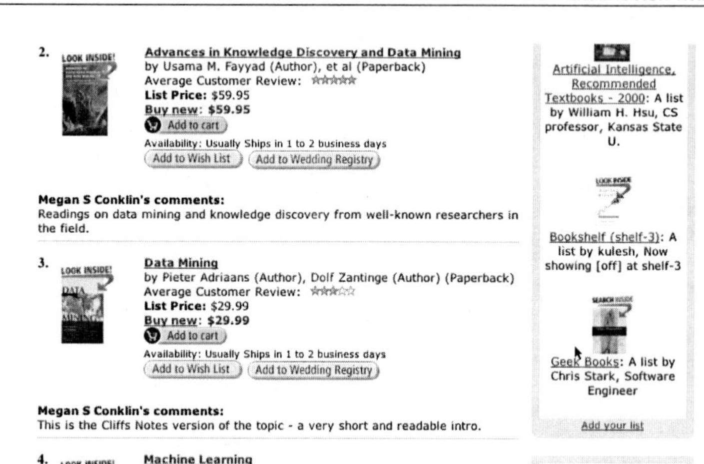

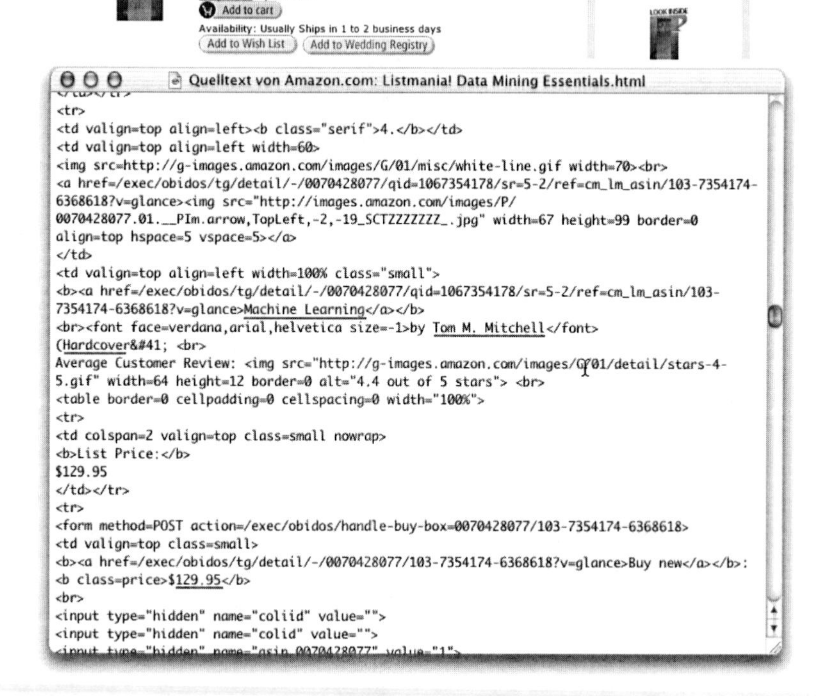

Figure 3.19.: amazon.com web page and *HTML* source

l_4 :`token([ttype=html,tag=b]), token([ttype=sym,value=$])`
r_4 :`token([ttype=html_end,tag=b]), token([ttype=html,tag=br])`

Simply collecting left and right surrounding tokens leads to an almost complete wrapper. Obviously this naive method using only one example extraction for determining delimiter patterns leads to a too restrictive wrapper model in most cases. So as already mentioned before the human wrapper constructor normally checks each pattern against all other delimiters observable from the complete set of given examples. As soon as one of the delimiter patterns is too restrictive, or in other terms does not match one of the preceding or succeeding token sequences of a specific slot filler, he starts to generalize the pattern according to his intuition based on his own logic. In fact, this process of generalization is what the later learning algorithms presented in Part II try to imitate. Nevertheless, this construction example wants to provide some insights into this time-consuming and inconvenient task. To complete the naive example wrapper model the slot filler patterns and intermediate patterns have to be located. For the `title`, `book type` and `price` slots some assumptions regarding the occurring text types can be made. For instance, a book title consists solely of tokens of `ttype = word`. Of course this is too specific for the general case but as a first attempt it suffices to describe the given extraction example. The same idea holds for the `price` slot, where a simple pattern like `token(ttype=float)` can be used, because a price is usually a floating number. For the author slot a roughly chosen pattern describing an arbitrary number of tokens of type `word` and `punct` is defined. Thus the following set of slot filler patterns might be a good first attempt:

s_1 :`+ token([ttype=word])`
s_2 :`token([ttype=word]) , + (token([ttype=word]) ; token([ttype=punct]))`
s_3 :`token([ttype=word])`
s_4 :`token([ttype=float])`

After having defined patterns for the slot fillers and their delimiters the final step is to combine the various slot patterns with fitting gap patterns. Actually in multi slot extraction these gap patterns can be a crucial point concerning the quality of a wrapper. If the text sequences appearing between two slots differ a lot in length and content type specifying appropriate patterns that do not tend to be too general is a difficult task. A second problematic point regarding gap patterns is their possible overlapping with left delimiter patterns or even worse with whole slot units (l_i, s_i, r_i). Such gap patterns cause permutations over slot fillers of different extraction tuples and therefore increase the number of false extractions and the match complexity. As already discussed in Section 3.3.2 one solution is to constrain gap patterns not to match sequences described by a left delimiter of a succeeding slot. In some cases it also suffices to simply use a maximum length restriction as a gap pattern. Hence for this illustrated case we chose a pattern matching arbitrarily many tokens `* token([ttype=X])` but with an upper length bound of n `maxlength(n,* token([ttype=X])`:

M_1 :`maxlength(1,* token([ttype=X])`
M_2 :`maxlength(0,* token([ttype=X])`
M_3 :`maxlength(28,* token([ttype=X])`

Obviously M_1 can be reduced to `token([ttype=X])` and M_2 can be simply dropped because

Quelltext von Amazon.com: Listmania! Data Mining Essentials.html

```
</td>
<td valign=top align=left width=100% class="small">
<b><a href=/exec/obidos/tg/detail/-/0201403803/qid=1067354178/sr=5-2/ref=cm_lm_asin/103-7354174-
6368618?v=glance>Data Mining</a></b>
<br><font face=verdana,arial,helvetica size=-1>by Pieter Adriaans (Author), Dolf Zantinge (Author)</font>
(Paperback&#41; <br>
Average Customer Review: <img src="http://g-images.amazon.com/images/G/01/detail/stars-3-0.gif" width=64
height=12 border=0 alt="3.2 out of 5 stars"> <br>
<table border=0 cellpadding=0 cellspacing=0 width="100%">
<tr>
<td colspan=2 valign=top class=small nowrap>
<b>List Price:</b>
$29.99
</td></tr>
<tr>
<form method=POST action=/exec/obidos/handle-buy-box=0201403803/103-7354174-6368618>
<td valign=top class=small>
<b><a href=/exec/obidos/tg/detail/-/0201403803/103-7354174-6368618?v=glance>Buy new</a></b>: <b
class=price>$29.99</b>
<br>
<input type="hidden" name="coliid" value="">
```

Figure 3.20.: a further example text to be extracted

of its zero length. Let's try to apply the wrapper to the partial html text shown in Figure 3.20 where the intended extractions are underlined in red.

The pattern $(l_1, s_1, r_1)M_1(l_2)$ perfectly fits on the first slot, the gap between slot 1 and slot 2 and the left delimiter of slot 2. Unfortunately the slot filler pattern s_2 and the succeeding pattern r_2 do not match in the intended way. The pattern s_2 is too restrictive since it does not allow matches containing symbols `token([ttype=sym,value='('])` to appear . Hence this pattern has to be generalized. Between two reasonable methods the human wrapper constructor might chose either to add a disjunctive pattern to loosen the pattern such that punctuation marks can be matched, or the pattern is generalized by replacing attribute values with variables. As a consequence of the second method the resulting pattern must be reduced to remove redundant sub patterns. The following token patterns s_2' and s_2'' show the resulting pattern after applying both methods to s_2:

```
s2' :token([ttype=word]), + ( token([ttype=word]) ;
                              token([ttype=punct]) ; token([ttype=sym]) )
s2'' :token([ttype=word]) , + token([ttype=X])
```

As one can easily see there are many possibilities how to generalize a token pattern according to the method of introducing variables. Unfortunately s_2'' now also matches the right delimiter described by the pattern r_2. So a more reasonable modification would have been to combine both methods:

```
s2''' :token([ttype=word]), + token([ttype=X]), ( token([ttype=word]) ;
                                                 token([ttype=sym]) )
```

Finally a first version of the AV-wrapper can be of the form:

$$((l_1, X_1 = s_1, r_1), M_1, (l_2, X_2 = s_2''', r_2), (l_3, X_3 = s_3, r_3), M_3, (l_4, X_4 = s_4, r_4))$$

So far it remains unclear if a generalized pattern like s_2''' or a length bounded most general pattern like gap patterns (e.g. `maxlength(n,+token([ttype=X]))` will provide better or worse results in practice than the patterns obtained by incremental generalization steps. From observations it appears to be the case that especially with highly varying slot filler texts it becomes very difficult for the wrapper constructor to apply the discussed generalization steps in a reasonable way by manual operation. Consequently algorithms for the automatic construction of wrappers are presented in the succeeding chapters of Part II.

3.4.4. Example RTD-wrapper

To illustrate the basic functionality of *RTD-wrappers* this section presents an example of a simple wrapper based on a subset of the previously introduced *TDOM predicates.* To ease the representation for reasons of readability several abbreviations for tokens and other notations are used and only expanded if needed for a better appreciation. Let us start with handcrafting a *RTD-wrapper* for the same extraction task as discussed in Section 3.4.3. Figure 3.21 shows a partial and simplified version of a *TDOM-tree* for the amazon web page displayed in Figure 3.19. We begin by choosing one of the intended extractions (`<Machine Learning, Tom M. Mitchell, Hardcover, 129.95>`) and start to examine its features with respect to the given *TDOM predicates.* The first step is to identify the *positions* of the slot fillers in the *TDOM-tree* representation of the document. Each filler is placed within a column of a table (`<td>`) and some of them additionally in environments included in each of these table columns. The filler `Machine Learning` is located within an anchor (`<a>`) environment, `Tom M. Mitchell` within an font environment, `Hardcover` within the table column environment and `129.95` within a bold face (`<b>`) environment contained in a column of a table contained in a table column environment. Hence every slot filler can be described by a `xspan` predicate determining the minimal span each slot filler is contained in:

```
Machine Learning:  xspan(1,([0,2,4,3,2,0,0],0,1),Title)
 Tom M. Mitchell:  xspan(1,([0,2,4,3,2,2],1,4),Author)
       Hardcover:  xspan(1,([0,2,4,3,2],4,4),Type)
          129.95:  xspan(1,([0,2,4,3,2,13,3,0,0,2],1,1),Price)
```

Now a very first version of a simple *RTD-wrapper* extraction rule can be constructed. Following Definition 3.3.4 an *RTD-wrapper model* $RTD(W)$ consists of normal program P_H with predicates H and a set of extraction rules R based on P_H. Consequently P_H is given by the implementation of the previously introduced *TDOM predicates.* Thus what is left to define is an extraction rule of the form $extract(D_{ID}, \vec{S}, \vec{W}) \leftarrow l_1, \ldots, l_m$ where $l_1, \ldots, l_m$ are *TDOM predicates.* For the presented example this means a very first version of the rule has the following form, where the document ID is `1`, $[\mathtt{S}_1, \ldots, \mathtt{S}_4]$ is a list of variables where each variable stands for one slot filler and $(\mathtt{N_i}, \mathtt{L_i}, \mathtt{R_i})$ represents a span including a slot filler.

$$\mathtt{extract}(1, [\mathtt{S}_1, \mathtt{S}_2, \mathtt{S}_3, \mathtt{S}_4], [(\mathtt{N}_1, \mathtt{L}_1, \mathtt{R}_1), (\mathtt{N}_2, \mathtt{L}_2, \mathtt{R}_2), (\mathtt{N}_3, \mathtt{L}_3, \mathtt{R}_3), (\mathtt{N}_4, \mathtt{L}_4, \mathtt{R}_4)]) \ :- \ \mathtt{true}. \tag{3.3}$$

Obviously this rule will always be true for documents with identifier 1 but will not extract anything. Therefore we add the discussed `xspan` predicates and additionally we take care that the correct variable bindings regarding the head and body literals are used:

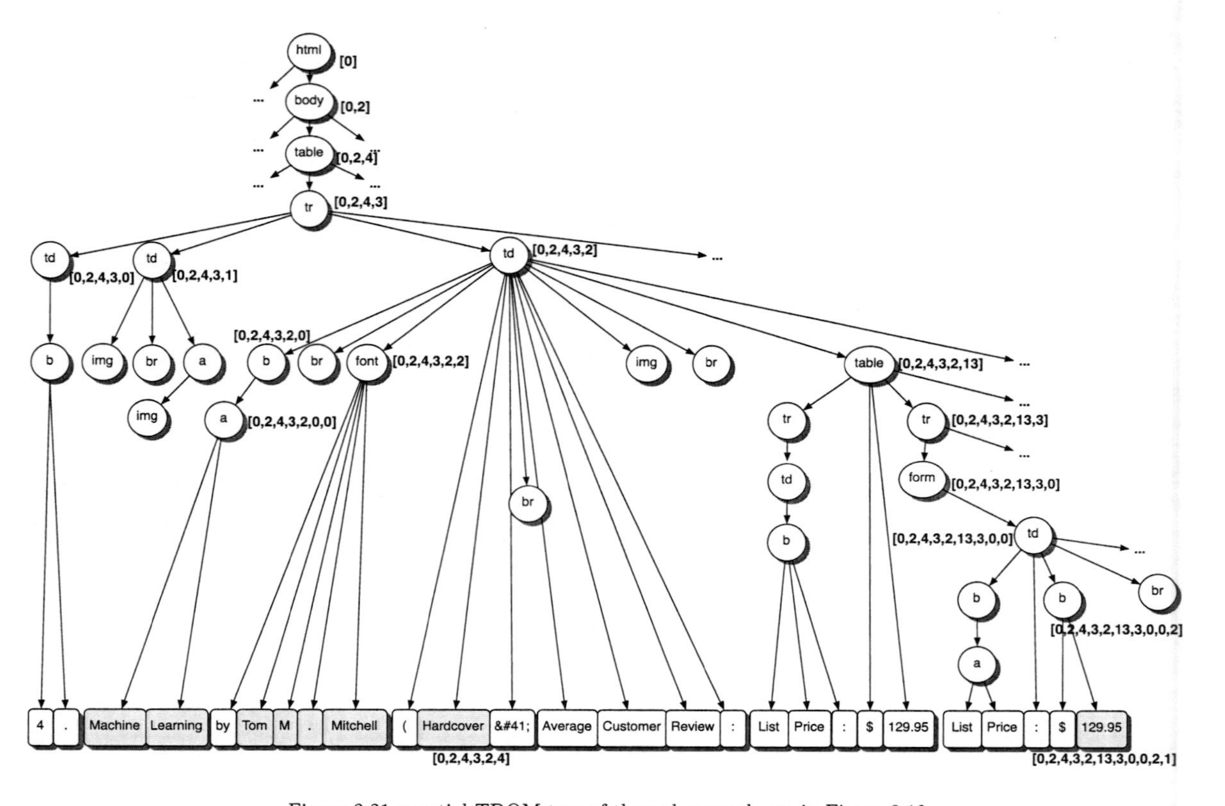

Figure 3.21.: partial TDOM tree of the web page shown in Figure 3.19

```
extract(1, [S1, S2, S3, S4], [([0, 2, 4, 3, 2, 0, 0], 0, 1), ([0, 2, 4, 3, 2, 2], 1, 4),
        ([0, 2, 4, 3, 2], 4, 4), ([0, 2, 4, 3, 2, 13, 3, 0, 0, 2], 1, 1)])  :-
            % title slot1
            xspan(1, ([0, 2, 4, 3, 2, 0, 0], 0, 1), S1),
            % author slot2
            xspan(1, ([0, 2, 4, 3, 2, 2], 1, 4), S2,
            % book typeslot3                                                    (3.4)
            xspan(1, ([0, 2, 4, 3, 2], 4, 4), S3),
            % price slot4
            xspan(1, ([0, 2, 4, 3, 2, 13, 3, 0, 0, 2], 1, 1), S4).
```

Rule 3.4 exactly extracts the example four-tuple but not any other product offers. Though this rule is not ground because of $S_1, \ldots, S_4$ it is still too restrictive and only covers the one investigated example. This is because the node identifiers and span parameters define unique text parts in the document. Since the intention is to build a wrapper model extracting all listed books the next step is to modify the rule such that it becomes more general. One possibility to do this is to generalize the node identifiers of each span such that the general structure is preserved but the extraction example shown in Figure 3.20 is also covered by this rule. Because this book description occurs before the *Machine Learning* book description a first attempt to generalize rule 3.4 is to replace those parts in the node identifier (`[0,2,4,3]`) by a variable which refer to the table row (`<tr>`) in which the offer is located (`[0,2,4,X]`):

```
extract(1, [S1, S2, S3, S4], [([0, 2, 4, X, 2, 0, 0], 0, 1), ([0, 2, 4, X, 2, 2], 1, 4),
        ([0, 2, 4, X, 2], 4, 4), ([0, 2, 4, X, 2, 13, 3, 0, 0, 2], 1, 1)])  :-
            % title slot1
            xspan(1, ([0, 2, 4, X, 2, 0, 0], 0, 1), S1),
            % author slot2
            xspan(1, ([0, 2, 4, X, 2, 2], 1, 4), S2,
            % book typeslot3                                                    (3.5)
            xspan(1, ([0, 2, 4, X, 2], 4, 4), S3),
            % price slot4
            xspan(1, ([0, 2, 4, X, 2, 13, 3, 0, 0, 2], 1, 1), S4).
```

Although the node identifier `[0,2,4,X]` refers to all direct children nodes of the table environment and the two investigated examples are children of this node, we can not be sure that all rows in this table contain book descriptions. If, for instance, the last row in this table contains something else than a book description but also has a similar structure regarding sub related spans, the extraction rule probably provides false extractions. So the problem is that after this generalization the relative `xspan` description probably becomes too general. One way to solve this problem is to introduce additional predicates describing the structure of each filler more detailed and not only based on relative node identifier positions. The `xpath` predicate is a reasonable extension to constrain the `xspan` predicate by additional information on the environment types each filler is located in.

$$
\begin{array}{l}
\mathtt{extract}(1,[\mathtt{S}_1,\mathtt{S}_2,\mathtt{S}_3,\mathtt{S}_4],[([0,2,4,\mathtt{X},2,0,0],0,1),([0,2,4,\mathtt{X},2,2],1,4),\\
\qquad\qquad ([0,2,4,\mathtt{X},2],4,4),([0,2,4,\mathtt{X},2,13,3,0,0,2],1,1)]) \quad :-\\
\quad \%\ \textit{title slot1}\\
\quad \mathtt{xpath}(1,[0,2,4,\mathtt{X},2,0,0],[\mathtt{html},\mathtt{body},\mathtt{table},\mathtt{tr},\mathtt{td},\mathtt{b},\mathtt{a}]),\\
\quad \mathtt{xspan}(1,([0,2,4,\mathtt{X},2,0,0],0,1),\mathtt{S}_1),\\
\quad \%\ \textit{author slot2}\\
\quad \mathtt{xpath}(1,[0,2,4,\mathtt{X},2,2],[\mathtt{html},\mathtt{body},\mathtt{table},\mathtt{tr},\mathtt{td},\mathtt{font}]),\\
\quad \mathtt{xspan}(1,([0,2,4,\mathtt{X},2,2],1,4),\mathtt{S}_2),\\
\quad \%\ \textit{book typeslot3}\\
\quad \mathtt{xpath}(1,[0,2,4,\mathtt{X},2],[\mathtt{html},\mathtt{body},\mathtt{table},\mathtt{tr},\mathtt{td}]),\\
\quad \mathtt{xspan}(1,([0,2,4,\mathtt{X},2],4,4),\mathtt{S}_3),\\
\quad \%\ \textit{price slot4}\\
\quad \mathtt{xpath}(1,[0,2,4,\mathtt{X},2,13,3,0,0,2],[\mathtt{html},\mathtt{body},\mathtt{table},\mathtt{tr},\mathtt{td},\mathtt{table},\mathtt{tr},\mathtt{form},\mathtt{td},\mathtt{b}]),\\
\quad \mathtt{xspan}(1,([0,2,4,\mathtt{X},2,13,3,0,0,2],1,1),\mathtt{S}_4).
\end{array}
\tag{3.6}
$$

Some short remarks concerning the notation: the token lists of the third argument of the *xpath* predicate are abbreviated such that only the value of the token feature `tag` is given. By Definition 3.3.4 the second argument of the *extract* predicate has to be a list of words of intended text to be extracted. Therefore it would be necessary to list all tokens of the third argument of the *xspan* predicate and to refer to the appropriate feature value containing the relevant words. For instance a correct notation for the title slot text `Machine Learning` is: `xspan(1,([0,2,4,3,2,0,0],0,1), [token([type=word,txt="Machine"]), token([type=word,txt="Learning"])`. We simplify the representation for this example in that we assume that $S_1, \ldots, S_4$ refer to the right feature values.

Though the modifications made in Rule 3.6 seem to be the right way to generalize and thus to extract more information more steps are needed. Comparing the *TDOM* structures of both examples shows that they are almost identical. The only difference is the number of text nodes (leaves) of the span containing the author information. Although in both examples the author information starts with the second child node (`[0,2,4,X,2,2,1]`) their number of nodes differ and hence also the node identifier of the last node differs. In fact this problem also exist for the title slot. But since the book title span only contains text to be extracted a solution for this case is easily achieved. The predicate `xcomplete_span` determines all nodes in the maximal span of a given node identifier. For the author span we have to use a different approach, because we have to omit the first word node (`by`) but have to extract all succeeding tokens of this span. An appropriate solution is the combination of the `xmax_child` and `xspan` predicate. This leads to the following refined extraction rule:

```
extract(1, [S1, S2, S3, S4], [([0, 2, 4, X, 2, 0, 0], Y, Z), ([0, 2, 4, X, 2, 2], 1, N),
        ([0, 2, 4, X, 2], 4, 4), ([0, 2, 4, X, 2, 13, 3, 0, 0, 2], 1, 1)])  :-
    % title slot1
    xpath(1, [0, 2, 4, X, 2, 0, 0], [html, body, table, tr, td, b, a]),
    xcomplete_span(1, ([0, 2, 4, X, 2, 0, 0], Y, Z), S1),
    % author slot2
    xpath(1, [0, 2, 4, X, 2, 2], [html, body, table, tr, td, font]),
    xmax_child(1, [0, 2, 4, X, 2, 2], N, _)
    xspan(1, ([0, 2, 4, X, 2, 2], 1, N), S2),                                        (3.7)
    % book typeslot3
    xpath(1, [0, 2, 4, X, 2], [html, body, table, tr, td]),
    xspan(1, ([0, 2, 4, X, 2], 4, 4), S3),
    % price slot4
    xpath(1, [0, 2, 4, X, 2, 13, 3, 0, 0, 2], [html, body, table, tr, td, table, tr, form, td, b]),
    xspan(1, ([0, 2, 4, X, 2, 13, 3, 0, 0, 2], 1, 1), S4).
```

Finally this *RTD-wrapper model* extracts exactly all the requested information about books presented on the example web page. It becomes apparent that handcrafting a wrapper is a tedious work to do. One can imagine how much effort it takes to build sets of wrappers for more than just a few documents. Though the *RTD-wrapper model* offers a nice declarative way to construct wrappers the manual construction can not honestly considered to be an adequate means for real world applications; not because of the quality of the wrappers but of the non acceptable time consumption and expert knowledge needed for construction. Without any doubt this example shows the urgent need for automatic construction techniques as presented in Part II of this thesis.

3.5. Logic Programming Based Implementation Techniques

This section is intended to give a sketch of the basic ideas how to use *logic programming* to implement and to apply the introduced wrapper models.

3.5.1. Implementing AV and CD-Wrappers in Prolog

In Section 3.4.1 the *token pattern* language has been introduced. This language is closely related to the attribute value representation of documents (Definition 2.3.1). Following strictly the language semantics definition (Definition 3.4.4) and the proposed *AV-representation* of documents, these two concepts can be integrated in a very convenient manner into a logic programming framework. As already described the *AV-representation* is by definition a set of ground unit clauses. The different pattern language operators (e.g. `*`,`times`) and the token unification (Definition 3.4.2) have to be implemented. Both programming tasks are very simple and solely need standard *Prolog* language features. Figure 3.22 shows the implementation of a token pattern matcher for the proposed *AV* and *CD-wrapper* language.

Some very brief details concerning the implementation have to be noted. The predicate `iterate(N,P,NP)` concatenates `N` times conjunctively the pattern `P` and instantiates the resulting pattern `NP`. The predicate `sequence(D,S,E,X)` instantiates `X` with the token sequence from document `D` starting at token number `S` and ending before token number `E`. Except with the token unification predicate `tunify(T1,T2)` the *Prolog* experienced reader should be familiar with all other predicates. Note that using a token pattern interpreter is only one alternative to implement *AV* and *CD-wrappers* in logic. Another possibility is to define a transformation

```
% maxlength(n,p) pattern
pattern(D,S,maxlength(N,P),E) :-
        pattern(D,S,P,E),
        sequence(D,S,E,X),
        length(N,X).

% not(p) pattern
pattern(D,S,not(P),E) :-
        not(pattern(D,S,P,E)).

% times(n,p) pattern
pattern(D,S,times(N,P),E) :-
        iterate(N,P,NP),
        pattern(D,S,NP,E).

% upto(n,p) pattern
pattern(D,S,upto(N,P),E) :-
        between(1,N,X),
        iterate(X,P,NP),
        pattern(D,S,NP,E).

% (X = p) pattern
pattern(D,S,X = P,E) :-
        pattern(D,S,P,E),
        sequence(D,S,E,X).

% ';' pattern
pattern(D,S,(P1;_P2),E) :-
        pattern(D,S,P1,E).
pattern(D,S,(_P1;P2),E) :-
        pattern(D,S,P2,E).

% ',' pattern
pattern(D,S,(P1,P2),E) :-
        pattern(D,S,P1,E1),
        pattern(D,E1,P2,E2).

% * pattern
pattern(_D,S,star(P),S).
pattern(D,S,star(P),E) :-
        pattern(D,S,P,E1),
        pattern(D,E1,star(P),E).

% + pattern
pattern(D,S,atleast1(P),E) :-
        pattern(D,S,P,E1),
        pattern(D,E1,star(P),E).

% ? pattern
pattern(_D,S,max1(P),S).
pattern(D,S,max1(P),E) :-
        pattern(D,S,P,E).

% atomic pattern
pattern(D,S,P,E) :-
        token(D,S,T),
        tunify(P,T),
        E is S+1.
```

Figure 3.22.: naive *Prolog* implementation of the token pattern matcher which maps token patterns into a set of *Prolog* rules which is in fact identical to implementing a compiler. Using the interpreter view provides the advantage of developing a uniform view on learning AV and *CD-wrappers* and *RTD-wrappers*.

Given an AV and *CD-wrapper* like the one presented in Section 3.4.3 the model can be completely implemented based on the presented pattern matcher. The basic idea is to define a predicate $extract(D_{ID}, \vec{X}, \vec{Pos})$, which is also called extraction rule in the following, with body literals $l_1, \ldots, l_n$ where each body literal represents one of the delimiter, slot or gap patterns of the AV and *CD-wrapper*. Once the wrapper model is implemented as an extraction rule it is used to compute extractions from arbitrary documents. The example wrapper from Section 3.4.3 is represented by the following extraction rule:

```
extract(D, [X1, X2, X3, X4], [(S1, E1), (S2, E2), (S3, E3), (S4, E4)])   :-
pattern(D, P1, (token([ttype = html, tag = b]), token([ttype = html, tag = a])), S1),
pattern(D, S1, X1 = atleast1(token([ttype = word])), E1),
pattern(D, E1, (token([ttype = html_end, tag = a]), token([ttype = html_end, tag = b])), P3),
pattern(D, P3, (maxlength(1, star(token([ttype = X])))), P4),
pattern(D, P4, (token([ttype = html, tag = font]), token([ttype = word, value = by])), S2),
pattern(D, S2, X2 = (token([ttype = word]), atleast1(token([ttype = X])),
                (token([ttype = word]); token([ttype = sym])))), E2),
pattern(D, E2, (token([ttype = html_end, tag = font]), token([ttype = sym, value =' (']))), P5),
pattern(D, P5, (token([ttype = html_end, tag = font]), token([ttype = sym, value =' (']))), S3),
pattern(D, S3, X3 = token([ttype = word]), E3),
pattern(D, E3, (token([ttype = special, value =' &#41;' ]),token([ttype = html, tag = br])), P6),
pattern(D, P6, (maxlength(28, star(token([ttype = X])))), P7),
pattern(D, P7, (token([ttype = html, tag = b]), token([ttype = sym, value = $])), S4),
pattern(D, S4, X4 = token([ttype = float]), E5),
pattern(D, E5, (token([ttype = html_end, tag = b]), token([ttype = html, tag = br])), P8).
```

(3.8)

Finally the computation of extractions from documents given an AV and *CD-wrapper model* works as follows: Let P_M be the logic program implementing the token matcher, R the extraction rule representation of an AV and *CD-wrapper* and $AV(D)$ the attribute value representation of an arbitrary document D. Computing answers for the query $extract(D_{ID}, \vec{X}, \vec{Pos})$ on D,P and R provides extractions from D as substitutions θ for $\vec{X}$: $AV(D) \cup P_M \cup R \vdash extract(D_{ID}, \vec{X}, \vec{Pos})\theta$.

3.5.2. Implementing RTD-Wrappers in Prolog

Implementing *RTD-wrapper models* in *Prolog* is almost an identical proceeding as the proposed implementation and application framework for AV and *CD-wrapper models*. Only the program clauses P of the token matcher have to be exchanged by an appropriate set of program clauses of *TDOM predicates* whereas the application or computation of extractions remains exactly the same. To be more precise an implementation of P_H from Definition 3.3.4 as *Prolog* rules has to be given. Additionally the attribute value document representation $AV(D)$ suited for AV and *CD-wrapper models* has to be replaced by an appropriate *TDOM* unit clause set $TD(D)$ (Definition 2.3.4). Without going into details of programming techniques we want to note that especially an elaborated implementation of *TDOM predicates* can have a significant impact on the runtime behaviour of a wrapper. Nevertheless, to derive extractions $\vec{X}\theta$ from a document D the same query scheme and answer computation as for the AV and *CD-wrapper models* is applied: $TD(D) \cup P_H \cup R \vdash extract(D_{ID}, \vec{X}, \vec{Spans})\theta$.

3.5.3. Implementing Wrappers as Elementary Formal Systems

A further method for the representation of *AV* and *CD-wrapper models* are *Elementary Formal Systems (EFS)* [Smullyan, 1961]. *EFS* have a certain kind of closeness to both Logic Programming and string matching. Roughly speaking they are logic programs with an extended unification, namely A_f*-unification* or word equations [Baader and Schulz, 1998]. The reasoning calculus is an adapted SLD respectively SLDNF-resolution [Lloyd, 1987] procedure.

They are of interest for the area of wrapper learning because much research has been done on the computational learnability of language learning [Zeugmann and Lange, 1995] with EFS systems [Arikawa *et al.*, 1992; Miyano *et al.*, 2000]. For instance [Grieser *et al.*, 2000] showed that *AEFS* [Lange *et al.*, 2003], which are *EFS* allowing negated literals in the body of a rule, implementing a subclass of *CD-wrapper models* called *island wrappers* [Thomas, 1999a] are learnable in the limit [Gold, 1967].

For the general case *EFS* and *AEFS* require the implementation of an unification algorithm to decide the word equation problem. Informally this means to decide if there exists a substitution for two arbitrary patterns consisting of variables and constants (terminals) such that they are syntactically equal. A typical example for such problems is to decide if appropriate substitutions for U, X, Y, Z from the set of words $\{a, b\}^*$ exist such that both strings $XaUZaU$ and $YZbXaabY$ are identical: $XaUZaU \stackrel{?}{=} YZbXaabY$.

Unfortunately it has been shown that this decision problem is NP-hard [Benanav *et al.*, 1985] and the time complexity of an existing algorithm is nondeterministic triple-exponential [Koscielski and Pacholski, 1990]. Of course it would be necessary to investigate if an AEFS implementation of *AV* and *CD-wrapper models* belong to this complexity class or perhaps only requires a restricted class of word equations. In the LExIKON project [LExIKON, 2000] *CD-wrapper models* represented as *AEFS* were transformed into *Prolog* programs and formative studies were carried out. Without any formal verification the success gives reason to speculate that the better choice is not to use *AEFS* for implementing *AV* and *CD-wrapper models.* Since the aim of this thesis is more orientated to combine the traditional fields of ILP and IE with the goal to represent solutions for practical applications it seemed more reasonable to make use of logic programming systems for the implementation. Because they are considerably better suited concerning existing efficient unification and automated reasoning algorithms.

3.6. Identifying Class Properties of Wrapper Models

After having introduced the three basic wrapper models which will be subject for the ongoing discussion throughout this thesis some further remarks concerning their wrapper class properties are necessary. One important point in the ongoing discussion on identifying the three wrapper models in the wrapper class hierarchy is the understanding of the concept of *modelling a wrapper* by one of the *wrapper models.* Disregarding the trivial cases that exclude the membership of a wrapper model to a wrapper class for the non trivial classes we always assume that the n extraction tuples of a wrapper W[2] are not represented by n trivial wrapper model rules r_i respectively R (see Definitions 3.3.1,3.3.2 and 3.3.4). A trivial rule is a rule that only

[2] Note that W denotes the set of available example extractions for which a wrapper model is to be learned and $\mathcal{W}$ denotes the theoretic target wrapper.

allows to detect one specific $e \in W$ and will fail for all other possible documents. In terms of logic we can speak of a grounded extraction rule yielding only one extraction for one specific document.

Why is this assumption important? It makes a great difference to exhaustively enumerate all n extraction tuples of W by one wrapper model consisting of n rules (extensional, see Section 2.2) or by only a few rules capturing general properties of the n extraction rules for correct identification (intensional). The overall motivation of this work is not to find a wrapper model solely describing the presented examples, instead the wrapper model has to have the ability to *operate* on future documents in the intended manner. Consequently a wrapper model should use an intensional description of elements in W, because they provide the opportunity to yield extractions from future data by generalized intensional rules. We call this the *intensional modelling assumption.* From this observation it follows that in fact the wrapper models have to be investigated concerning their quality to represent intensional descriptions.

For instance, the *CD-wrapper model* has a serious shortcomming based on the definition of constraints. The intention of a slot filler constraint c_i^s is to disallow the slot filler to include the right delimiter d_i^r. But if the intended slot filler text designedly contains text identical to the right delimiter this will contradict the put up constraint. During the wrapper construction phase this can be eluded by defining appropriate slot and delimiter patterns (intensional) or by using for each extraction one highly specific pattern rule (extensional). Nevertheless, for some future documents the idea of constraint delimiters will hinder the *CD-wrapper model* to provide the desired extractions, because the actual slot fillers are not predictable for future documents and thus the use of slot and delimiter constraints bears the risk of missing correct extractions.

From a practical perspective another point is important to discuss, which is the number of false extractions and how they depend on the definition of the wrapper model. Strictly following the definition and thereafter deciding the class a model belongs to leaves out the aspect of covering false extractions. The presented wrapper model definitions demand that all elements of a wrapper W are described or derivable from the wrapper model, but they do not define what is not to be covered by the model and thus not to be extracted. Due to the fact that a wrapper model shall be applied to unseen documents this definition fits perfectly into the basic idea since we cannot know in advance what exactly is to be excluded. This causes the strange behaviour that if a wrapper model M is a model for a wrapper W of class X it does not necessarily has to mean that the model will provide just correct extractions. This leads to the concept of a *partial wrapper model* as already discussed in Section 3.3.4. So, under this assumptions it follows that if a wrapper model M is said to be a model for a wrapper W of class X it is meant that M is a partial model for W.

3.6.1. Class Properties of AV and CD-wrapper Models

The first striking property easily observable from the definition of AV and *CD-wrapper models* is the requirement that the given order of slot fillers of an example does not have to represent the occurrence order within a document. For instance, if slot filler 2 occurs before 3 for one extraction example and for another example filler 2 occurs after slot filler 3 within the document then this is representable by AV and *CD wrapper models*, either by one rule with suitable delimiter patterns or by several rules each describing one alternative occurrence order. Nevertheless the presented models do not support to describe that a slot varies in its occurrence order. So, no general AV and *CD* rule can be defined in an intensional way. Expectably, the wrapper results will be limited on future data with unseen filler occurrence orders.

If the chosen description language $\mathcal{L}$ allows to describe the empty token sequence by some $p \in \mathcal{L}$ both the AV and *CD wrapper model* can represent instances of the ϵ-wrapper class by setting a slot pattern s_i to match the empty token sequence.

For the ϵ^+ class the AV*-wrapper model* in theory is also able to represent this class. Choosing a left and right delimiter and slot pattern to match an empty sequence models the structure of an ϵ^+-wrapper as defined. And also the *CD-wrapper model* can represent this class. All that has to be done is to use constraint patterns not including the empty token sequence to meet the constraints conditions.

In general there is a crucial problem regarding the usage of AV and *CD-wrapper models* that use delimiter patterns which are intended to describe two different delimiter sequences, which is the case when representing ϵ^+-wrappers (see Definition 3.1.4). Because in almost all cases there is a relationship between the left and right delimiter of a slot, which means that depending on the left delimiter a specific pattern for the right delimiter must be chosen to yield the correct extraction. In this case the naive approach to use a token pattern like `?p`, where `p` matches the non-empty slot or respectively delimiter is not sufficient, since the introduced token pattern language does not offer conditional language constructs. A solution to this problem is the use of an AV and *CD-wrapper model* with two token pattern rules r_1 and r_2, explicitly modelling the delimiter pattern dependencies. Concluding from this discussion obviously both wrapper models are able to model wrappers of the $\vee$ and $\wedge$-wrapper class.

Under the intensional modelling assumption only a subset of the x*-wrapper* class can be represented by AV and *CD-wrapper models.* The major problem is that in general the border between two directly adjacent slot fillers is difficult to detect. Especially if no further assumptions about the type, length etc. of the potential slot fillers are observable. In this case no significant general pattern or rule for the separation of the two slot fillers is possible. From a theoretical point of view a large disjunctive pattern of the presented slot fillers will suffice to model W. But this would again lead to a similar extensional enumeration as discussed before. There is the possibility to define an AV and *CD-wrapper model* for x*-wrapper* extraction tasks if slot filler patterns can be defined that are significant enough. Therefore all patterns between two slot fillers are defined to match the empty sequence and reasonable slot filler patterns have to be defined. Thus it depends on the character of the slot filler text if a model can successfully be constructed in this manner. If, for example, the token pattern language is used and the slot fillers have different values for the attribute `type` it is quite reasonable that the wrapper will provide correct extractions. Solely using slot filler patterns obviously requires a very attribute rich tokenization and not too strong varying slot fillers texts. Unfortunately slot fillers vary too much as this approach works reliable for all cases. These considerations clarify the problems to be expected when constructing x-class wrappers in practice. For this reason, where in theory a wrapper model with extensional parts but no appropriate intensional part can be constructed, the $\subset$ symbol is used in Table 3.4. Because not for all wrappers of this class a reasonable model (intensional) can be constructed.

inc-wrappers cannot be represented by AV and *CD-wrapper models* since they require the use of nested patterns (i.e. slot filler pattern i must be a sub pattern of a slot filler pattern j) which obviously contradicts the sequential order of slot filler patterns as defined for AV and *CD-wrapper models.*

model	ϵ	ϵ^+	MV	$\vee$	$\wedge$	x	single	linear	non-linear	inc	nested
AV	✓	✓	✓	✓	✓	$\subset$	✓	✓	$\subset$	-	-
CD	✓	✓	✓	✓	✓	$\subset$	✓	✓	$\subset$	-	$\subset$
RTD	✓	✓	✓	✓	✓	$\subset$	✓	✓	✓	✓	✓

Table 3.4.: wrapper models and classes

Roughly speaking an *MV-wrapper* is characterized by a list of values represented as such in the document for a certain slot. A general scheme matching the definition of an *MV-wrapper* can be abstractly described as: $v_{slot_{i-1}}\ s\ v1_{slot_i}\ d\ v2_{slot_i}\ d\ \ldots\ d\ vn_{slot_i}\ e$. Where s and e are delimiter patterns for the start and end of the value list, d is the delimiter pattern separating list elements, v_j are the slot fillers. Adapting an *AV* and *CD-wrapper model* to be consistent with such a definition can be achieved by setting up three rules:

$r_1 = \quad (\ldots(l_{i-1}, s_{i-1}, r_{i-1})\ s\ (\epsilon, s_i, d)g_i(l_{i+1}, s_{i+1}, r_{i+1})\ldots)$
$r_2 = \quad (\ldots(l_{i-1}, s_{i-1}, r_{i-1})g_{i-1}(d, s_i, d)g_i(l_{i+1}, s_{i+1}, r_{i+1})\ldots)$
$r_3 = \quad (\ldots(l_{i-1}, s_{i-1}, r_{i-1})g_{i-1}(d, s_i, e)g_i(l_{i+1}, s_{i+1}, r_{i+1})\ldots)$

Obviously this scheme models an *MV-wrapper* with appropriate definitions of gap patterns. The gap patterns can be simple length restricted token patterns matching any token e.g. `maxlength(n,*token(ttype=X))` where the length n is the longest gap observable from the given extractions of W. Logically a much better wrapper model can be defined by a *CD-wrapper model* constraining the gap patterns g_i to not contain l_{i+1}.

For the class of *nested wrappers* assume A, B and C are delimiter and slot pattern triples used to construct an *AV* and *CD-wrapper model*. A basic scheme for constructing a *CD-wrapper model* for a *nested wrapper* can be: $\ldots A(g_A, *A)B(g_b, *B)C\ldots$ This can be read as follows: the first slot filler is detected by matching the patterns of triple $A = (l, s, r)$, then an appropriate gap pattern g_A is matched, which is not allowed to match the slot description A, namely given by the pattern (l, s, r). This gap pattern can be interpreted to shift the patterns B for recognizing the next slot filler and all further occurring slot fillers for slot B within the nested structure of A. Without the use of constraint delimiters the boundary between the nested structure introduced by the different fillers for the top-level slot A will be ignored. As a result wrong subordinated slot fillers would be extracted. Trying to model a *nested-wrapper* with an *AV-wrapper model* will apparently result in a very weak partial model with very many false extractions and without the use of conditions or constraints there is no possibility to decrease this number. Therefore we argue that *AV-wrapper models* cannot satisfactorily represent *nested-wrappers*.

3.6.2. Class Properties of RTD-wrapper Models

RTD-wrapper models can represent *linear wrappers* and *non-linear wrappers*, because no fixed order regarding the occurrence of slot fillers in a document is stated by its definition. Additionally a relational description allows to describe relative position relations between slot fillers without explicitly stating an occurrence order.

The class of *inc-wrappers* can be modeled by *RTD-wrapper models*. Figure 3.18 gives an illustrative example how the $span\_in\_span(D_{ID}, s_i, s_j)$, predicate can be used to describe such slot filler relationships. Nevertheless, the quality of such an inclusion wrapper depends on the existence of significant delimiters or span borders for the included span resp. slot filler.

The insights drawn from the previous discussion on the *x-wrapper* class and *AV* and *CD-wrapper models* can directly be adapted to *RTD-wrappers.* As long as the slot filler text properties and additional information like token attributes are rich and significant enough *RTD-wrappers* are capable of representing at least a subset of *x-wrappers.*

AV and *CD-wrapper models* based on the introduced token pattern language have the ability to use disjunctive patterns to identify for example two different delimiters. So far *RTD-wrapper models* based on the presented *TDOM* predicates do not allow to use a disjunction of predicates (which are comparable to the role of patterns) in the body of an extraction rule. This is because an extraction rule is defined as an definite program clause (see Definition 3.3.4 and 2.1.5). Instead of introducing the concept of disjunction of literals in body of a rule e.g. `p :- c, a ; b` (where `;` stands for the connective $\vee$), the plausible semantical equivalent transformation is used `p :- c,a` and `p :- c,b`. Consequently two example extractions with strongly varying descriptions that cannot be reasonably described with non ground (generalized) *TDOM predicates,* are representable by a set of extraction rules. Hence *RTD-wrapper models* are able to represent $\wedge$ and $\vee$-wrappers.

Two intuitive modelling techniques demonstrate that *RTD-wrapper models* belong to the class of ϵ and ϵ^+-wrappers. For the first case, assume a left and right delimiter of a slot filler s_i are described by some *TDOM predicates* such that they provide references to *node identifiers* of the left and right boundary of a span predicate used for extracting the slot filler.

This scheme with variable boundary references accepts empty slot fillers (e.g. if the left and right reference refer to the same node identifier). For ϵ^+- wrappers two extraction rules are defined, one describing examples where the slot in focus does not occur with an empty filler and a second rule where the empty slot is treated as not present. Hence different relationships among the slots are identifiable. Since in this discussion we are interested in demonstrating how *partial wrapper models* can be constructed, this approach suffices.

A small MV example

- slot1 : v1, v2, **v3**, v4 ,**b5**, v6
- slot2 : n1, **n2**, n3

Figure 3.23.: MV example

RTD-wrapper models can extract multiple values per attribute. Two cases have to be distinguished. First, each multiple value list is contained in a different subtree as shown in Figure 3.23, where `<slot1,v2>` `<slot1,v3>` are some intended example extractions. For this case, where each *MV* list is located in a different subtree (here because of the `<li>` tag) node identifier patterns in conjunction with span predicates and successor resp. predecessor predicates for determining the list element delimiter (`,`) build a simple but efficient rule for modelling *MV-wrappers.* Second, if two or more *MV* lists are contained within one subtree on the same level the same technique as discussed for *AV* and *CD-wrapper models* can be applied.

The general idea of *RTD-wrapper models* strongly depends on the used document type. By definition as well semi-structured as "flat" documents, containing no annotational or tagged text parts, can be represented as a *TDOM.* In particular semi-structured documents allow to use the whole set of *TDOM predicates* for the description of relevant extraction parts.

Especially those predicates describing inclusions and sub related span relationships. Nested structures can easily be represented with those predicates. For example, Figure 3.17 depicts a similar nested structure as discussed in Section 3.1.7. For flat documents, where concepts like sub related spans do not occur similar techniques like those applied for *CD-wrapper models* can be applied. Because we focus on semi-structured documents the circumstances associated with flat documents can be disregarded.

4. Summary Part One

In the first part of the thesis the basic idea, formal concepts and practical issues of *wrappers* have been introduced and discussed. Motivated by practical issues arising from different extraction tasks reported in several publications [Muslea, 1998; Muslea, 1999; Kushmerick and Thomas, 2003] the basic concept of a wrapper has been formally separated into classes. Independent of a special class well known qualitative and quantitative measurements have been adapted to the concept of wrappers to provide measures for their evaluation.

Based on two different document representations, namely the *AV* representation interpreting documents as a sequence of attribute value terms and the *TDOM* representation interpreting documents as hierarchical tree structures over attribute value terms, three models for representing wrappers have been introduced.

These three models, *AV-wrapper*, *CD-wrapper* and *RTD-wrapper model*, represent general structural schemes independent of a specific modelling language. They are merely restricted in the assumed document representation. Progressively pursuing the goal to present extraction procedures founded on a formally defined basis two languages for the description of wrapper models were presented. For *AV* and *CD-wrapper models* a unification and regular expression based pattern language, namely the *token pattern language* was introduced. A more extensive version of the token pattern language (token-templates) was published in [Thomas, 1999a; Thomas, 2000]. For the learning algorithms presented in Part II it is sufficient to consider the subset presented here.

For the modelling of *RTD-wrappers* the concept of *TDOM predicates* was established. In the discussion on *RTD-wrappers* it was emphasized that *TDOM predicates* can be any reasonable set of predicates to describe properties of nodes, spans and sub-trees of a *TDOM-representation*.

A sketch how to implement and to use wrapper models in a logic programming system was presented in Section 3.5. An important outcome of this brief discussions is the fact that in both cases as well for *AV* and *CD-wrapper models* as for *RTD-wrapper models* identical queries and identicial program structures can be used. This means that each program can be divided into the chosen document representation D, a logic program P implementing either a token pattern language interpreter (resp. compiler) or a set of *TDOM predicates* and the wrapper model itself, which is determined by a set of extraction rules R.

To summarize, starting from a theoretical set based definition of a wrapper as given in the Preliminaries 2.1.1 we presented an approach to transfer the general wrapper concept to a general logic programming based concept of wrappers.

The main intention of Section 3.6 was to demonstrate which of the defined wrapper models can be used in theory to tackle a certain extraction problem. Under the assumption of *partial wrapper models*, a wrapper model is said to model a wrapper of a specific wrapper class if it approximates the wrapper. How good or bad such a model is, can be evaluated by the introduced metrics. It should have become clear that the presented class properties are meant to be indicators for a reasonable practical application of a certain wrapper model and not theoretically proven properties.

Thesis Contributions of Part One

The following list summarizes the contributions of Part One of the thesis:

- a formal definition of a general wrapper concept (Section 2.1.1)
- a uniform and formal definition of wrapper classes which have been informally discussed in previous publications (Section 3.1)
- three general wrapper models independent of a specific wrapper language and implementation technique (Section 3.3)
- two wrapper languages: a pattern language based on feature structures and unification techniques (Section 3.4.1) using a sequential token view on documents (Section 2.3.2), and a predicate logic language (Section 3.4.2) using a logical document representation strongly related to DOM trees (Section 2.3.3)
- a pure logic programming framework for wrapper based information extraction computing extractions by means of answer computations (Section 3.5) with interchangeable wrapper models and document representations

Part II.

Inductive Logic Programming Based Wrapper Learning

5. Information Extraction and Inductive Logic Programming

Part I illustrated and discussed the basic concept of a wrapper and how it is transferred into a pure logic programming (LP) framework. One of the outcomes of Part I is the insight that different wrapper models and different document representations can be conveniently be represented as logic programs.

One can argue that the mere representation is not a convincing argument to chose the advocated LP approach, since any other suitable representation could be chosen. But in fact the LP approach has one eminent advantage. Consider the general idea of wrapper representation and application presented in Section 3.5. It consists of D a unit clause representation of a document, a logic program P modeling either a wrapper language interpreter or a set of *TDOM predicates* and the wrapper model R given by a predicate definition `extract(D,X,S)`.

Modifying one or all of these three components leaves the actual computation, namely the derivation or answer computation, untouched. Thus new wrapper concepts including new document representations and background knowledge can be solely modeled in a declarative logical manner. This offers a certain degree of freedom for the application to various different problem domains. It has the advantage, in practice, that the approach can be applied on different document formats and that depending on the extraction task different wrapper models can be used, easily modified or introduced. Especially the introduced concept of a *wrapper clause template* (Section 5.4) demonstrates that new wrapper classes can be easily developed and evaluated by defining them in a declarative manner in terms of logic programming rules.

In the following we call P the *background knowledge*, since once the wrapper model and document representation is chosen, it is fixed and assumed to be given. In the context of machine learning the wrapper language is referred to as *hypothesis language*. Hence in the discussed context the *background knowledge* is the implementation of the *hypothesis language*.

So far, the construction of a wrapper was assumed to be done in manual operation. Where the picture one has in mind is that some example extractions $(w_1, \ldots, w_n)$ from a document D are given, which follow a certain semantics (e.g. addresses, product descriptions).

So, in fact the problem is to find a suitable extraction procedure (set of rules R) that calculates all of the intended extractions from D. This task, which we call the *wrapper learning task*, can be formulated as: $D \cup P \cup R \vdash E$ where E is the set of extractions we expect to be drawn from D, namely $extract(D, [w_1, \ldots, w_n], S)$.

Now the question arises if this set of rules R (the wrapper) can be learned such that all $e \in E$ can be computed from $D \cup P \cup R$ and no false extractions are derivable. And if so is it possible to learn R solely from a subset of E automatically without any further user interaction? And will the learned wrapper be general enough to be applicable to future documents yielding correct extractions?

Because the research field of *Inductive Logic Programming* has produced auspicious methods and techniques to synthesize logic programs in comparable settings like the presented wrapper learning task it is more than reasonable to assume that those techniques can help to find good solutions.

In this chapter we introduce some of the basic concepts and notations of *Inductive Logic Programming* (ILP). Albeit it is not intended to give an extensive introduction and overview of this active machine learning research area. Thus we assume the reader to be familiar with the area of ILP and LP. The less advanced reader is referred for exhaustive discussions on ILP in general, its history, application areas and existing systems to [Bergadano and Gunetti, 1996; de Raedt, 1996; Lavrac and Dzeroski, 1994; Lavrac and Dzeroski, 2001; Wrobel, 1996]. Additional readings can be found in the *Inductive Logic Programming* conference proceedings published in the Springer *Lecture Notes in Artificial Intelligence* series.

5.1. The ILP Problem

The origins of *Inductive Logic Programming*, which has become an active research field in the area of Machine Learning and a subfield of Artificial Intelligence, can be found in early works by Plotkin [Plotkin, 1970; Plotkin, 1971b] and Shapiro [Shapiro, 1981]. Plotkin studied properties of least general generalization (lgg) operations of first-order clauses. It is to date one of the essential operation used in many existing ILP systems. Shapiro was one of the first who presented an approach to inductive synthesis of logic programs. His *Model Inference System (MIS)* induces a finite set of Horn clauses (the axiomatization of a unknown model M) from false and true ground units.

The term *Inductive Logic Programming* was coined by Muggleton [Muggleton, 1991] and a widely accepted definition can be found in [Muggleton and Raedt, 1994]. The *standard ILP setting* consists in general of three parameters

B the background knowledge. Usually a first-order logic (FOL) formula. Normally this is a clause set theory, definite logic program or just a set of unit clauses for which a completion (regarding the correct description of examples) is searched. In most cases it implements predicates used in the hypothesis language. But in some settings B can also be empty.

E the set of positive and negative examples denoted by E^+ and E^-. As with B according to different learning settings E can consist of anything from single ground units to FOL formulae.

Given this standard ILP setting, the general *ILP problem* consists of finding a hypothesis H such that:

- $\forall e \in E^+ : H \wedge B \models e$
- $\forall e \in E^- : H \wedge B \not\models e$

Depending on restrictions and required properties regarding $H, E^{+/-}$ and B several different *ILP settings* and *semantics* have been defined. Usually, a *ILP semantics* is described by some formulae, which define requirements regarding $H, E^{+/-}$ and B. The *prior satisfiability* and *necessity* criteria state what requirements regarding B and $E^{+/-}$ have to hold regardlessly of H. *Posterior satisfiability* and *sufficiency* criteria are defined to describe the requirements concerning the hypothesis H, which is to be induced. Not all semantics use all four criteria and some use different notions but in general the properties are defined by means of logical consequence or model theoretic statements.

In the remainder of this section the most relevant *ILP semantics* and derivatives are briefly introduced. This deals as basis for the discussion in Section 5.2 and definition of an adequate semantics for inducing logic programs for information extraction tasks.

5.1.1. ILP Settings

Following the notions of [Muggleton and Raedt, 1994] the *standard ILP setting* plus some further requirements regarding B, H and E is defined as the *normal semantics* setting.

Definition 5.1.1 (ILP normal semantics) *Let B, E^+ and E^- be clause theories. Under a* normal semantics *a clause theory H (hypothesis) has to suffice the following criteria regarding B, E^+ and E^-:*

prior satisfiability $B \wedge E^- \not\models \Box$

posterior satisfiability $B \wedge E^- \wedge H \not\models \Box$

prior necessity $B \not\models E^+$

posterior sufficiency $B \wedge H \models E^+$

□

Some remarks concerning prior and posterior criteria: The *prior necessity* states that E^+ is not allowed to be logically entailed by the background knowledge B. If this would be the case we were already done, because B would explain E^+ sufficiently well enough. The *prior satisfiability* assumes that E^- together with B is not inconsistent. Usually we assume that a given background knowledge B is consistent, hence in this setting it is also required that E^- is not inconsistent and $E^- \wedge B$ has a model. The hypothesis also has to suffice certain criteria, which are summarized as *posterior criterions*. For the *normal semantics* the *posterior sufficiency* requires H to be a description for all $e \in E^+$. This is also known as *completeness* regarding E^+. The *posterior satisfiability* criterion assures that B, E^- and H are *consistent*.

There have been numerous publications on other ILP learning settings[1] [Flach, 1992; Helft, 1989; Raedt, 1997; Muggleton and Raedt, 1994; Wrobel and Dzeroski, 1995] based on restricting the background knowledge, hypothesis and example properties of subclasses of full first-order logic formulae. In the following we only mention those learning settings relevant for the upcoming discussion.

De Raedt proposes in [Raedt, 1997] the *learning from entailment* setting, which is a slightly restricted version of the *normal semantics*. In this setting each example is restricted to be exactly one clause. Further H is required to cover the *sufficiency* criterion and a background knowledge B is omitted. Wrobel et al. [Wrobel and Dzeroski, 1995] describe a modified version of the *normal semantics* called *ILP prediction learning problem* where the *prior satisfiability* is extended such that the positive and negative examples are to be consistent: $E^+ \wedge E^- \wedge B \not\models \Box$. Additionally they define the *posterior sufficiency* such that no $e \in E^-$ logically follows from $B \wedge H$, which correlates with the *standard ILP setting*. The basis for the most widely applied semantics is presented by Muggleton under the name *definite semantics*.

[1] In the remainder of the thesis the notion ILP setting and ILP semantics are used in an identical sense.

Definition 5.1.2 (ILP definite semantics) *Let B be definite clause sets. Let E^+ and E^- be clause theories. Further let $\mathcal{M}(T)$ denote the minimal Herbrand model of a clause theory T, which is unique for definite clause theories. Under the definite semantics a definite clause set H (hypothesis) has to suffice the following criteria regarding B,E^+ and E^-:*

prior satisfiability *all $e \in E^-$ are false in $\mathcal{M}(B)$*

posterior satisfiability *all $e \in E^-$ are false in $\mathcal{M}(B \wedge H)$*

prior necessity *some $e \in E^+$ are false in $\mathcal{M}(B)$*

posterior sufficiency *all $e \in E^+$ are true in $\mathcal{M}(B \wedge H)$*

□

Further restrictions of the *definite semantics* such that examples solely consist of ground unit clauses leads to the *example setting* [Muggleton and Raedt, 1994], which is de facto the main setting of actual ILP approaches and systems. Similar to the *example setting* the *learning from interpretations setting* [Raedt, 1997] also requests examples to be ground unit clauses. This semantics is discussed in Section 5.2.

To complete the overview of basic ILP learning settings, a third major semantics, the *non-monotonic* semantics [Helft, 1989; Flach, 1992], is presented. The basic idea differs from the *standard ILP setting* since it assumes that solely a background theory as definite clause theory is given. It is assumed that all examples are contained in this background theory. Whereas the positive examples are considered to be part of this theory, the negative examples are derived implicitly under a closed world assumption. This leads to the following definition taken from Muggleton [Muggleton and Raedt, 1994].

Definition 5.1.3 (ILP non-monotonic semantics) *Let B be a clause set. Under non-monotonic semantics a clause set H (hypothesis) has to suffice the following criteria:*

validity *all $h \in H$ are true in $\mathcal{M}(B)$*

completness *if a clause g is true in $\mathcal{M}(B)$ then $H \models g$*

minimality *there is no proper subset G of H which is valid and complete*

□

In this setting the minimal Herbrand model of B contains only those examples which are positive ones that are entailed by B. Everything not contained in the minimal model is considered to be false, and thus interpreted to be a negative example. Hence if there is a $h \in H$ which is not true in $\mathcal{M}(B)$ it covers something not contained in the minimal model and thus covers some false examples. Or in other terms, only what follows from B is considered to be a positive example, everything else is a negative example. Because of its similarity to the *closed world assumption* this semantics is called non-monotonic because no negative examples are explicitly given. Instead they are implicitly given.

The validity requirement states that the hypothesis H must be an explanation for information represented by B. Vice versa the completeness axiom requires that each clause in B has to be a logical consequence of H. In fact, this is the most conservative semantics of all, since it does not allow for *inductive leaps*.

Example 5.1.1 (Inductive Leap) *(taken from [Muggleton and Raedt, 1994])*
For instance, let $B = \{\{bird(tweety)\}, \{bird(oliver)\}\}$ *and* $E^+ = \{flies(tweety)\}$*. In the example setting a valid hypothesis is* $h = flies(X) \leftarrow bird(X)$*, because* $\mathcal{M}(B \wedge h) = \{bird(tweety), flies(tweety), bird(oliver), flies(oliver)\}$ *and* $flies(tweety)$ *is true in this model. Note that though we have not said that oliver can fly it is a logical consequence of* $B \wedge h$ *which is known as an inductive leap. Now assume we use the non-monotonic semantics and* $B' = B \cup E^+$*. Then* $\mathcal{M}(B') = B'$ *and* h *is false in this model. In general, the non-monotonic setting is more conservative in that it allows less general hypothesis than the normal setting.* ⌟

A further generalized definition of the *non-monotonic semantics* is proposed in [Dzeroski, 1995] which extends the definition of B to positive E^+ and negative E^- datasets of interest and corresponding background theory as definite clause theory B. *Validity* is then defined as $\forall D \in E^+, H$ is true in $\mathcal{M}(D \wedge B)$ and $\forall D \in E^-, H$ is false in $\mathcal{M}(D \wedge B)$.

Setting up different learning settings by restricting the expressiveness for the description of examples has some pro's and con's. On one hand, restrictions regarding example descriptions and hypotheses lead to loss in expressiveness but on the other hand there is a gain in computational efficiency. For instance, a positive example like $child(X, X)$ stating that no-one is their own child is not allowed in the *learning from interpretations* but in the *learning from entailment* setting. In the *learning from interpretations* setting de Raedt proposes to test the coverage of hypotheses by testing if $e \in E^+$ is a model for H, which is obviously less complex than testing for logically entailment of $H \models e$ as in the *normal semantics.*

Hence for some problem domains it is reasonable to reduce the expressiveness for efficiency reasons. Particularly in many application scenarios where the full standard ILP setting is not needed. And secondly more and more ILP systems [ILP2Net, 2000] have been developed over the last years especially tailored for certain ILP learning settings motivated by practical tasks and experiences.

5.2. The IE-ILP Setting

Before starting to clarify which of the previously discussed ILP semantics is best suited for ILP based wrapper learning it is necessary to determine how learning examples (example extractions) are to be represented. Two reasonable representations can be chosen. 1) Each example is represented as one ground unit clause describing a relationship between a document, text tuples and some additional information regarding the chosen wrapper model (i.e. spans, word positions). 2) Each example is represented as a wrapper consisting of one extraction rule (i.e. ground clause) as exemplified in Section 3.5. The actual example representation is explained in Section 5.5. For the remainder of this section the exact representation is not important. Furthermore it is of interest to determine the role in the context of an ILP learning task and which representation is to be favored.

Representing an example by a ground wrapper model (i.e. ground extraction rule) has a significant advantage over the ground unit representation. Since the overall goal is to induce a set of clauses (hypothesis) providing correct predictions, or in terms of wrapper learning, to extract intended text tuples from so far unseen documents, a sensible shortcut for finding good hypotheses is to use ground extraction rules as a starting point. If they are interpreted as initial most specific hypotheses they deal as some sort of clause templates and semantic guidance in a bottom-up search process. This reduces the search space dramatically, because unnecessary tests for rule body literals can be left out during the learning phase. A reasonable learning algorithm then has to find a good generalization of this initial very specific clause set. This is of course only expected to work if an extraction rule includes all valid hypothesis language literals regarding the given example extraction. Undoubtedly, in this context the learning task is to find a hypothesis (set of extraction rules) that explains the given examples by clauses consisting of literals from the hypothesis language.

To point it out more clearly, the set of examples still consists of the ground unit clauses (i.e. *extract*/3), but as a starting point for the ILP task the observable properties of each example (e.g. relational descriptions or delimiters) are represented by a ground extraction rule. For instance, given an example extraction $(D, (w_1, \ldots, w_n))$ then its unit clause representation is $extract(D, [w_1, \ldots, w_n], S)$ and an appropriate ground extraction rule is of the form $extract(D, [w_1, \ldots, w_n], S) \leftarrow l_1, \ldots, l_n$ with $l_i \in \mathcal{H}$ where $\mathcal{H}$ is the chosen hypothesis language.

These extraction rules can be calculated in a preprocessing step of the learning algorithm. This is discussed in Section 5.4. Ground extraction rules thus serve as a set h' for which a suitable algorithm has to be developed to generate a hypothesis h from h' with good predictability. So, this gives a sketch how an algorithm can start to induce reasonable hypotheses. It should be clear now, that learning examples are ground unit clauses and that examples represented by ground extraction rules are the first step in a *specific-to-general* learning method.

The setting discussed so far, corresponds to the *definite semantics* because the task is to find a hypothesis h such that each example extraction is a logical consequence of h and the background theory. Since the examples consist solely of ground unit clauses this setting corresponds to the *example setting.* Additionally in this context of wrapper learning, hypotheses consist solely of Horn clauses (Section 3.5) with identical positive literals, namely the *target predicate extract*/3.

The *learning from interpretations* setting [Raedt, 1997] is also based on the assumption that examples are ground unit clauses but plus some additional criteria which are discussed in the following. For the *learning from interpretations* settings there exist some very attractive learning systems (e.g. [Blockeel and Raedt, 1998; Blockeel *et al.*, 2000]). Based on the underlying semantics these methods offer the advantage that the validity of a current hypothesis (i.e. coverage) does not have to be tested globally. Thereby [Raedt, 1997] uses the notion that a hypothesis h covers an example e if e is a model for h. Global coverage testing means that to test if a hypothesis covers one example it requires the whole set of examples and background theory to be considered. In most ILP systems this is a crucial bottleneck for the system's performance. Instead, under *learning from interpretations* it is sufficient to perform local coverage tests on one example or certain subsets of examples (see [de Raedt *et al.*, 1998]).

Unfortunately the proposed learning setting by [Blockeel and Raedt, 1998] and [Blockeel *et al.*, 2000] is not well suited for the wrapper learning task investigated in this thesis. The *learning from interpretations* setting uses following components: a set of classes C (nullary predicates), a set of examples E with $(e, c) \in E$ so called classified examples with e a set of ground units and $c \in C$ a class label and a background theory B. De Raedt formalizes the learning task as finding a hypothesis h such that for all $(e, c) \in E$ it holds that $h \wedge e \wedge B \models c$ and $\forall c' \in C - \{c\} : h \wedge e \wedge B \not\models c'$. Within this setting the basic idea is to construct a hypothesis consisting of a set of clauses of the form $c \leftarrow l_1, \ldots, l_n$ with $c \in C$ and l_i literals from E or predicates from B. In a broader sense the intention in this setting is to *classify* or to predict to which class a new example belongs. Of course the hypothesis also yields some compact or generalized description of the examples to allow predictability of class memberships. But in contrast to the *definite* or *normal semantics* it tries to find an explanation for the distinctive features of the examples, namely their classes ($h \wedge B \wedge e \models c$), whereas in the *definite semantics* the examples are to be explained ($h \wedge B \models e$).

This is a somewhat different intention than the one associated with the wrapper learning task, since we do not want to present text tuples for classification. Our intention is that text tuples are computed or in terms of logic are derived from the learned hypothesis and background theory consisting of the document representation and the implementation of hypothesis language predicates. This observation is independent of the question how to represent examples or what to chose as initial hypotheses as discussed at the beginning of this section.

5.2.1. Learning From Positive Examples Only

So far it has not been discussed what negative examples are in the context of wrapper learning and how they can be provided to one of the presented ILP settings.

In general it is quite easy to determine what should be extracted from a document, but to determine in a reasonable fashion what is not to be extracted by an implicit description instead of using an explicit enumeration is not trivial. Apparently the most convenient way to learn a wrapper for some documents is simply to present a handful of positive examples. Unfortunately early learning theory results [Gold, 1967] show, that even regular languages are not learnable in the limit from positive examples alone.

Drawing a pessimistic conclusion would result in resigning from the idea of learning from positive examples alone. But there are a lot of practical applications where a user cannot be annoyed by giving examples of what he is not interested in. Imagine a system that aims at learning price comparison wrappers on the fly while the user is crawling the web. It is probably acceptable for the user to mark the products he is interested in, but his acceptance towards the system will certainly decrease if he has to mark also things he is not interested in. So, despite the theoretical learning results the investigation and development of techniques for learning from sparse positive data alone is of great practical interest.

In Part I, Section 3.3.4 we illustrated how the absence of explicitly given negative examples can be overcome by the exhaustive enumeration of positive examples regarding a given document. The basic idea behind this is to a certain degree comparable to the notion of the *closed world assumption (CWA)*. Which means, roughly speaking, everything not stated or not derivable, is assumed to be false. The presented *non-monotonic semantics* is also based on this *CWA*. This does not mean that the *non-monotonic semantics* is to be favored over the *example setting*. From a theoretical point of view the *example setting* can also be used in conjunction with a *CWA* based derivation of negative example. Therefore the set of negative examples is given by the complement of the minimal Herbrand model $E^- = \mathcal{M}^-(B \wedge E^+)$. To compute $\mathcal{M}^-$ the set difference of the Herbrand base of $B \wedge E^+$ and the minimal Herbrand model of $B \wedge E^+$ is computed. It should be clear that this set of negative examples can become dramatically huge and in the worst case infinite. Especially if we think of large documents and a background theory based on a hypothesis language like the one used for *RTD-wrapper models* as presented in Part I, Section 3.4.2. In the case where a positive example is a ground unit clause, the set E^- can be reasonably reduced so that only atoms with the same functor as those appearing in E^+ are considered.

Nevertheless, this set can be intractably huge for practical applications. Assume n extraction examples for a k-slot wrapper are given that are drawn from n different documents where each document has on average w words. Because the position of the word within the document is relevant for the extraction task, identical words also have to be incorporated. Thus there are $(w^k) - 1$ possible negative examples per document. In terms of logic for one document there are at least w^k atoms $extract(d, [w_1, w_2, w_3], S)$ in the Herbrand base, leaving out the number of combinations depending on the additional information S (e.g. word positions, spans). Hence the total number of *extract* atoms in the Herbrand base is $n^2 * (w^k)$. For a normal scenario of a 3-slot wrapper and 10 positive examples drawn from 10 different documents where each document has in average 1000 words, this results in $10^2 * (1000^3) = 10^{11}$ atoms. Taking the unrealistic assumption that one atom can be stored using only one byte of memory then there are still 93 giga-bytes needed to store all atoms. Of course this is a naive calculation, because almost no automated reasoning algorithm would compute the whole Herbrand base. Consequently it is much more reasonable in the wrapper learning context to use an implicit representation of negative examples instead of computing explicitly the set of negative examples by the proposed method. So the naive idea to compute the inverse Herbrand model seems not to be feasible in practice, but it serves as a theoretical model for the exhaustive enumeration idea.

Though the *non-monotonic semantics* seems to be well suited regarding the absence and implicitly determined negative examples, it lacks important features required by the wrapper learning task. Assuming the background theory B consists of a logic program P_{RTD} implementing the hypothesis language for *RTD-wrappers*, all extraction examples E, either a set of ground unit clauses or ground extraction rules, and for each example the associated clause set representing the document it belongs to. Obviously the *completness* requirement, stating that every clause g that is true in B must follow from H is not very helpful for inducing a hypothesis representing a set of extraction rules.

For illustration, consider the following scenario in which a document contains at least one *span-in-span* relation (see Section 3.4.2 Table 3.3). Consequently the corresponding clause $g : span\_in\_span(D_{ID}, s_i, s_j) \leftarrow l_1, \ldots, l_n$ contained in B is also true in $\mathcal{M}(B)$. Because per definition only hypotheses are allowed which are a model for this clause the hypothesis (extraction rule) has to model g, $H \models g$. Additionally assume none of the presented examples can be described by a *span-in-span* relation, then the learned wrapper H does not model g. Obviously this semantics does not meet the intended wrapper learning semantics, since a learned wrapper is sufficiently good enough if it models all examples with respect to a hypothesis language and given document representation. It is not necessary that it models clauses contained in the background theory as required by the completeness criterion. Consequently, in this example a sensible extraction rule H would not contain a span-in-span literal and therefore g would not follow from H. Hence an extraction rule describing solely the observable features of an example extraction is not a valid respectively complete hypothesis. On the other hand extending H to meet the *completeness* criterion involves computational overhead not needed from a mere wrapper learning task perspective.

So far, none of the presented most commonly used ILP semantics fit perfectly on the *wrapper learning task*. The best suited one is the *example setting* though until now it remains unclear how to efficiently handle the problem of missing negative examples. Finally, we introduce the *IE-ILP setting* in Definition 5.2.1.

Definition 5.2.1 (IE-ILP semantics) *Given a background theory B consisting of a normal program P (e.g. implementing a wrapper hypothesis language) and a set of ground unit clauses*

D, the logical representation of documents. Let E be an exhaustive example set regarding D of positive examples given by a set of ground units where each atom has the same predicate symbol e and arity n. Let $E^+ \subset E$ be the set of positive learning examples. Furthermore, p denotes an n-ary atom with predicate symbol e and n arguments of disjoint variables. Under the IE-ILP semantics *the definite clause set H (hypothesis) has to satisfy the following criteria regarding E^+* and *B:*

prior satisfiability & necessity *there is no substitution θ such that $B \models p\theta$*

posterior sufficiency $\forall e \in E^+ : B \wedge H \models e$

posterior satisfiability *for all substitutions θ such that $B \wedge H \models p\theta$ it holds that $p\theta \in E$*

□

The *IE-ILP semantics* differs in three minor points from the *example setting.* The *IE-ILP prior necessity* and *satisfiability* requirement is stronger, since it does not allow any positive examples to be logically entailed by B. Secondly, though the hypothesis in both semantics are definite clauses the background theory in the *IE-ILP semantics* is allowed to be a normal program. This reflects the applicatory character to understand the background theory in a procedural or logic programming sense such that it offers a maximal range for implementing extraction system components (e.g. *RTD hypothesis languages*). Thirdly, the *prior satisfiability* and *posterior satisfiability* criteria are formulated without computing E^- in the previously mentioned costly manner.

Hypotheses can be tested much more efficiently under these reformulations. Instead of using the theoretical approach of computing the complete set of negative examples based on $\mathcal{M}^-$ an iterative computation of answers (i.e. extractions) can be used. As soon as one violating answer substitution (example extraction not in E) is computed it follows that the current hypothesis does not satisfy the satisfiability criterion.

This can be easily checked by computing an answer substitution θ for p. If $p\theta$ is not ground then the satisfiability condition is violated since it is more general than any $e \in E$. If $p\theta$ is ground and not contained in the set E then $p\theta$ represents an negative example under the strong exhaustive example enumeration assumption. This test has to be repeated until a violating answer is computed. If no such answer is found the satisfiability requirement holds and the set of computed answers are used to check the *posterior sufficiency.* Obviously the *IE-ILP semantics* represents the proposed wrapper learning task to learn from positive (exhaustively enumerated) examples only under a closed world assumption.

The *posterior satisfiability* is of great relevance for the application of learned wrappers. It guarantees that a learned wrapper solely computes grounded answer substitutions. From the *posterior satisfiability* requirement it follows that only range restricted[2] hypotheses are allowed in the *IE-ILP setting.* If for instance non range restricted hypothesis clauses are learned then in the worst case a hypothesis consisting of only one unit clause like $extract(d_{ID}, X, I)$ with X and I variables would fulfill the *posterior sufficiency* criterion under certain logical calculi (e.g. SLDNF-resolution). In this cases non grounded substitutions for X are computable which yield no reasonable extractions.

[2] A clause is range restricted if every variable occurring in the head of a program clause also occurs in the body of the clause.

Not *range restricted* clauses are only one reason for non ground answer substitutions for the query $extract(d_{ID}, X, I)$. A second possible reason is the implementation of hypothesis predicates. If they are defined such that for some predicate l from a hypothesis language L a non grounding answer substitution θ for the query $P_L \wedge D \vdash l\theta$ is computable with P_L the implementation of L and D a logical representation of a document, then P_L is to be rejected. Otherwise range restricted hypothesis clauses could lead to non ground answers.

The *IE-ILP semantics* can be best compared to the assumptions taken in the ILP system *FOIDL* [Mooney and Califf, 1995] and the related successor system *CLOG* [Manandhar *et al.*, 1998]. Both systems are modifications of the *FOIL* system [Quinlan, 1990] and learn *first-order decision lists* which are ordered lists of clauses ending with a cut. Or in other terms *Prolog* rules where the order of rules is important and each rule is terminated by the *cut* operator. *FOIDL* was developed because existing systems like *GOLEM* [Muggleton and Feng, 1992] or *FOIL* were not well suited for the tackled task of learning the English past tense. This task and the wrapper learning task share two basic assumptions: 1) Background knowledge is represented as a logic program and not as ground units. 2) No negative examples are explicitly given. In contrast to *FOIDL* the *IE-ILP setting* does not require *cut* terminated and ordered rules as hypotheses, but both aim at learning one *target predicate*.

The most important similarity between these two settings is the treatment of missing explicit negative examples. [Mooney and Califf, 1995] use the notion of *output completness* to overcome the intractably large or infinite computation of negative examples in the discussed model theoretic based *CWA* approach. *Output completeness* states that the training set includes all of the correct outputs (answers) in form of so called mode declarations. With mode declarations regarding a target predicate the *FOIDL* system can determine if the output of the learned program yields non intended negative examples. Roughly speaking, mode declarations state for which arguments of a target predicate the positive example set is exhaustively enumerated. For example $append(-,-,+)$ means that all possible lists of pairs that can be appended to build a list are included in the training set (e.g. with respect to the third argument $append([], [a,b], [a,b]), append([a], [b], [a,b]), append([a,b], [], [a,b])$). Obviously the negative example treatment used in the *IE-ILP semantics* is a special case of the setting used by the *FOIDL* or *CLOG* systems. Other ILP systems capable of learning from positive examples only, but following different techniques, are for example *PROGOL4.2* [Muggleton, 1996] and *MERLIN2.0* [Boström, 1998].

5.3. Basic ILP Methods

Independent of the problem related *ILP semantics*, a method for the construction of hypotheses is needed. Such construction or learning methods are divided into incremental and non incremental methods. Up to now almost all existing ILP systems use incremental learning methods. The reason is simply found in the worse quality of a hypothesis obtained from a learning method with a missing evaluation step. Such an evaluation step allows incremental methods to refine a current hypothesis and consequently to increase the hypothesis's quality. Of course such an evaluation step would not be necessary if the non incremental method, also called *one step learning*, used a perfect or at least a sufficiently good learning operator. Unfortunately so far such a *one step learning* operator does not exist in general.

A *one step learning* method computes a hypothesis from all provided examples and background knowledge without any verification of its hypothesis. One advantage of this method lies in faster learning times in comparison to incremental approaches. In contrast, an *incremental learning technique* starts with an initial hypothesis which is refined in a loop of modification and evaluation steps using in each round some of the examples until a certain quality threshold is reached. This results in most cases in better hypotheses, allows to bias the hypothesis construction by feedback taken from the evaluation phase, but the trade off to pay is the increased learning time.

Especially, the evaluation of intermediate hypotheses during the construction is a crucial point in ILP based learning methods. In the worst case when hypotheses and background knowledge are allowed to be full first-order formulae, the testing of hypotheses can be undecidable (depending on the learning setting). This again demonstrates the importance of correctly classifying the problem task and its associated *learning setting.* Besides the correct choice of a learning calculus and specific restrictions on the hypothesis language there also exist approaches to optimize the deductive component of an ILP system. For instance [Blockeel *et al.*, 2002] propose, roughly speaking, to minimize the number of hypothesis tests (queries) by building one elaborated query by adding disjunctions and control operators (e.g. the *cut*). By intelligent *packing* of several queries for testing the validity of a hypothesis, the proof time can be significantly reduced and many redundant sub queries can be omitted.

5.3.1. Bottom-Up and Top-Down Methods

Learning methods either start from a most general hypothesis which is specialized by some operations based on the information derivable from the given examples and background knowledge, or they start with a most specific hypothesis, which is generalized until certain criteria are fulfilled. Methods starting from the examples or most specific hypothesis are called *bottom-up* methods. Starting with the most general hypothesis and trying to specialize it until no negative examples are covered any more is called *top-down* method. Constructing or learning a sufficiently good hypothesis can also be considered to be a search problem [Mitchell, 1982]. Therefore the search space is the set of all hypotheses and a goal state of this search problem is a hypothesis satisfying the criteria regarding the given ILP semantics. Consequently a bi-directional search through this space is also applicable which actually can make use of an ordered search space. For instance, such an ordering on hypotheses is given by a lattice (see Section 5.3.2) with respect to *θ-subsumption* on hypotheses (*clauses*). From this it follows that it is also reasonable to combine top-down and bottom-up ILP techniques [Mooney *et al.*, 1994].

It is hard to advocate one method over the other, since in practice it depends often on the given problem task. Bottom-up methods have the advantage that the positive examples are entailed by the initial hypothesis, because the initial hypothesis is in almost all techniques the set of examples. The arising difficulty is then to apply generalization operators such that the resulting generalized hypothesis does not become too general, hence does not cover any negative examples or becomes inconsistent. So, bottom-up methods strongly depend on the available operators used for generalization. The construction of a compressed representation of the examples regarding a background knowledge, as which a hypothesis can be understood, is solely achieved by iterative application of these operators. In general no additional search operators are applied.

In top-down approaches the crucial point is the choice of reasonable specializations for the initial most general hypothesis. For instance, if we want to learn a description for the concept $father(X,Y)$ we start with a hypothesis $father(X,Y) \leftarrow$. In fact, any literal contained in the hypothesis space is one potential candidate to extend the body of this clause to gain a more specific hypothesis. One can imagine that this space is considerably huge in certain problem domains and mere *generate and test* methods are intractable for practical applications. Consequently restrictions concerning the possible body literals [Feng and Muggleton, 1990; Dzeroski *et al.*, 1992] and general structure of the rules to be learned [Bergadano and Giordana, 1988; Kietz and Wrobel, 1991] or heuristics concerning the search through the hypothesis space are made. All these attempts to guide the search for a sufficiently good hypothesis by specific assumptions and restrictions are summarized under the notion of *inductive bias* [Mitchell, 1990]. In general two types of biases are considered, the *declarative bias*, which defines the hypothesis space, and the *preference bias* guiding the search for hypotheses. The declarative bias is distinguished into *syntactic* and *semantic* biases. The latter sets up restrictions on the behaviour or meaning of hypotheses. The *declarative bias* is also called *language bias* since it restricts the form of the hypotheses, e.g. hypotheses with rules where at most two body literals are considered.

5.3.2. ILP Operators

Two basic types of operators are commonly used in bottom-up and top-down approaches, *specalization* and *generalization operators.* A *specialization operator* is a mapping from a clause theory T onto a set of maximal specializations S of T, where maximal specialization means that no specialization S' of T exists such that S is a specialization of S'.

A *generalization operator* is a mapping from a clause theory T onto a set of minimal generalizations S of T where minimal generalization means that no generalization S' of T exists such that S is a generalization of S'.

It is fairly straightforward to see that the major task in building a good ILP algorithm reduces to determining efficient operators. Most existing ILP techniques are based on three basic operations, namely *θ-subsumption*, *inverse resolution* and *inverting implication.* A lot of research work has been put into the adaptation and modification of these three basic learning operators. Because in this thesis the presented learning algorithms are based on Plotkin's *θ-subsumption* (see Definition 2.1.15), we discuss the other two techniques only in brief.

Inverse resolution, first studied in [Muggleton and Buntine, 1988], captures the idea of inverting the resolution operator [Robinson, 1965] which is the basic component of *Prolog* and hence most successful logic programming systems based on *SLD/SLDNF - resolution* [Kowalski and Kuehner, 1971; Lloyd, 1987]. Because it has been proven that *resolution* is a complete deductive inference rule, its inverse should also be complete. So, if a (learning) example can be deductively derived under the resolution calculus from a background knowledge plus an additional clause theory (the hypothesis) then it should also be possible to construct this hypothesis from the background knowledge plus the examples and an inverted resolution operator.

More precisely, the key idea of inverting resolution is as follows: given two clauses e and c_1, where e is considered to be the resolvent of c_1, and a unknown clause c_2. Under this assumption we can find a clause c_2 such that resolving c_1 and c_2 results in e. The problem one is confronted with is that there is not a unique parent clause in general. For instance, let $e = p(a) \leftarrow r(a), s(a)$. and $c_1 = q(a) \leftarrow s(a)$. Then two possible solutions for c_2 are $p(a) \leftarrow q(a), r(a)$ or $p(X) \leftarrow q(X), r(X)$. Besides the problem to decide how to introduce or instantiate variables or turn terms into variables in the parent clause there is a second problem. There are more than these two solutions for c_2, if the assumption of disjoint body literals in c_1 and c_2 is dropped. For instance $c_2 = p(a) \leftarrow q(a), r(a), s(a)$ is such a solution. In fact, this basic idea was elaborated and four inference rules were introduced, *absorption*, *identification*, *intra-construction* and *inter-construction*. Roughly speaking they stem from observations taken from resolution proof trees, where *absorption* and *identification* describe the inverse of one resolution step and *intra* and *inter-construction* the combination of two resolution steps. Important results regarding the learnability with inverted resolution operators have been stated by [Lapointe and Matwin, 1992]. They show that inverse resolution is not capable of reversing SLD-derivations containing self-resolution steps.

Another attempt at inverting a deductive inference rule is the approach to define an *inverted implication* operator. Theorem 2.1.1 says that for any clause c_2 that is *θ-subsumed* by a clause c_1 it also holds that c_2 is a logical consequence of c_1, that is c_2 is true in all models of c_1. Unfortunately the reverse, that every c_2 which is a logical consequence from a clause c_1 is also θ-subsumed by c_1, does not hold in general. A simple counter example is, $c_1 = p(f(X)) \leftarrow p(X)$ and $c_2 = p(f(f(X)) \leftarrow p(X)$ [Plotkin, 1970]. Obviously c_1 implies c_2, because c_2 is the resolvent obtained by self-resolution of c_1. But there is no substitution such that c_1 θ-subsumes c_2. Obviously subsumption is weaker than implication and therefore *θ-subsumption* cannot be a complete learning operator. Despite these facts, depending on certain domains and biases this operator can yield promising results as we demonstrate in Part III.

However, there has been extensive research work on constructing inverted implication operators. Short introductions and discussions are given in [Muggleton and Raedt, 1994; Bergadano and Gunetti, 1996]. The major problem in inverting logical implication is, that it is undecidable for full first-order predicate logic [Schmidt-Schauss, 1988]. Even worse, in [Marcinkowski and Pacholski, 1992] it is also shown that even for Horn clauses, the problem of clause implication is undecidable. Nevertheless the study of inverse implication is essential for building self-recursive predicates. Since it is the essential concept in practical logic programming. Despite the theoretical boundaries of decidability there are a few approaches which are all based on a similar idea to invert logical implication. As a starting point they all try to identify structural regularities of terms among clauses based on observations drawn from resolution and unification properties. Such terms are indicators for clauses to be generalized. We leave out a more detailed discussion, because almost all of these approaches aim at learning recursive predicates, which is not needed in the context of wrapper learning as presented in this thesis. More details on these approaches are published in [Lapointe and Matwin, 1992; Aha *et al.*, 1994; Muggleton, 1992].

LGG-Operators

The *least general generalization* operator under *θ-subsumption* was first discussed in broader detail by Plotkin [Plotkin, 1970; Plotkin, 1971b; Plotkin, 1971a]. This operator forms one basis for both bottom-up ILP systems like *GOLEM* [Muggleton and Feng, 1992] and for top-down [Muggleton, 1995] and hybrid approaches as presented in [Zelle and Mooney, 1996]. In the following some of the important properties of *θ-subsumption* (Definition 2.1.15) are discussed. We already demonstrated that θ-subsumption is not complete with respect to logical implication among clauses, in the sense that there is a clause c which is a logical consequence of P but there is no substitution θ such that $P\theta \subset c$. Nevertheless the converse always holds, which is a helpful property for learning hypotheses. For instance (taken from [Bergadano and Gunetti, 1996]), let two clauses $C_1 = win(c_1) \leftarrow occ(p_1, x, c_1), occ(p_2, o, c_1)$ and $C_2 = win(c_2) \leftarrow occ(p_1, x, c_2), occ(p_2, x, c_2)$ be given, which represent some winning configurations in a certain board game.

Under *θ-subsumption* we can find a clause C such that $C \preceq C_1$ and $C \preceq C_2$. For instance, $C = win(X) \leftarrow occ(p1, x, X), occ(p2, Y, X)$. Obviously this is only one out of many possible clauses that hold under θ-subsumption for C_1 and C_2 and in fact this is the least general clause.

So far, it has not been discussed how to construct subsuming clauses C and how to chose the most reasonable one. A quite sufficient solution to the latter question is already given by the notion of *least general generalization* of clauses (Definition 2.1.18) and its computation on basis of the *lgg of literals* (Definition 2.1.17). The computation of an *lgg* of two clauses according to Definition 2.1.19 mainly consists of two basic steps. First, build the *lgg* of every literal in clause one with every literal in clause two. Second, remove all redundant literals with respect to θ-subsumption from the calculated set of *lgg* literals. Because of step One the maximal length of a *clause lgg* is $|C_1| \times |C_2|$ literals. Hence iterative application of the *clause lgg* can increase the length of the learned clauses in each iteration step. This is of course not intended, because a more general description normally is also a shorter one. Therefore redundant literals are removed in a second step and also literals not occuring in both clauses are abolished. Nevertheless depending on the clauses to be generalized the increasing number of literals can be one of the shortcomings.

Example 5.3.1 exemplifies a *clause lgg* computation. According to Definition 2.1.19 we have to compute the *lgg* of each literal in clause c_1 and each literal in clause c_2. Whereby it is important that identical variables are introduced for the *lgg* of identical terms. For instance, given the two clauses $\{p(a, b), \neg q(e, f(b))\}$ and $\{p(a, c), \neg q(e, f(c))\}$ then the calculated *lgg* of the subterms (e.g. $lgg(b, c) = X$) in $p(a, b)$ and $p(a, c)$ has to be applied to the subterms $f(b)$ and $f(c)$. Hence the resulting *clause lgg* is $\{p(a, X), \neg q(e, f(X))\}$.

Example 5.3.1 (Clause LGG) *Let C_1 be $win(c_1) \leftarrow occ(p_1, x, c_1), occ(p_2, o, c_1)$ and C_2 be $win(c_2) \leftarrow occ(p_1, x, c_2), occ(p_2, x, c_2)$. For the two example clauses the computation is as follows, where $(c_1, c_2)/X$ denotes the replacement of term c_1 and c_2 by the variable X:*

$$\begin{aligned}
lgg(win(c_1), win(c_2)) &= win(X) \;\; (c_1, c_2)/X \\
lgg(occ(p_1, x, c_1), occ(p_1, x, c_2)) &= occ(p_1, x, X) \;\; (c_1, c_2)/X \\
lgg(occ(p_1, x, c_1), occ(p_2, x, c_2)) &= occ(Y, x, X) \;\; (p_1, p_2)/Y, (c_1, c_2)/X \\
lgg(occ(p_2, o, c_1), occ(p_1, x, c_2)) &= occ(Y, Z, X) \;\; (o, x)/Z, (p_1, p_2)/Y, (c_1, c_2)/X \\
lgg(occ(p_2, o, c_1), occ(p_2, x, c_2)) &= occ(p2, Z, X) \;\; (o, x)/Z, (p_1, p_2)/Y, (c_1, c_2)/X
\end{aligned}$$

As one can easily observe there are a few literals subsuming each other and since we are interested in the least general clause it is necessary to reduce the obtained set of literals. The proposed algorithm by Plotkin is NP-complete, which led to further studies on the removal of redundancies from clauses under θ-subsumption by [Gottlob and Fermüller, 1983]. In this example the third literal subsumes the second and the fourth one subsumes the second and the fifth. Thus both literals are removed and the resulting clause is: $C = win(X) \leftarrow occ(p1, x, X), occ(p2, Y, X)$. ⌟

Though the *clause lgg* seems to provide a reasonable generalization of the given examples in many cases the calculated subsuming clause does not meet an intended generalization. Example 5.3.2 illustrates the weakness of naive application of the *lgg*-operator.

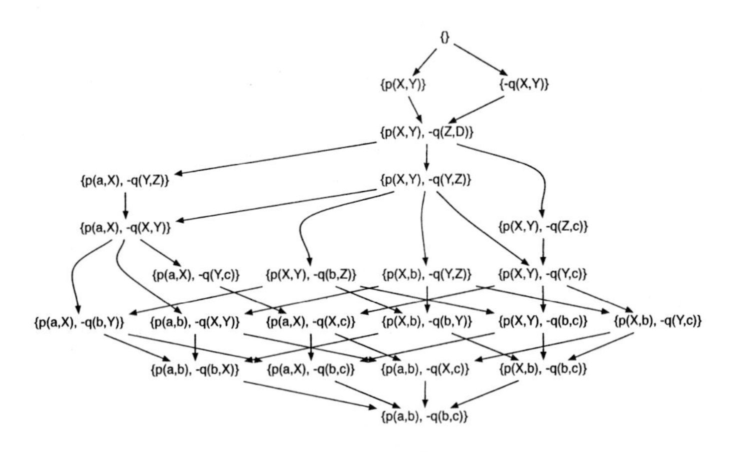

Figure 5.1.: θ-subsumption lattice for $p(a,b) \leftarrow q(b,c)$

Example 5.3.2 (Overly general LGG) *Let C_1 be $uncle(b,n) \leftarrow brother(b,k), child(k,n)$ and C_2 be $uncle(b,m) \leftarrow sister(b,j), child(j,m)$ clauses, representing knowledge about family relationships. Calculating the clause* lgg *of C_1 and C_2 results in:*

$$\begin{aligned} lgg(uncle(b,n), uncle(b,m)) &= uncle(b,X) \;\; (n,m)/X \\ lgg(child(k,n), child(j,m)) &= child(Y,X) \;\; (k,j)/Y, (n,m)/X \end{aligned}$$

The resulting clause lgg *of C_1 and C_2 is $uncle(b,X) \leftarrow child(Y,X)$. Unfortunately this clause is not a very helpful generalization, because it says that a person X is an uncle of b if X is a child of some person Y. This is of course not an intended general description for the uncle concept.* ⌟

Though this observation shows that the *clause lgg* can be calculated in a very uncomplicated straightforward manner, it is also very limited in its quality of generalization. As soon as the clauses do not contain identical predicates and don't have a simple structure it tends to build overly general clauses. In particular, if we are interested in computing the *lgg* of sets of clauses the fact that no unique *lgg* exists for sets of clauses, can become a serious problem.

There are other important properties of θ-subsumption and *clause lgg* as for example the existence of infinitely many clauses d for a clause c such that $c\theta \subseteq d$. Depending on the learning algorithm this can be a very important point. Furthermore θ-subsumption defines equivalence classes of clauses. For instance, $\{p(X), \neg q(Y)\}$, $\{p(X), \neg q(Y), \neg q(Z)\}$ θ-subsume each other and thus belong to one equivalence class. Indeed this can be a very helpful fact, since all elements of one class are logical equivalent (see Theorem 2.1.1).

One of the most important properties of θ-subsumption is the lattice order introduced on redundant free clauses. This lattice allows a structured search through the space of reasonable generalizations of a clause. Consider Figure 5.1 showing the lattice given under *θ-subsumption* regarding the clause $p(a,b) \leftarrow q(b,c)$. Finding the *least general generalization* of two clauses corresponds to identifying identical clauses c_1' and c_2' (with respect to variable renaming) in the θ-subsumption lattices of clause c_1 and c_2 such that there are no other clauses less general than c_1' and c_2' θ-subsuming c_1 and c_2. Or in other words given a set of clauses and taking θ-subsumption as ordering a lattice of clauses is defined in which the least upper bound is the *clause lgg* of c_1 and c_2.

It is important to note that if the assumption to consider only non-redundant clauses is dropped then the least generalization of a clause under θ-subsumption can be of infinite length. Assume the following clause $d : \{p(X,X), \neg q(X,X)\}$ a reasonable most specific generalization is a clause with as many literals in the body as possible. Hence $\{p(Z,W), \neg q(W,W), \neg q(Z,Z), \neg q(W,Z), \neg q(Z,W), \neg q(Z,K), \neg q(K,M) \ldots\}$ is of infinite length but obviously it θ-subsumes d because for each variable occurring in this clause a substitution can be determined that maps it onto X. The computation of the *lgg* for terms, literals and clauses as defined by Plotkin's rules (Definitions 2.1.16, 2.1.17, 2.1.18) are inductively defined over the term construction and thus do not run into infinite loops. Nevertheless it is also necessary for the *lgg* computation of clauses to remove redundant literals. Unfortunately in the general case there is not always a unique *lgg* for a set of clauses.

To conclude this brief discussion on *θ-subsumption* and the computation of *lgg* of clauses, we want to point out two important observations relevant for learning wrappers in the context introduced so far. First, if the structure of extraction rules is restricted in a reasonable way such that non-intended generalizations are omitted (see Example 5.3.2), the *lgg* operator provides promising results. Second, if such a rule (clause) representation is given, learning of wrappers (i.e. sets of generalized extractions rules) can be performed by computing the *lgg* of example extraction rules. Consequently a pure bottom-up learning method solely based on the standard *clause lgg* can be used to synthesize wrappers (i.e. logic programs). This idea is illustrated and discussed in the following sections.

5.4. Representing Wrapper Models for the IE-ILP Setting

Following strictly the definition of *RTD-Wrappers* (Definition 3.3.4) it is quite obvious how to define a reasonable *language bias* for the *IE-ILP semantics*. The language bias is defined by the notion of an *extraction rule* from Definition 3.3.4. From this definition it follows that hypotheses (wrappers) are constrained to consist of exactly one predicate (set of Horn clauses where each clause has the same positive literal). Each literal in the body of these definite or Horn program clauses belongs to a given relational description language, the set of *TDOM predicates* (see Definition 3.3.3). This hypothesis language $\mathcal{L}_{RTD}$ for learning *RTD-wrappers* is given by the set of *TDOM predicates*. The background theory consists of a normal logic program implementing the *TDOM predicates*. The *language bias* is determined by non-recursive Horn clauses consisting of one predicate. Recursion is excluded, because the head literal of an extraction rule is not included in the set of *TDOM predicates*. Accordingly, preparing and representing *RTD-wrappers* to be learned in the proposed ILP setting is straightforward.

To limit the length of hypothesis clauses we restrict each clause to consist of at most $n \times$ (|delimiter predicates|+|content predicates|+|structural predicates|) + $\frac{n(n-1)}{2} \times$|span predicates| literals taken from $\mathcal{L}_{RTD}$ with n the number of slots of the wrapper to be learned. For empty slots no literals from $\mathcal{L}_{RTD}$ are added to the clause. Further, every variable occurring in the head of an extraction rule has to be linked with a body literal (*range restricted*) and at least one constant or variable of each body literal must occur or be linked with an constant or variable in the head literal. A clause meeting these criteria is called *RTD-wrapper clause template.*

The *clause templates* used in this framework differ somewhat from those used in systems like *ML-SMART* [Bergadano *et al.*, 1989] or *MOBAL* [Kietz and Wrobel, 1991], because their *rule schemata* can be interpreted as second order rules guiding the number and order of possible body literals within a program rule. So, the clause templates used in this work can be considered to be a subset of those.

The adaptation of *AV* and *CD-wrapper models* for ILP tasks is not that straightforward. Though in Section 3.5 a naive and sketchy implementation of *AV-wrappers* was presented, some additional modifications are necessary for adapting it to the *IE-ILP setting.* Generally speaking, *AV* and *CD-wrappers* are sets of rules where each rule consists of tuples of patterns for slot fillers, delimiter texts and gaps. In general, patterns are defined by an arbitrary language capable of matching token sequences. Within this thesis the *token pattern language* is used for *AV* and *CD-wrappers.* Consequently, this token pattern language serves as basis for the hypothesis language in the context of ILP based *AV* and *CD-wrapper* construction. But in contrast to the *RTD-wrapper* hypothesis language this language cannot be used without a suitable transformation within a logic programming setting. Two possible ILP representations for the *AV* and *CD-wrapper* hypothesis language are conceivable. Firstly, each of the token pattern operators is transformed into a predicate with a document reference argument, an *atomic token pattern list,* arguments determining start and end matching positions, and additional arguments required by the operator. Secondly, an interpreter based approach as presented in Section 3.5 is used, where complex token patterns appear as arguments of particular predicates that (e.g. $pattern(Doc, Start, Pattern, End)$) implement a token pattern interpreter (e.g. matcher).

Transforming every token pattern operator into a predicate definition causes some problems. Nested patterns like `(a; (b,c)),d` cannot be represented by one definite clause by the proposed transformation, because this requires operator predicates to accept non-atomic token patterns as arguments. But this contradicts the idea not to use the interpreter approach. An alternative to represent disjunctive patterns is to use disjunctions in the body of a rule (e.g. $p \leftarrow (a; (b, c)), d$). But since the *IE-ILP semantics* requires definite clause hypotheses this results in transforming disjunctive token patterns into sets of definite clauses (e.g. $p \leftarrow a, c$, $p \leftarrow b, c, d$). All nested patterns (not only disjunctive ones) can be unfolded in this way. Obviously this leads to hypotheses larger in the number of clauses, but guarantees a simpler (function-free) representation.

Since token patterns are to be built from a finite small set of examples in the wrapper learning context, an approximated more general description for these patterns can be given by lists of atomic token patterns. For instance, given three extraction examples with right delimiters $\mathtt{r_1} = \mathtt{html}, \mathtt{int}, \mathtt{html}, \mathtt{html}$, $\mathtt{r_2} = \mathtt{html}, \mathtt{html}, \mathtt{int}, \mathtt{html}$, and $\mathtt{r_3} = \mathtt{html}, \mathtt{int}, \mathtt{int}, \mathtt{html}$. For these delimiters it seems reasonable to use a delimiter pattern like $\mathtt{r} := \mathtt{html}, (\mathtt{int}; \mathtt{html}), (\mathtt{int}; \mathtt{html}), \mathtt{html}$ for an *AV-wrapper* with maximal delimiter length of $mdl = 4$. An approximation in the proposed sense for this pattern is given by a predicate $\mathtt{rightdelimiter}(\mathtt{D_{ID}}, [\mathtt{html}, \mathtt{X_1}, \mathtt{X_2}, \mathtt{html}], \mathtt{S}, \mathtt{E})$. This predicate is true if the given atomic token sequence $[\mathtt{html}, \mathtt{X_1}, \mathtt{X_2}, \mathtt{html}]$ matches on $\mathtt{D_{ID}}$, returning start and end matching positions $\mathtt{S}$ and $\mathtt{E}$. Note, that if sequences of arbitrary length and tokens shall be matched this can be represented by a *Prolog* list with an open tail (e.g. $[\mathtt{html}|\mathtt{X}]$) as for example for representing slot filler patterns where the length is not fixed.

Obviously this approximation is more general then the original token pattern, but it still keeps the *greatest common sub pattern* (i.e. a match starting and ending with a `html` token) for length restricted patterns.

Using this representation in a bottom-up approach has the advantage that no special learning operators for token patterns have to be invented. Normally, in a bottom-up approach we would start with a set of token sequences and would have to construct a reasonable token pattern from the given example sequences. This involves to observe significant common features and a generalization step, which results in a set of operators for inventing and selecting different token pattern operators. Since we investigate in this thesis solely bottom-up learning techniques based on *lgg* operations, this representation allows us to use the same learning techniques for *AV* and *CD* and *RTD-wrappers*. The modified and restricted hypothesis language for *AV* and *CD-wrappers* used in Chapter 6 and 7 is shown in Table 5.1.

`leftdelimiter(D,P,S,E)` the predicate holds if the reverse list of atomic token patterns `P` matches (token unifies) on document `D` starting at token position `S` ending at token position `E`. This is equivalent to a token pattern $P' = p_n, p_{n-1}, \ldots, p_1$ with `P'` the conjunction of atomic token patterns p_i in $P = [p_1, p_2, \ldots, p_n]$.
`rightdelimiter(D,P,S,E)` the predicate holds if the list of atomic token patterns `P` matches (token unify) on document `D` starting at token position `S` ending at token position `E`. This is equivalent to a token pattern $P' = p_1, p_2, \ldots, p_n$ with `P'` the conjunction of atomic token patterns p_i in $P = [p_1, p_2, \ldots, p_n]$.
`filler(D,X,P,S,E)` the predicate holds if the list[3] of atomic token patterns `P` matches (token unifies) on document `D` starting at token position `S` ending at token position `E` with `X` the matched text.
`maxlength(P,N)` the predicate holds if the list `P` of atomic token patterns is at most of length `N`. This predicate in conjunction with the `filler` predicate and same pattern list `P` occurring within one clause is equivalent to the token pattern `maxlength(N,P')`.
`gap(D,P,S,E)` the predicate holds if the list of atomic token patterns `P` matches (token unifies) on document `D` starting at token position `S` ending at token position `E`.
`notcontains(S,P)` the predicate holds if the list of atomic token patterns `P` does not token unify with any sub-list of the token sequence `S`.

Table 5.1.: logical hypothesis language for *AV* and *CD-wrappers*

For *AV* and *CD-wrapper* learning, a language bias determined by clause templates is defined in Definitions 5.4.1 and 5.4.2. The similarity between the *AV-wrapper* definition and the logical program clause template representation is apparent. The clause template models exactly a rule scheme of an *AV-wrapper* (i.e. $((L_1, F_1, R_1), G_1, (L_2, F_2, R_2), \ldots))$. The only important difference between the proposed *AV-wrapper model* and the clause template representation of *AV-wrapper models* lies in the chosen wrapper language. The clause representation misses equivalent predicates for the token pattern operators `times(n,P)`, `upto(n,P)` and `not(P)`. The iteration operators (e.g. `?,*,+`) can be approximated by sets of clauses in the discussed manner as long as an upper length for matched sequences is given. This always holds for delimiters because of the *maximal delimiter length* criterion (see Part I Section 3.3.1). For slot filler sequences and gap fillers the upper bound is derived from the presented examples.

Definition 5.4.1 (AV-wrapper clause template) *An* AV-wrapper clause template *for an* n-slot wrapper W *is of the form:*

$$\texttt{extract}(\texttt{D}, [\texttt{X}_1, \ldots, \texttt{X}_n], [(\texttt{S}_1, \texttt{E}_1), \ldots, (\texttt{S}_n, \texttt{E}_n)]) \leftarrow B_1, G_1, B_2, G_2, \ldots, G_{n-1}, B_n$$

where every B_i (filler block) *consists of the conjunction of literals:*

```
leftdelimiter(D, L, LS, FS),
filler(D, X, F, FS, FE), maxlength(F, N),
rightdelimiter(D, R, FE, RE),
```

and every G_i (gap block) *consists of the conjunction of literals:*

```
gap(D, G, GS, GE), maxlength(G, M),
```

For every $\texttt{X}_i$ *in* $[\texttt{X}_1, \ldots, \texttt{X}_n]$ *and corresponding* $(\texttt{S}_i, \texttt{E}_i)$ *in* $[(\texttt{S}_1, \texttt{E}_1), \ldots, (\texttt{S}_n, \texttt{E}_n)]$ *there exists exactly one* filler block B_m *such that for variables* $(\texttt{X}, \texttt{FS}, \texttt{FE})$ *in* B_m *it holds that:* $\texttt{X} = \texttt{X}_i$, $\texttt{FS} = \texttt{S}_i$ *and* $\texttt{FE} = \texttt{E}_i$. *Every* B_m *is related to exactly one* $\texttt{X}_i$ *in* $[\texttt{X}_1, \ldots, \texttt{X}_n]$. *For variables* $(\texttt{RE})$ *in* B_i, $(\texttt{GS}, \texttt{GE})$ *in* G_i *and* $(\texttt{LS})$ *in* B_{i+1} *it holds that:* $\texttt{RE} = \texttt{GS}$, $\texttt{GE} = \texttt{LS}$.

□

Only a slight modification is necessary to build a clause template for *CD-wrappers* from an *AV-wrapper clause template.* Therefore every filler block and gap block is extended by a `notcontains` literal as stated in Definition 5.4.2.

Definition 5.4.2 (CD-wrapper clause template) *An* CD-wrapper clause template *for an* n-slot wrapper W *is of the form:*

$$\texttt{extract}(\texttt{D}, [\texttt{X}_1, \ldots, \texttt{X}_n], [(\texttt{S}_1, \texttt{E}_1), \ldots, (\texttt{S}_n, \texttt{E}_n)]) \leftarrow B_1, G_1, B_2, G_2, \ldots, G_{n-1}, B_n$$

where every B_i (filler block) *consists of the conjunction of literals:*

```
leftdelimiter(D, L, LS, FS),
filler(D, X, F, FS, FE), maxlength(F, N),
notcontains(F, R),
rightdelimiter(D, R, FE, RE),
```

and every G_i (gap block) *consists of the conjunction of literals:*

```
gap(D, G, GS, GE), maxlength(G, M), notcontains(G, NL),
```

For every $\texttt{X}_i$ *in* $[\texttt{X}_1, \ldots, \texttt{X}_n]$ *and corresponding* $(\texttt{S}_i, \texttt{E}_i)$ *in* $[(\texttt{S}_1, \texttt{E}_1), \ldots, (\texttt{S}_n, \texttt{E}_n)]$ *there exists exactly one* filler block B_m *such that for variables* $(\texttt{X}, \texttt{FS}, \texttt{FE})$ *in* B_m *it holds that:* $\texttt{X} = \texttt{X}_i$, $\texttt{FS} = \texttt{S}_i$ *and* $\texttt{FE} = \texttt{E}_i$. *Every* B_m *is related to exactly one* $\texttt{X}_i$ *in* $[\texttt{X}_1, \ldots, \texttt{X}_n]$. *For variables* $(\texttt{RE})$ *in* B_i, $(\texttt{GS}, \texttt{GE}, \texttt{NL})$ *in* G_i *and* $(\texttt{LS}, \texttt{L})$ *in* B_{i+1} *it holds that:* $\texttt{RE} = \texttt{GS}$, $\texttt{GE} = \texttt{LS}$, $\texttt{NL} = \texttt{L}$.

□

The template representation of a *CD-wrapper* differs from the definition of a *CD-wrapper model* in that no constraints on the left and right delimiters are stated. This was justified by observations taken from experiments on HTML documents. Based on the assumption that delimiters mostly contain HTML tags it is observable that if slot fillers contain HTML tags it happens that left delimiter constraints become too restrictive after generalization of slot filler and delimiter patterns. This results in extraction failures. Hence, less constrained wrapper models are investigated. Nevertheless, the clause template representation still conforms with the *CD-wrapper model* definition and using empty constraint patterns has no consequences for the presented learning algorithms.

5.5. Preparing and Representing Examples

The preceding sections introduced and discussed an appropriate learning setting for the automatic construction of *AV*, *CD* and *RTD-wrappers* in an ILP context. What remains to de clarified is how example extractions are modfied to fit into the proposed *IE-ILP setting*. We already mentioned that the *IE-ILP setting* expects positive examples to be units (e.g. ground unit clauses) all having the same predicate symbol. For the following the predicate symbol `extract` is used. The overall goal of the learning task is to learn a definition (e.g. hypothesis) for this target predicate `extract`. In Definition 2.1.2 and 2.2.1 an example (extraction tuple) is determined by a tuple of words $\vec{w}$ from a document D. For *AV*, *CD* and *RTD-wrapper* learning it is necessary to find a suitable representation for each example. Since *AV* and *CD-wrappers* work with a linear document representation ($AV(D)$, Definition 2.3.1), each argument $w.i$[4] of $\vec{w}$ is determined by its start and end position in the linear ordered set $AV(D)$. Hence, every example is represented by an instance of the target predicate in the following way: $\mathtt{extract}(\mathtt{D_{ID}}, [\mathtt{w_1}, \mathtt{w_2}, \ldots, \mathtt{w_n}], [(\mathtt{s_1}, \mathtt{e_1}), (\mathtt{s_2}, \mathtt{e_2}), \ldots, (\mathtt{s_n}, \mathtt{e_n})]$ with $\mathtt{s_i}$ the starting token position and $\mathtt{e_i}$ the end position of $Tok(w_i)$ in $AV(D_{ID})$. This representation is called *AV* and *CD extraction example*. In a similar manner *RTD extraction examples* are constructed. Instead of containing start and end token positions the last argument of the target predicate consists of a list containing minimal spans $[(\mathtt{s_1}, \mathtt{l_1}, \mathtt{r_1}), (\mathtt{s_2}, \mathtt{l_2}, \mathtt{r_2}), \ldots, (\mathtt{s_n}, \mathtt{l_n}, \mathtt{r_n})]$ determining the position of each $w.i$ in the *TDOM-representation* of D. If w_i is the empty word (i.e. empty slot represented by a *Prolog* empty atom `''`) then the corresponding start and end positions (spans) are represented by constants (e.g. *empty*). The representation of wrapper examples $(D, w) \in \mathcal{W}$ as instances of the target predicate is called *logical example representation.* By $E_M(E^+)$ the set of *logical example representations* of examples in E^+ regarding the wrapper model M is denoted.

In fact the mere token positions and span positions for each example are only a fraction of the information that can be observed in advance before starting to learn. Finally, each example can be represented by a ground clause template. This idea of *example description clauses* is summarized by the Definitions 5.5.1 and 5.5.2.

Definition 5.5.1 (AV and CD Example Description Clause) *Let W be an n-slot wrapper and $(D, (w_1, w_2, \ldots, w_n)) \in W$ an example extraction from D with document identifier D_{ID}. Further let $I = [(s_1, e_1), (s_2, e_2), \ldots, (s_n, e_n)]$ be the list of tuples containing start and end positions for each w_i regarding $AV(D)$. Let mdl be the maximal delimiter length. Additionally, C is an* AV *or* CD-wrapper *clause template and P a normal logic program implementing predicates occurring in the body of C (hypothesis language $\mathcal{L}_{AV}$ and $\mathcal{L}_{CD}$ respectively).*

[4]In the actual implementation of the learning system *LIPX* each $w.i$ is represented by a white-space separated *Prolog* list of words. For instance, let $w.i$ be "Hello World" then $w.i$ is represented as `['Hello','World']`. Because each word in $w.i$ occurs as one token in $AV(D)$ and thus has to be referable as an argument within the word list.

Let $\theta = \{D/D_{ID}, \vec{X}/(w_1, w_2, \ldots, w_n), \vec{S}/((s_1, e_1), (s_2, e_2), \ldots, (s_n, e_n))\}$ *be a substitution for document identifier* D*, slot variables* $\vec{X}$ *and position variables* $\vec{S}$ *occurring in the head literal of* C*. Then a clause* $R = C\theta\theta'$ *is called* AV and CD example description clause *iff:*

- θ' *is ground*
- $P \wedge AV(D) \wedge C\theta\theta' \models extract(D_{ID}, [w_1, w_2, \ldots, w_n], I)$
- *every pattern of any* `leftdelimiter` *and* `rightdelimiter` *predicate in* $C\theta\theta'$ *is at most of length mdl.*

By $cd_P^C(e, I) = R$ *with* $e \in W$ *the* example description clause *for example e regarding program P implementing* $\mathcal{L}_{AV}$ *and* $\mathcal{L}_{CD}$ *respectively, position information I and clause template C is denoted.* □

Not for every example a *CD example description clause* exists. A simple example illustrates the incompleteness of *CD example description clauses*. Assume an example for a 1-slot wrapper is given. The example text contains the symbol : and the minimal delimiter length is 1. Immediately succeeding the slot filler text, the symbol : appears in the document. Hence : functions as right delimiter for the slot. Consequently, no substitution θ can be found such that the requirements are fulfilled, because the predicate `notcontains`$('\ldots:\ldots', ':')$[5] will always be false.

Definition 5.5.2 (RTD Example Description Clause) *Let* W *be an n-slot wrapper and* $(D, (w_1, w_2, \ldots, w_n)) \in W$ *an example extraction from D with identifier* D_{ID}*. Further let* $I = [(n_1, l_1, r_1), (n_2, l_2, r_2), \ldots, (n_n, l_n, r_n)]$ *be the list of minimal spans for each* $w_1, \ldots, w_n$ *regarding* $TD(D)$*. Let* $\mathcal{L}_{RTD}$ *be a hypothesis language for* RTD-wrapper models *and P a normal logic program implementing all predicates in* $\mathcal{L}_{RTD}$*. Further let d be the context distance and C be a* RTD-wrapper clause template *regarding* $\mathcal{L}_{RTD}$*.*

Let $\theta = \{D/D_{ID}, \vec{X}/(w_1, w_2, \ldots, w_n), \vec{S}/((n_1, s_1, e_1), (n_2, s_2, e_2), \ldots, (n_n, s_n, e_n))\}$ *be a substitution for document identifier D, slot variables* $\vec{X}$ *and span variables* $\vec{S}$ *occurring in the head literal of C. Then a clause* $R = C\theta\theta'$ *is called* RTD example description clause *iff:*

- θ' *is ground*
- $P \wedge TD(D) \wedge C\theta\theta' \models extract(D_{ID}, [w_1, w_2, \ldots, w_n], I)$
- *there is no other* RTD-wrapper clause template C' *such that* C' *has more literals than* C*.*
- *every context distance related predicate has at most context distance d.*

By $cd_P^C(e, I) = R$ *with* $e \in W$ *the* example description clause *for example e regarding program P implementing* $\mathcal{L}_{RTD}$*, position information I and clause template C is denoted.* □

Given a set E^+ of examples then $CD_M(E^+)$ denotes the set of *example description clauses* for the wrapper model M and examples in E^+.

RTD example description clauses can be easily computed with the help of a *Prolog* system. Therefore a *Prolog* program P for $\mathcal{L}_{RTD}$ has to be implemented. The *RTD* document representation is already in *Prolog* syntax and thus can be used without changes. Each example is given

[5] For ease of notation here we use strings and not the token representation

by $e = (D, (w_1, w_2, \ldots, w_n), (s_1, s_2, \ldots, s_n))$ consisting of a document D with identifier D_{ID}, a text tuple w and tuple of minimal spans s for w. The set of body literals for a *RTD example description rule* regarding e is given by $Body(e)$ the set of ground literals $l\theta\sigma$ derivable from $P \wedge TD(D) \vdash l\theta\sigma$ with $l \in \mathcal{L}_{RTD}$, σ a computed ground answer and θ a ground substitution according to the semantics of l. This means that for each example e the corresponding D_{ID}, w_i and s_i is used to partially instantiate l to compute a description for e. Here $\vdash$ denotes the logical derivation operator and we assume a standard logical calculus (e.g. SLDNF-Resolution [Lloyd, 1987]). The conjunction atoms in $Body(e)$ plus the ground target predicate `extract` with respect to e forms the *RTD exmaple description clause* for e.

Computing an *AV* or *CD example description clause* is easily achieved. Therefore a *Prolog* program P for $\mathcal{L}_{AV}$ or $\mathcal{L}_{CD}$ has to be implemented. Similar to computing the set of body literals for *RTD example descriptions* the instantiations for `filler`, `gap` and `maxlength` predicates of the *AV* and *CD-wrapper clause templates* are computed from $P \wedge AV(D)$. The computed answer substitutions are applied to the whole clause template. Missing start and end positions for the delimiter predicates are computed on basis of the given *maximal delimiter length.* Finally, grounding answer substitutions for the partially instantiated delimiter predicates are computed and the appropriate order of body literals (filler and gap blocks) with the according dependencies are estimated according to the slot filler's text position. This process results in a ground clause template with respect to the given example.

6. One Step Learning of Wrappers

In this chapter a supervised, off-line, one step ILP learning approach for synthesizing wrappers is presented. In off-line learning the learner is presented a set of pre-classified examples, from which without any teacher interaction a hypothesis is computed. *One step learning* (*OSL*) means, that all examples are used only once for the construction of a hypothesis. This means hypotheses are not constructed in an iterative process of evaluation and refinement based on example subsets, like in incremental learning systems. In general, one step systems do not yield as good results regarding the hypothesis quality as incremental systems, because of the missing evaluation of hypotheses during the learning phase. On the other hand a *one step learning* system needs less learning time. This can be of great practical relevance, if fast learning is required and higher error rates are tolerated. For example, this is the case in application scenarios where a user revises extraction results and can sort out false or erroneous extractions.

Intuitively, a one step system only makes sense if the crucial components of a learning scenario, which are the size of the hypothesis space, the hypothesis structure (i.e. clause structure), are reduced. *OSL* learning of *AV* and *CD-wrapper models* seems to be attractive for one major reason: both wrapper models have a fixed structure, for which reasonable clause templates are given in Definition 5.4.1 and 5.4.2. Clause templates impose a very strong language bias and hence reduce the search space for good hypotheses significantly.

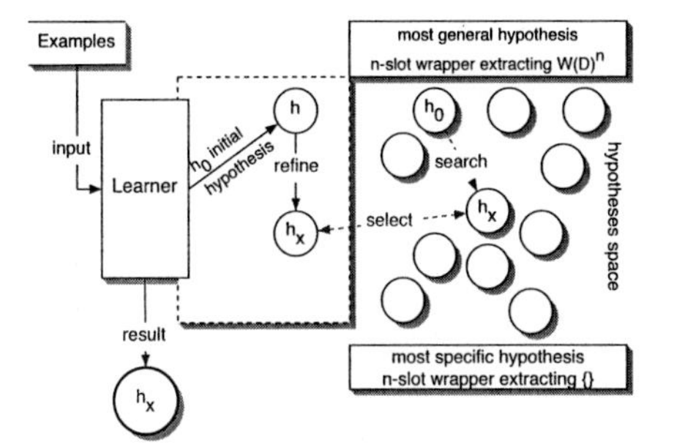

Figure 6.1.: One Step Learning strategy as one search step hypothesis selection

This wrapper learning task consists of finding a sensible partial instantiation of a clause template such that all examples are derivable from the instantiated template union the program implementing the template literals. Because the space of instantiations can be intractably large, it is more feasible to determine instantiations by generalizing from a set of ground instantiated clause templates than to determine these suitable substitutions in a top down manner.

In the case of *RTD-wrappers* the clause templates may vary in its number of literals. This underlines that the idea to start from grounded clause templates seems to be a promising idea, because this saves the learning algorithm from finding suitable body literals, variable bindings and constants in a general-to-specific search, for which it seems unclear how to be done in a one-step manner. In contrary to *AV* and *CD-wrappers*, the assumption to minimize the hypothesis space by strong constraints on the structure of a hypothesis clause is a lot weaker (see Section 5.4). This results in a requirement for the one-step learning operator not only to find suitable generalizing substitutions (i.e. terms are substituted by variables) but also to adapt the structure of the hypothesis clauses by elimination of conflicting literals from the clause templates.

6.1. LGG based Bottom-Up Learning

The idea of starting from grounded instantiated example clause templates to construct a generalized example description fits perfectly into the concept of using *least general generalization* operators and *example description clauses* for the bottom-up based construction of wrappers. In Section 5.3.2 we discussed several properties of *lgg* operators. One crucial point of the *clause lgg* operator is its affinity to produce overly general results, based on certain structural criteria of clauses. Two reasons can be identified for explaining some over-generalizations:

literal elimination Given two clauses c_1 and c_2 containing literals l_1 and l_2 with different predicate symbols then l_1 and l_2 are eliminated per *clause lgg* definition from the resulting *lgg* of c_1 and c_2.

unintended compression Given two program clauses c_1 and c_2 with multiple occurrences of a literal p in c_1 and c_2 where each p is linked to one different argument in the head of the clause. For instance, given two reduced grounded clause templates:

```
c1 : extract(D, ['book','5'], [(10,11), (11,12)]) ←
                filler(D, 'book', [word], (10,11)),
                filler(D, '5', [int], (11,12)).
c2 : extract(D, ['12.9',z3], [(20,21), (21,22)]) ←
                filler(D, '12.9', [real], (20,21)),
                filler(D, z3, [alphanum], (21,22)).
```

The clause-lgg of c_1 and c_2 before the removal of redundant literals is:

$$\mathtt{extract}(\mathtt{D},[\mathtt{X}_1,\mathtt{X}_2],[(\mathtt{S}_1,\mathtt{E}_1),(\mathtt{S}_2,\mathtt{E}_2)]) \leftarrow \mathtt{filler}(\mathtt{D},\mathtt{X}_1,\mathtt{T}_1,(\mathtt{S}_1,\mathtt{E}_1)),\mathtt{filler}(\mathtt{D},\mathtt{X}_3,\mathtt{T}_2,(\mathtt{S}_3,\mathtt{E}_3)), \mathtt{filler}(\mathtt{D},\mathtt{X}_4,\mathtt{T}_3,(\mathtt{S}_4,\mathtt{E}_4)),\mathtt{filler}(\mathtt{D},\mathtt{X}_2,\mathtt{T}_4,(\mathtt{S}_2,\mathtt{E}_2)).$$

And after the removal of redundant literals:

$$\mathtt{extract}(\mathtt{D},[\mathtt{X}_1,\mathtt{X}_2],[(\mathtt{S}_1,\mathtt{E}_1),(\mathtt{S}_2,\mathtt{E}_2)]) \leftarrow \mathtt{filler}(\mathtt{D},\mathtt{X}_1,\mathtt{T}_1,(\mathtt{S}_1,\mathtt{E}_1)).$$

If a hypothesis clause is understood as a concept description, then every literal in the body of this rule describes a certain property of the concept. Hence *literal elimination* can be interpreted to determine the largest common set of property describing predicates. Indeed, this effect is helpful to reduce *RTD-wrapper example description clauses* to find a unifying representing description.

Another observable property of the *clause lgg* operation is, what we call *unintended compression*. Consider the given example in which two reduced versions of example description clauses are generalized under the *clause lgg* operation. The basic idea of all logical representations of wrapper models is that a subset of body literals in each rule describe a certain slot filler text which is appearing as argument in the head literal of the rule. Consequently generalizations among such clauses are expected to generalize only those slot descriptions related to each other and not any description with any other. And furthermore even if the resulting set of generalized literals contains redundant literals, it must be granted that for each slot argument (variable) in the head literal a linked body literal exists. This requirement is also known under the notion of *range restricted clauses*. It states that every variable occurring in the head of a program clause must occur in the body of the clause. Obviously the standard definition of a *clause lgg* does not suffice this desired requirement for the special case of learning wrappers from *example description clauses.*

Without changing the definition of the *clause lgg*, there are two possible solutions to this problem. The first one consists in transforming all example description clauses into a set of clauses with different predicate definitions. Then this set of clauses is partitioned according to the predicates and generalized under the *clause lgg* operation. For instance, given a example description clause $e \leftarrow block_1, block_2$. where $block_i$ is a conjunction of literals describing a slot filler i. This clause is transformed into an equivalent clause set: $e \leftarrow s_1, s_2$ with s_1 and s_2 having different predicate symbols and two program clauses $s_1 \leftarrow block_1$ and $s_2 \leftarrow block_2$. Obviously generalizing each set of s_i clauses separately will overcome the illustrated problem.

A second solution simply renames the predicate symbols in the body of each example description rule according to the slot fillers they describe. This is called *prefix protection* [Thomas, 2003], because by adding a prefix to each predicate symbol of a body literal, they are protected against unintended compression. The prefix is chosen according to the slot the literal is related to. For instance, the clause c_1 from the unintended compression example is transformed into:

```
c1 : extract(D, ['book','5'], [(10,11), (11,12)]) ←
                slot1_filler(D, 'book', [word], (10,11)),
                slot2_filler(D, '5', [int], (11,12)).
```

The *prefix protection* can be applied before calculating the clause lgg. Or alternatively the background theory is modified such that prefix protected predicates are added to it (e.g. `slot1_filler` ← `filler`). In the following the *prefix protection* is used and an according modification of the background theory is assumed.

6.2. One Step Learning Algorithm

Next the basic *one step learning* algorithm is presented (*basic-OSL*). Assume a set of positive examples $E^+ \subset \mathcal{W}$ is given, from which the corresponding *logical example representation* $E_M(E^+)$ is calculated. $E_M(E^+)$ is the set of instances of the target predicate with respect to the wrapper model M (see Section 5.5). Then the *example description clause set* $CD_M(E^+)$ for E^+ with M the wrapper model (e.g. *AV*,*CD* or *RTD*) is calculated. Computing CD_M also requires a normal logic program P_M, implementing the hypothesis language of wrapper model M.

The learning of a logic program H, that implements a wrapper model M for a wrapper $\mathcal{W}$ from examples E^+, reduces to the mere computation of the *clause lgg* of a set of *example description clauses* $CD_M(E^+)$. Algorithm 1 shows the *clause lgg* based *one step learning* algorithm.

The *lgg* of a set of clauses (Definition 6.2.1) is easily computed by iterative application of the *clause lgg* of two clauses as shown in Function 2. Because the binary *clause lgg* operator (Definition 2.1.19) is commutative and associative the selection order of clauses to be generalized is unimportant.

Definition 6.2.1 (Clause Set LGG) *Let C be an arbitrary set of clauses.* The clause set lgg of C *is defined as:*

$$clause\_set\_lgg(C) = \begin{cases} undef. & : \; C = \emptyset \\ c & : \; C = \{c\} \\ clause\_lgg(c_1, c_2) & : \; C = \{c_1, c_2\} \\ clause\_lgg(c', clause\_set\_lgg(C \setminus \{c'\})) & : \; |C| > 2 \end{cases}$$

□

---

input : $E^+ \subset \mathcal{W}$ a set of examples for a wrapper $\mathcal{W}$ to be learned
M the wrapper model to be learned
$R \leftarrow CD_M(E^+)$
$c \leftarrow clause\_set\_lgg(R)$
return c

Algorithm 1: *OSL*

---

function $clause\_set\_lgg(C)$
input : C a set of clauses
$lgg \in C$
$C \leftarrow C \setminus \{lgg\}$
while $C \neq \emptyset$ **do**
  $c \in C$
  $C \leftarrow C \setminus \{c\}$
  $lgg \leftarrow clause\_lgg(lgg, c)$
return lgg

Function 2: clause set lgg

At first glance the question arises if this very simple learning technique can provide reasonable synthesized wrappers. Therefore it is important to recall some important points. The use of clause templates in the case of *AV* and *CD-wrapper models* impose a fixed structure of clauses that also holds under the *clause set lgg* operator, since all clause description clauses have an identical set of literals. Hence effects like *literal elimination* do not occur. Furthermore, if the clause template is chosen in the right way, that means if it models the structural properties of the wrapper to be learned in a sufficient way, then the expectable quality of the learned wrapper solely depends on the quality of generalized atomic token patterns. Taking the assumption that a teacher has a good understanding of determining clause templates and wrapper models, then the crucial point remaining is the generalization of delimiter patterns. So, if delimiters and hence atomic token patterns of the example description clauses differ too much, the simple application of the *clause lgg* operator will obviously result in overly general delimiter patterns. In fact, this is a general problem showing the overall limitation of a pure delimiter based wrapper concept. Luckily, experiments (Chapter 10) show that for common test sets used in the *IE-ML* research community these concerns do not become true. These concerns also hold for *OSL* learning of *RTD-wrappers*, because *literal elimination* is to happen with high probability. Nevertheless, as for *AV* and *CD-wrappers* it also strongly depends on the heterogeneity of examples regarding the observable relational properties like ancestor nodes, brother nodes etc. Regardless of these issues several other entrapments of the *one step learning technique* can be identified. *Inconsistency* problems regarding the coverage of learning examples are discussed in Section 6.4. A second entrapment is an observable increasing number of false extractions, if learning examples contain *empty slot fillers*, or in general, if ϵ and ϵ^+*-class wrappers* are to be learned. In the following Section a refinement for this problem is presented.

6.3. Refinements

Using the *lgg* operator for generalizing strongly differing token patterns results typically in too general patterns. The natural way to overcome this problem is to introduce disjunctive patterns. This would require either to modify the computation of the *clause set lgg* to identify *problematic terms* and to provide extended *lgg* operations or problematic terms can be identified in advance and the example sets are partitioned. Then each of the partitioned example set can be generalized independently. The second approach has the advantage that the standard *lgg* operator remains unmodified and therefore no new semantics and properties have to be introduced and proven. Though this idea is still conform with the *one step learning* scenario, because it does not use any hypotheses test and refinement cycles, it yields a set of learned clauses instead of a single clause as defined in the basic version.

Especially when learning wrappers belonging to the ϵ and ϵ^+*-class* (Section 3.1) the examples differ strongly regarding their slot filler texts and delimiters. Intuitively it is not very sensible to generalize the slot filler patterns of an empty slot example with an non empty one. For AV and *CD-models* this results in the most general filler pattern allowing to match every token sequence up to the length of the non empty slot filler. For *RTD-models* this results in elimination of all predicates describing the slot text of the non empty slot example, which leads to a generalized rule describing no property regarding the specific slot (i.e. no filler ever).

Since all introduced wrapper models allow by definition to consist of several *extraction rules* (i.e. a target predicate definition of more than one program clause) it is reasonable to detect heterogeneity among examples regarding empty slots before starting one step learning. Algorithm 3 shows the extended version of the *one step learning* algorithm, called ϵ*-OSL*.

input : $E^+ \subset \mathcal{W}$ a set of examples for a wrapper $\mathcal{W}$ to be learned
M the wrapper model to be learned

$LearnedRules \leftarrow \emptyset$
$S \leftarrow partition(E^+)$
while $S \neq \emptyset$ **do**
  $E \in S$
  $R \leftarrow CD_M(E)$
  $c \leftarrow clause\_set\_lgg(R)$
  $LearnedRules \leftarrow LearnedRules \cup \{c\}$
  $S \leftarrow S \setminus \{E\}$
return $LearnedRules$

Algorithm 3: ϵ-*OSL*

The function $partition(E^+)$ calculates the set S of non empty subsets E_i of E^+ regarding examples in E^+ having an identical number and positions of empty slots such that E^+ is the disjunctive union of all $E_i \in S$. The consistency results for the *basic-OSL* (see Section 6.4) also hold for the extended version, since generalization of one partitioned example set is an application of the *basic-OSL* algorithm. But, by reducing the probability of false extractions because of using less generalized patterns and preventing the elimination of predicates respectively, the resulting wrapper (i.e. clause set) can become too specific. This means, the learned wrapper covers only the presented examples and no intended extractions from future documents. This effect takes place if all examples differ in its number and positions of empty slots. Or in other terms, given E^+ containing n examples and the partitioning of E^+ yields n partitions, then the proposed learning algorithm demotes to simply storing the example description clauses of each provided example, which abolishes the ability to predict. Depending on the quality of examples, this does not have to be a disadvantage, because in contrast to the basic algorithm the partitioning intuitively results in increased precision rates and decreased recall rates of the learned wrapper. Because one step generalization of the complete example set leads to over-generalization (*AV* and *CD-wrapper*) and clearly false extractions with missing slot fillers (*RTD-wrappers*), ϵ-*OSL* seems to be an preferable step to raise the overall quality of the learned wrapper although yielding decreased recall values.

6.4. Properties of OSL

In Section 5.5 it is discussed that a *CD example description clauses* does not always exist due to the underlying assumed structure of a *CD wrapper* and the defined clause template. The question arises if *OSL* in general learns consistent wrappers, i.e. wrappers that cover all learning examples (see Definition 2.2.4 and 3.2.3). Note that the notion of consistency for wrappers (Definition 2.2.4) and logical formulae (theories) (Section 2.1.2) are not identical concepts.

The following Theorems 6.4.1, 6.4.3 and 6.4.5 summarize the most important consistency issues regarding *OSL* learning of *AV*,*CD* and *RTD-wrappers* with respect to the ϵ and ϵ^+ wrapper classes.

Two reasons are identified, that causes *OSL* to become inconsistent in some settings. Firstly, the elimination of certain literals under *clause lgg* operation can yield inconsistent wrappers as discussed in Section 6.3. Secondly, the problem of *floundering* in the context of *SLDNF-resolution* [Lloyd, 1987]. Roughly speaking, *floundering* describes the problem that different orders of body literals in a program clause result in different computed answer sets. *Floundering* can only occur in logic programs using negation (negated body literals). From a mere logic perspective the order of literals within a clause is irrelevant, therefore *floundering* causes some unintended results in logic programing systems. It is typically present in standard *Prolog* systems, like the one used for implementing the algorithms presented in this thesis. In the remainder we denote by $\vdash_f$ a floundering logical calculus.

Before we investigate these two claims more detailed Theorem 6.4.1 and the following proof shows that under a model theoretic semantics *OSL* learning of *AV* and *CD* wrappers is always consistent.

Theorem 6.4.1 OSL *learned AV and CD-wrappers are consistent under logical consequence* ($\models$).

Proof 6.4.2 (Theorem 6.4.1) *Proof Assumption: Let P be an appropriate clause set definition of the hypothesis language for a wrapper model M (e.g. AV,CD, or RTD). Let E^+ be a set of examples such that for all $e \in E^+$ an example description clause $cd_P^C(e, I)$ regarding M and P exists. Further let D be the set of appropriate document representations of D' occurring in E^+ with respect to the wrapper model M (e.g. $AV(D')$ and $TD(D')$).*

Proof Assertion: For E^+ with $S = CD_M(E^+)$ regarding wrapper model M (e.g. AV or CD) there exists a clause C with $C = clause\_set\_lgg(S)$ which is an explanation for E^+ such that $\forall e \in E_M(E^+) : P \wedge D \wedge C \models e$ holds.

Proof by induction on the number of examples in $E_M(E^+)$
Base cases:

1. *if $|E_M(E^+)| = 1$ with $E_M(E^+) = \{e_1\}$ then the clause set lgg of $CD_M(E^+) = \{c_1\}$ contains a single clause c_1 given by $cd_P^C(e, I)$ and from the proof assumption and Definition 5.5.1 it follows that $P \wedge D \wedge c_1 \models e$.*

2. *if $|E_M(E^+)| = 2$ with $E_M(E^+) = \{e_1, e_2\}$ and corresponding example description clauses $c_1 = cd_P^C(e_1, I_1)$ and $c_2 = cd_P^C(e_2, I_2)$ then the clause set lgg c of $CD_M(E^+) = \{c_1, c_2\}$ is given by the clause $c = clause\_lgg(c_1, c_2)$. From Definition 5.5.1 it follows that $e_1 \subseteq c_1$ and $e_2 \subseteq c_2$ and therefore $lgg(e_1, e_2)$ is in c. From Theorem 2.1.1 it follows that $c \models c_1$ and $c \models c_2$ and therefore it also holds that $c \models e_1$ and $c \models e_2$. Clearly, $P \wedge D \wedge c \models e_1$ and $P \wedge D \wedge c \models e_2$ hold, because c is the only predicate definition for the target predicate and clauses P and D do not contain the target predicate.*

Induction Hypothesis: For $n > 0$ with $|E_M(E^+)| = n$ it holds that $\forall e \in E_M(E^+) : P \wedge D \wedge C \models e$ with $C = clause\_set\_lgg(CD_M(E^+))$.

Induction Step: We show that for $|E_M(E^+)| = n + 1$ it also holds that all $e \in E_M(E^+)$ are logical consequences from $P \wedge D \wedge C'$ with $C' = clause\_set\_lgg(CD_M(E^+))$, $E_M(E^+) = \{e_1, e_2, \ldots, e_n, e_{n+1}\}$, and $CD_M(E^+) = \{c_1, c_2, \ldots, c_n, c_{n+1}\}$, that is

$$\forall e \in E_M(E^+) : P \wedge D \wedge clause\_set\_lgg(\{c_1, c_2, \ldots, c_n, c_{n+1}\}) \models e$$

$$P \wedge D \wedge clause\_set\_lgg(\{c_1, c_2, \ldots, c_n, c_{n+1}\}) \models e_1 \wedge e_2 \wedge \ldots \wedge e_n \wedge e_{n+1}$$

From the inductive definition of the clause set lgg operation (Definition 6.2.1) it follows that:

$$P \wedge D \wedge clause\_lgg(c_{n+1}, clause\_set\_lgg(\{c_1, c_2, \ldots, c_n\})) \models e_1 \wedge e_2 \wedge \ldots \wedge e_n \wedge e_{n+1}$$

Let $c' = clause\_set\_lgg(\{c_1, c_2, \ldots, c_n\})$ and $c'' = clause\_lgg(c_{n+1}, c')$. From base case 2 we can conclude that $c'' \models e_{n+1}$ and from the induction hypothesis it follows that: $P \wedge D \wedge c' \models e_1 \wedge e_2 \wedge \ldots \wedge e_n$. By Definition 5.5.1 every example description clause c_i contains a target predicate e_i with $e_i \in E_M(E^+)$. Hence, for the clause lgg c'' it holds that $\forall e \in E_M(E^+)\ \exists$ a substitution $\theta : e \in c''\theta$. From this fact and from Definition 2.1.15 it follows that $c''\theta \subseteq c'$ and $c'' \models e_1 \wedge e_2 \wedge \ldots \wedge e_n$. Thus it follows that for every $|E_M(E^+)| > 0$ it holds that all $e \in E_M(E^+)$ are logical consequences from $P \wedge D \wedge C'$ with $C' = clause\_set\_lgg(CD_M(E^+))$. Hence OSL of AV and CD-wrappers is consistent under logical consequence. q.e.d.

Theorem 6.4.3 OSL *learned CD-wrappers are not necessarily consistent under a floundering logical calculus* ($\vdash_f$).

Proof 6.4.4 (Theorem 6.4.3)

Proof Assertion: For every set of examples E^+ *representable as a set of* CD *example description clause* $S = CD_{CD}(E^+)$ *there exists a* $C = clause\_set\_lgg(S)$ *such that under floundering* SLDNF-resolution ($\vdash_f$) *it holds that:* $\forall e \in E_{CD}(E^+) : P \wedge D \wedge C \vdash_f e$ *with* P *an appropriate logic program implementing the* CD hypothesis language *and* D *the* $AV(D')$ *document representations of* D' *occurring in* E^+.

Counter Example: Given two examples e_1 *and* e_2 *such that the slot filler* n *of* e_1 *and* e_2 *are* `'a;b'`*and* `'c-b'`. *The appropriate atomic token patterns used in example description clauses for* e_1 *and* e_2 *are given by the token sequences* t_1 *and* t_2. *The right delimiters with* $mdl = 1$ *are given by* r_1 *and* r_2*:*
$t_1 = [[\mathtt{ttype} = \mathtt{word}, \mathtt{value} =' \mathrm{a}'], [\mathtt{ttype} = \mathtt{punct}, \mathtt{value} =';'], [\mathtt{ttype} = \mathtt{word}, \mathtt{value} =' \mathrm{b}']]$
$r_1 = [[\mathtt{ttype} = \mathtt{punct}, \mathtt{value} =':']]$
$t_2 = [[\mathtt{ttype} = \mathtt{word}, \mathtt{value} =' \mathrm{c}'], [\mathtt{ttype} = \mathtt{punct}, \mathtt{value} =' -'], [\mathtt{ttype} = \mathtt{word}, \mathtt{value} =' \mathrm{b}']]$.
$r_2 = [[\mathtt{ttype} = \mathtt{punct}, \mathtt{value} =' .']]$
Then the lgg *of slot filler patterns and right delimiter patterns are:*
$\mathtt{LGG}_{\mathtt{t_1,t_2}} : [[\mathtt{ttype} = \mathtt{word}, \mathtt{value} = \mathtt{X}], [\mathtt{ttype} = \mathtt{punct}, \mathtt{value} = \mathtt{Y}], [\mathtt{ttype} = \mathtt{punct}, \mathtt{value} =' \mathrm{b}']]$[1]
$\mathtt{LGG}_{\mathtt{r_1,r_2}} : [[\mathtt{ttype} = \mathtt{punct}, \mathtt{value} = \mathtt{Z}]]$.
This results in a generalized example description clause containing two literals: $\mathtt{filler(D, F, LGG_{t_1,t_2}, S, E)}$ *and* $\mathtt{notcontains(LGG_{t_1,t_2}, LGG_{r_1,r_2})}$. *Obviously* LGG_{r_1,r_2} *token-unifies with a sublist of* LGG_{t_1,t_2} *for any instantiations of* X *and* Y *as long as* LGG_{r_1,r_2} *is not ground. Hence notcontains does not hold thus no* $e \in E_{CD}(E^+)$ *is derivable from* P,D, *and the learned clause. Consequently the learned wrapper is inconsistent.*

q.e.d.

The proof is based on the assumption that the right delimiter pattern LGG_{r_1,r_2} is not grounded before checking if the constraint predicate holds. This clearly depends on the chosen logical calculus. Any calculus proving that predicate `notcontains` holds before having proven that `rightdelimiter` holds will be inconsistent in the discussed sense. Because the right delimiter pattern will not be instantiated and therefore it token unifies with a subset of the grounded slot filler token pattern of the counter examples.

Since the general aim of this thesis is to synthesize *Prolog* programs in the framework of logic programming which uses a *SLDNF-resolution* calculus the literal order plays an important role in learning extraction predicates. Hence, modifying the order of literals (e.g. placing `notcontains` after `rightdelimiter`) yields a refutation for the counter example given in the proof.

The discussed problem is also known in the context of *SLDNF-resolution* under the notion of *floundering* [Lloyd, 1987]. In more detail, if a logic programming system like *Prolog* does not ensure to use a computation (selection) function[2] that selects only grounded negative literals, the *safeness condition* [Lloyd, 1987] is violated and the *SLDNF-resolution* is not sound.

Obviously, this discussion could have been omitted if a modified version of the *CD-wrapper clause template* is used, a flounder free program clause. But it demonstrates the important

[1]The *lgg* of lists is computed pairwise (i.e. element i in list one is only generalized with element i in list two).
[2]A function deciding which body literal of the program clause is selected to be proven next.

point, that investigation results based on the logical consequence operator ($\models$) have to be reconsidered, if they are applied under a specific logical calculus (proof procedure) ($\vdash$) as for example in a *Prolog* system. So one solution to the floundering *CD-wrapper clause template* problem is to use modified slot filler and gap blocks as shown in Definition 6.4.1. Another possibility is the usage of a logic programming system that has features as delaying of sub-goals or reordering of sub goals to ensure that the *safeness condition* is fulfilled.

Definition 6.4.1 (Flounder free CD-wrapper clause template) *A* CD-wrapper clause template *for an* n-slot wrapper *W is of the form:*

$$\texttt{extract(D,}[\texttt{X}_1,\ldots,\texttt{X}_\texttt{n}],[(\texttt{S}_1,\texttt{E}_1),\ldots,(\texttt{S}_\texttt{n},\texttt{E}_\texttt{n})]) \leftarrow B_1, G_1, B_2, G_2, \ldots, G_{n-1}, B_n$$

*where every B_i (*filler block*) consists of the conjunction of literals:*

```
leftdelimiter(D,L,LS,FS),
filler(D,X,F,FS,FE), maxlength(F,N),
rightdelimiter(D,R,FE,RE),
notcontains(F,R),
```

*and every G_i (*gap block*) consists of the conjunction of literals:*

```
gap(D,G,GS,GE), maxlength(G,M),
leftdelimiter(D,NL,GE,NF),
notcontains(G,NL),
```

For every $\texttt{X}_\texttt{i}$ in $[\texttt{X}_1,\ldots,\texttt{X}_\texttt{n}]$ and corresponding $(\texttt{S}_\texttt{i},\texttt{E}_\texttt{i})$ in $[(\texttt{S}_1,\texttt{E}_1),\ldots,(\texttt{S}_\texttt{n},\texttt{E}_\texttt{n})]$ there exists exactly one filler block *B_m such that for variables $(\texttt{X},\texttt{FS},\texttt{FE})$ in B_m it holds that: $\texttt{X} = \texttt{X}_\texttt{i}$, $\texttt{FS} = \texttt{S}_\texttt{i}$ and $\texttt{FE} = \texttt{E}_\texttt{i}$. Every B_m is related to exactly one $\texttt{X}_\texttt{i}$ in $[\texttt{X}_1,\ldots,\texttt{X}_\texttt{n}]$. For variables $(\texttt{RE})$ in B_i, $(\texttt{GS},\texttt{GE},\texttt{NL},\texttt{NF})$ in G_i, and $(\texttt{LS},\texttt{L},\texttt{FS})$ in B_{i+1} it holds that: $\texttt{RE} = \texttt{GS}$, $\texttt{GE} = \texttt{LS}$, $\texttt{NL} = \texttt{L}$, $\texttt{NF} = \texttt{FS}$.* □

Theorem 6.4.5 OSL *learned RTD-wrappers for ϵ and ϵ^+ class wrappers are not necessarily consistent under SLD and SLDNF-resolution ($\vdash$).*

For Theorem 6.4.5 the notion of *consistency* has to be reconsidered. In the *IE-ILP* setting the *posterior sufficiency* criterion requires that $\forall e \in E^+ : B \wedge H \models e$. This means, whenever all examples logically follow from the background knowledge and the learned hypothesis, then the wrapper is said to be consistent. In the context of a logic programming system like *Prolog*, where the logical consequence operator is replaced by the calculus depending derivation operator, this leads to a problem.

Since the overall goal is to synthesize a logic program that computes extractions, we are interested in the computable answers and not the testing if a certain query is satisfiable, i.e. extraction is derivable. This is of great importance for practical application. We already discussed that generalizing *RTD example description clauses* in the case of *ϵ-class wrappers* can lead to not *range restricted clauses*[3]. In the worst case, this leads to a wrapper H consisting solely of one unit clause, namely the literal `extract(d,X,I)`. A query test like $B \wedge H \models$

[3] A program clause is range restricted if all variables in the head of the clause also occur in the body of the clause

$extract(d, a, b)$ obviously holds for $extract(d, a, b) \in E^+$. Under SLDNF-resolution the ground query $extract(d, a, b)$ also follows from the logic program $P \cup H$, because one simple resolution step yields a refutation.

From the theoretical point of view there is no problem concerning the consistency of the learned wrapper, since all queries with $e \in E^+$ are true regarding P and H. But in practice this wrapper is absolutely useless, since it does not compute any extraction (i.e. grounded substitution) for the discussed non ground query. Hence for this example, in a *Prolog* system the query $P \cup H \vdash extract(d, X, I)$ to calculate extractions X from d results in non grounding substitutions for X and I. Apparently, non ground answer substitutions are useless in the context of information extraction.

In fact a learning algorithm following strictly the *posterior satisfiability* criterion of the *IE-ILP setting* is protected against the discussed inconsistency, because the *posterior satisfiability* criterion excludes such hypotheses from the set of possible hypotheses. So, *consistency* in the *IE-ILP setting* means, that the set of computed answers for the query $extract(d_{ID}, X, I)$ has to be at least equal to the *set of logical example representation* $E_M(E^+)$ of the learning examples and a subset of the exhaustive example enumeration set regarding the used training documents. Proof 6.4.6 shows that one step learning of *RTD ϵ-class wrappers* is not consistent in general.

Proof 6.4.6 (Theorem 6.4.5)

Proof Assumption: Given a set of examples E^+ for an ϵ or ϵ^+ wrapper representable as a set of RTD *example description clauses that contains at least two examples $e_i, e_j \in E^+$ such that e_i and e_j have different number or positions of empty slots.*

Proof Assertion: For every set of examples E^+ meeting the proof assumption with $S = CD_{RTD}(E^+)$ and $C = clause\_set\_lgg(S)$ there exists an $e \in E_{RTD}(E^+)$ such that there is no grounding answer substituion θ for $P \wedge D \wedge C \vdash extract(D_{ID}, X, I)\theta$ and $extract(D_{ID}, X, I)\theta = e$. P is an appropriate logic program implementing the RTD *hypothesis language and $D = TD(D')$ the* TDOM *document representation of documents D' occurring in E^+.*

Proof: From the proof assumption it follows that there must be a n such that either slot $e_i.n$ or $e_j.n$ is empty. Because RTD *example description clauses (Section 5.4) do not contain literals for the description of empty slots there is a subset B of literals regarding the description of slot n either in example description clause $c_1 = cd_P^C(e_i, I_i)$ or $c_2 = cd_P^C(e_j, I_j)$. From the clause lgg definition (Definition 2.1.19) under the usage of prefix protection, we can conclude that there is no substitution θ such that $B'\theta \subset lgg(c_1, c_2)$ with $B'\theta' = B$. Hence, $C = lgg(c_1, c_2)$ contains no literals for the description of $e_i.n$ and $e_j.n$ and C contains a target predicate (head literal) with a non linked variable. Hence C is not range restricted and thus either the slot filler text for $e_i.n$ or $e_j.n$ is not computed. Finally, there is no ground answer substitution θ for the query: for $P \wedge D \wedge C \vdash extract(D_{ID}, X, I)\theta$ and therefore either e_i or e_j is not in the set of computed answers.* q.e.d.

To summarize the results, as long as a standard floundering logic programming system is used, clause templates have to be defined flounder free. For AV and CD wrapper models this is easily established as shown (Definition 6.4.1). For *RTD-wrapper models* this can be assured if body literals of *RTD example description clauses* are not linked (share same variables) among each other, then the order of body literals is irrelevant. But this holds only for the initial *example description clauses* where every literal of the proposed *RTD hypothesis language* describes some property of an example independently of another hypothesis literal. For a generalized set of description clauses it is very likely that body literals share variables as a consequence of the generalization process. This depends on the way the *prefix protection* is used. A better requirement to keep *RTD wrapper models* flounder free is the requirement not to use negative literals in the body of an *RTD wrapper description clause*.

It should have become clear that ϵ-*OSL* is a solution to the inconsistency problem stated by Theorem 6.4.6 with respect to *basic-OSL* and thus is an improvement in terms of fulfilling the *IE-ILP setting* criteria.

The most important outcome is the realization, that whenever the *clause lgg* of two clauses yields a non range restricted clause it can be discarded, because it yields non consistent wrappers and false extractions (i.e. missing slot fillers). This is an important insight for the later incremental learning of wrappers to reduce the complexity of evaluating the quality of hypotheses.

6.5. OSL Observations

The ϵ-*OSL* algorithm has been tested on various extraction tasks, reported in numerous ML based IE publications (see Chapter 10). A comparison and discussion of the proposed learning methods and wrapper models is given in Part III. Anyhow, two of the basic claims regarding the advocated wrapper models and *one step learning* are confirmed by empirical results in this section.

Maximal Delimiter and Context Length Observations

One basic idea all wrapper models have in common is the assumption that length bounded delimiters are a reasonable restriction. Figure 6.2 shows for four different *maximal delimiters lengths* (configuration 1 to 4 with mdl : $3, 5, 7, 9$) the precision and recall graphs of several test runs with increasing number of examples ($5 - 20$). The task was to learn an *CD-wrapper model* belonging to the *non-linear* ϵ^+*-wrapper class* (RISE problem: InternetAddressFinder). As observable from the graphs a small $mdl = 3$ results in very low precision values. For all four test configurations it is observable that an increasing number of training examples also results in increased recall values.

This meets the expected behavior, because the more examples given the more general the resulting clause set lgg of the example description clauses becomes. Hence, the intuition to use longer *mdls* can help to slow down the effect of over-generalization and low precision rates. The graph of *test configuration 3* with $mdl = 7$ displays this effect clearly, the maximum recall rate is reduced by 10% but the median precision rate is increased by $\sim 40\%$ regarding *test configuration 1* with $mdl = 3$. Similar to the problem of overfitting, choosing a too large *mdl* results in wrappers that generalize poorly to new data. Apparently, a $mdl = 9$ as used in *test configuration 4* yields for example sets with maximal 20 examples too specific extraction rules. In general we can not determine in advance how different delimiter texts or *TDOM* structures are with respect to the given examples. Therefore determining a good *mdl* in advance is difficult. But estimating an approximately good *mdl* can be done by some iterative approximation. A naive method consists of training-test cycles with increasing *mdl* values until a specific threshold regarding recall, precision or F1 rate is reached.

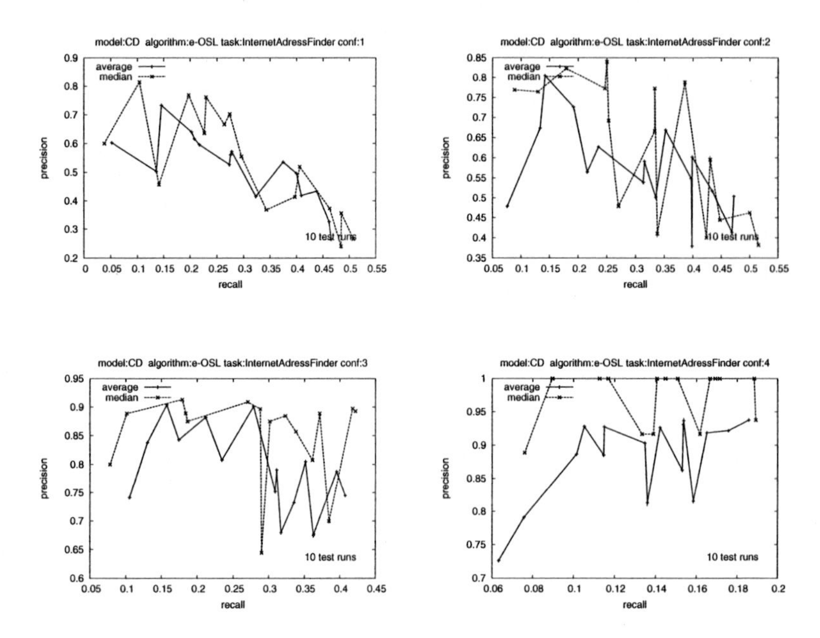

Figure 6.2.: recall vs. precision graphs ϵ-OSL of *CD* ϵ^+*-class wrappers* with $mdl : 3, 5, 7, 9$

ϵ-OSL Observations

ϵ-*OSL* claims to yield better precision rates, because the partition of example sets prevents from useless generalization of examples with too different slot fillers and missing delimiters. For non ϵ-based wrapper classes ϵ-*OSL* behaves like *basic-OSL*. Figure 6.3 shows the average and median number of learned rules for the previously discussed learning of an *AV* ϵ^+*-wrapper class model*. Ten times 20 randomly drawn examples belonged in average to 5 different example type classes (i.e. different number and position of empty slots). It is obvious that partitioning improves the quality of a wrapper, because partitioning and construction of more than one extraction rule partially suffices the need for disjunctive token patterns or the need for disjunction of literals in the body of a *RTD clause* as discussed in Section 5.4 under the notion of unfolding disjunctions.

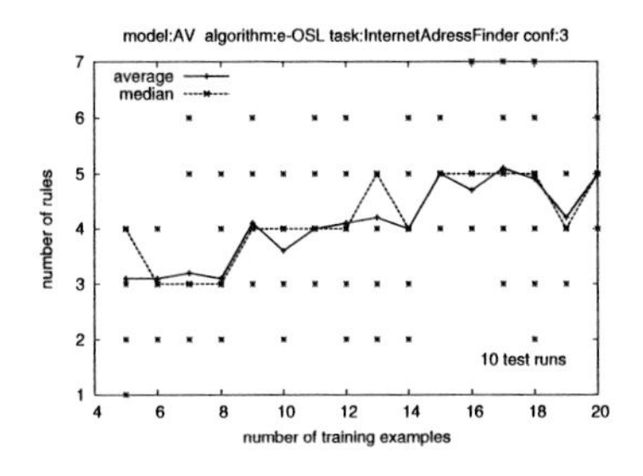

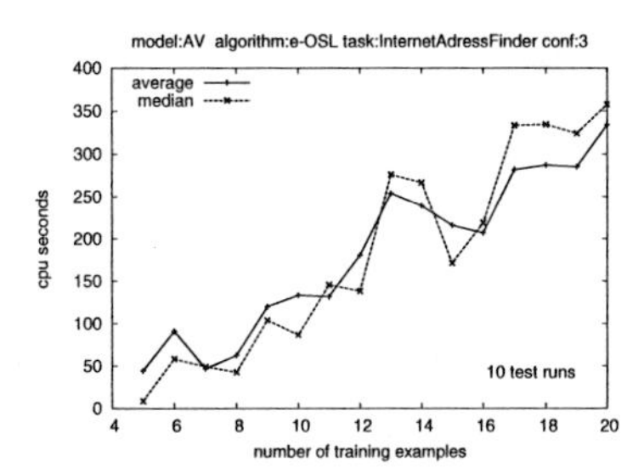

Figure 6.3.: number of learned rules and extraction times of ϵ-*OSL* learned wrapper

One consequence of ϵ-*OSL* is the increasing number of learned rules with respect to the number of training examples (upto a certain threshold, where no more different examples wrt. empty slots are presented). Learning more rules in this setting does not effect the overall learning time. But it leads to longer extraction times. Because all learned rules are tested on a given document. This is justified by the fact, that different extractions of different types can occur in a document. If for instance, rule 1 yields extractions from a document, this does not exclude the necessity to test if rule n also provides intended extractions. Hence, all rules have to be applied. Figure 6.3 depicts the testing times[4] of the discussed wrapper. The graph shows that a wrapper a trained on 20 examples needs significantly more time for extraction than a wrapper b that was trained only on 5 examples. There are two reasons: 1) wrapper a is much more general, hence more sub patterns matches and more combinations have to be tested (i.e. via backtracking instantiations of the body literals have to be computed). 2) because wrapper a was trained on more examples the probability to be trained on different example types is higher and therefore the set of learned rules is greater. Both reasons dependent on the number of examples and thus can not be viewed separately.

Despite this observation, the fact that in average the processing of one document takes circa 6 minutes for this problem class is not very satisfying. Although the F1 rate for the discussed *AV wrapper* is about 56% which is quite acceptable for this problem class, it seems to be not very optimized due to fast extraction times. In fact all test runs were limited such that the application of a learned wrapper (extraction phase) for each document was at most 7 cpu minutes. Although results for ϵ^+-*wrapper class* (e.g. Rise:QuoteServer) and *linear-$\wedge$-wrapper class* (e.g. Rise:BigBook) are promising, as show in Figure 6.4, the *non-linear wrapper class* seems to be a problem for *one step learning* (Figure 6.2).

[4]Tests were run on a Intel Pentium 4, 2.4 GHz, 512 MB memory machine, running LINUX Suse 8.0. The *LIPX* wrapper induction system is implemented using the ECLIPSE system [ECLiPSe, 2004].

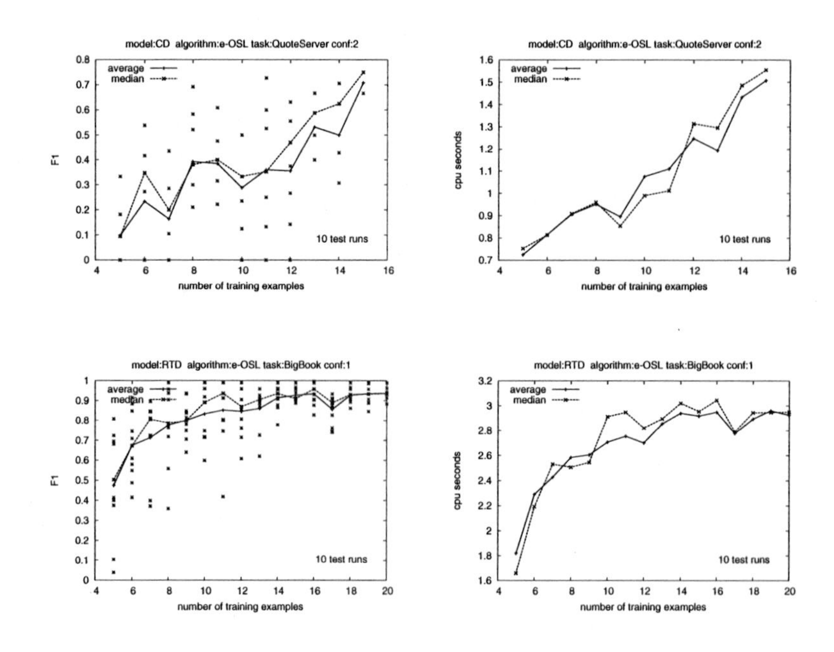

Figure 6.4.: ϵ-*OSL* learned *CD* ϵ-*class* and *RTD linear-x-*ϵ^+*-class wrapper* with $mdl : 5, 3$

Learning Time Observations

At the beginning of this chapter we motivated the use for one shot learning techniques by the argument that *OSL* has significant lower learning times than incremental systems. Indeed, experiments showed that learning times in comparison to the incremental techniques discussed in this thesis are significantly lower and that it is linear in the number of learning examples. Figure 6.5 illustrates this fact. The differing learning times for the investigated problem classes are based on the number of slots, the size of the documents and the number of documents used for learning, since examples were randomly selected from a given set of possible learning documents. These factors lead to different preprocessing and generalization times.

Noisy Data

Obviously *basic-OSL* and ϵ-*OSL* are not robust against noisy training data, because both methods do not use any hypothesis evaluation steps during learning. Additionally the training data, except the examples with different occurrences of empty slot fillers, is treated equally during the learning process by both approaches. Consequently already one false positive example (e.g. having completely different delimiters than all true positive examples) can strongly bias the hypothesis construction towards overly general hypothesis.

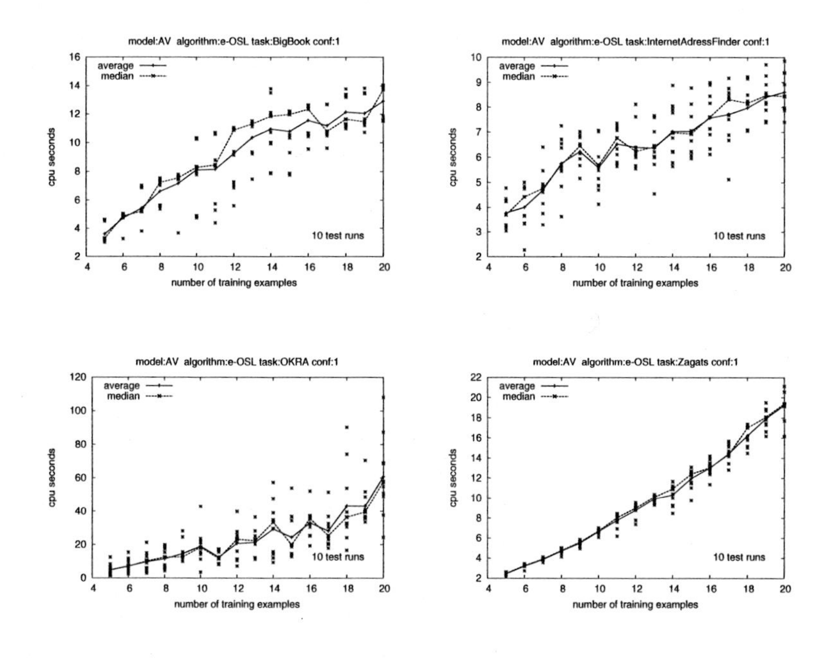

Figure 6.5.: learning times of *AV-wrapper models* under *ϵ-OSL*

Complexity Issues

At a first glance, *ϵ-OSL* seems to need longer learning time than *basic-OSL*. But in fact in the best case the number of *lgg* computations is reduced. For an example set of n examples *basic-OSL* has to compute the *clause set lgg* wich requires $n-1$ clause lgg computations. Given a partition for E^+ consisting of sets E_i such that each partition E_i contains only two elements then there exists $\frac{n}{2}$ partitions, whereas only the non trivial case is considered, where every partition at least yields a generalized clause[5]. Calculating the *clause set lgg* for each E_i then requires one *clause lgg* computation. Hence the total number of *clause lgg* computations is $\frac{n}{2}$. Thus *ϵ-OSL* has to compute half as much clause lggs than *basic-OSL* in the best case. In the worst case it does not find a partition and its effort is identical to that of *basic-BFOIL*. Partitioning E^+ has in the worst case order of n^2 time complexity in the number of examples n and the complexity to determine if two examples have the same empty slots is linear in the length of the examples. In the average case the partitioning can be done by sorting the

[5]Otherwise the best case consists of n partitions each containing one single clause and no learning (i.e. generalization) is needed.

example clause set (e.g. Quicksort). Then the partitions can be easily obtained. This has a complexity of $n * log(n)$. In the best case we need $n - 1$ tests to check if the set is ordered and to construct the partitions.

So, in contrary to the complexity of computing the *lgg* of two clauses, which for general clauses is NP-complete, partitioning is significantly less complex. Though there are more efficient *θ-subsumption* algorithms [Kietz and Lübbe, 1994] for restricted clauses (e.g. determinate clauses, k-local clauses), which are polynomially computable, the testing if a clause is reduced is co-NP-complete [Gottlob and Fermüller, 1983]. That means finding the reduction of a clause C requires at least $|C|$ linear number of calls to an NP oracle (i.e. NP-complete θ-subsumption). Consequently it is of great interest to minimize the number of *clause lgg* computations.

7. Bottom-Up Inductive Learning of Wrappers

Humans usually learn in cycles of construction, testing and refinement of hypotheses. In daily live one has to *train* to improve the ability to reach a goal in the best manner. So, learning is usually considered to be an incremental process, which can be understood as a search through the hypotheses space as illustrated in Figure 7.1, where a hypothesis is refined and evaluated iteratively.

Compared to the *one step learning* approach presented in Chapter 6 it appears to be obvious, that the presented *OSL* methods have to be extended to be able to applied to more difficult extraction tasks. Here, more difficult tasks denote those extraction problems where delimiter patterns and conjunctions of relational description predicates do not suffice anymore to represent the significant properties of the slot filler surrounding environments with solely one extraction rule. In other terms, descriptions based on the proposed wrapper languages start to become too general if a certain number and heterogeneity of examples is given. An incremental method that searches for appropriate subsets of examples to be generalized and which learns sets of rules seems to promise better results regarding the precision of learned wrappers. In this chapter an extended and incremental version of the *basic-OSL* algorithm is presented, the *BFOIL* (bottom-up first-order inductive learning) algorithm. Based on analysis of the results of *basic-OSL* and ϵ-*OSL*, several refinements of the *basic-BFOIL* algorithm are introduced and discussed in Section 7.3 and Chapter 8.

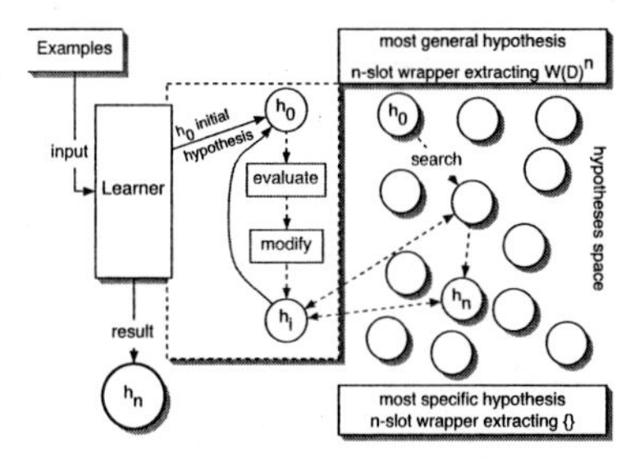

Figure 7.1.: Incremental Learning strategy as search through the hypotheses space

7.1. The BFOIL Algorithm

One shortcoming of the *basic-OSL* algorithm is the assumption that the *clause set lgg* of the whole example set can provide a reasonable generalized extraction rule. As we have shown in Section 6.5 for very simple wrapper classe with almost no structural differences like *linear* and $\wedge$-*wrappers*, this idea yields quite satisfying results. But as soon as the structure of a wrapper is more complex as in the case of *non-linear*,ϵ, $\vee$ or *nested-wrappers*, the quality of learned wrappers is significantly reduced.

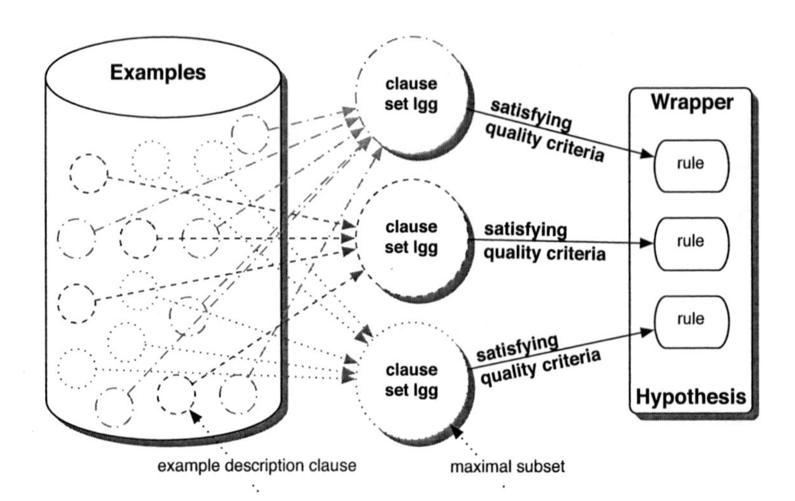

Figure 7.2.: BFOIL idea

This is due to the strong language bias of *clause templates.* For instance, the linear order of slot fillers of *AV* and *CD-wrappers*, which is imposed by the clause templates, simply disallows to yield high precision extractions with a non-linear order of slot fillers by one generalized extraction rule. Another reason is the *blind* application of the *clause lgg* operator regardless of observable properties of the learning examples.

A first attempt to improve this weak spot, is to test if the generalization of two example description clauses yields reasonable results or not. More elaborated solutions are presented in Section 7.3. In the basic variant of *BFOIL* a reasonable result is considered to be a hypothesis providing no false extractions from the training documents.

Since the *lgg* of a set of clauses is defined as an iterative application of computing *clause lggs* (Section 6.2 Function 2) this test can be easily achieved. Whenever a *clause lgg* is calculated in the generalization loop, that does not meet the predefined quality criteria, the last generalization step is revoked and the chosen example description clause is removed from the starting set and put aside. The process terminates when no more clauses are left to be generalized. The generalized rule represents one learned clause and the process is repeated with the remaining set of examples which had been put aside. This results in a set of clauses defining the target predicate, i.e. the wrapper. This is the basic *BFOIL* learning idea as shown in Figure 7.2. Since this is only a sketch the missing details are presented now.

What separates *BFOIL* from *OSL* is the test or evaluation of generalized clauses. The *IE-ILP setting* defines if a hypothesis is acceptable, i.e. if it covers the *posterior satisfiability* and *posterior sufficiency.* If a generalized example description clause does not violate the *posterior satisfiability,* then it does not provide any false extractions from the so far presented training documents and it is reasonable to proceed with it for further generalizations.

Additionally, from the results of the discussion on *OSL* properties, it is known that the *posterior satisfiability* test has not to be applied for *not range restricted clauses.* It suffices to show that at least one violating computed answer (i.e. false extraction) exists to terminate the hypothesis evaluation with a negative result. In fact *basic-BFOIL* partitions the set of learning examples such that each partition is maximal and fulfills the *posterior satisfiability* criterion of the *IE-ILP setting.* A partition M is maximal if any additional example added to M results in a *clause set lgg* of M no longer satisfying the required criteria anymore.

Testing the quality of intermediate constructed extraction rules during the partition and generalization process involves the use of a validation set. Either a validation set consists of positive and negative examples or the validation set is an exhaustive set of solely positive examples (Definition 2.2.3) as discussed in Section 3.3.4, 3.6 and 5.2.1. *BFOIL* requires exhaustively labeled documents, which means for each training document the complete set of extractions of this document must be given. The definition of the *basic-BFOIL* algorithm is shown in Algorithm 4.

The algorithm uses the Function $check(R, P, D, VS)$ (Algorithm 5) to compute the number of false extractions with respect to the validation set $VS = E_M(E)$. If the number of false positives is greater than zero the *posterior satisfiability* does not hold. The notation used (`check.attribute`) is interpreted in an object-oriented programming language style to refer to certain properties of R, e.g. the number of false positives calculated by `check`.

---

input : $E \subset \mathcal{W}$ an exhaustive example set E for a wrapper $\mathcal{W}$
$E^+ \subset E$ training examples
M the wrapper model to be learned
P a logic program implementing the hypothesis language of M
D logical document representation of documents in E

$VS \leftarrow E_M(E)$
$Pool \leftarrow CD_M(E^+)$
$LearnedRules \leftarrow \emptyset$
while $Pool \neq \emptyset$ **do**
  $Rule \in Pool$
  $Pool \leftarrow Pool \setminus \{Rule\}$
  $Remains \leftarrow \emptyset$
  while $Pool \neq \emptyset$ **do**
    $X \in Pool$
    $R \leftarrow clause\_lgg(Rule, X)$
    if $range\_restricted(R) \wedge check(R, P, D, VS).PosteriorSatisfiability$ **then**
      $Rule \leftarrow R$
    else
      $Remains \leftarrow Remains \cup \{X\}$
    $Pool \leftarrow Pool \setminus \{X\}$
  $LearnedRules \leftarrow LearnedRules \cup \{Rule\}$
  $Pool \leftarrow Remains$
return $LearnedRules$

Algorithm 4: *basic-BFOIL*

---

function *check*(R, P, D, VS);

input : R :=rule ; P = logic program ; D = documents rep. VS = validation set

$PosteriorSatisfiability \leftarrow false$
$Extractions \leftarrow \{R_{head}\theta | P \cup D \cup R \vdash R_{head}\theta$ with θ answer substitution$\}$
$FalsePositives \leftarrow |Extractions \setminus (VS \cap Extractions)|$
if $FalsePositives = 0$ **then**
$\quad PosteriorSatisfiability \leftarrow true$

Function 5: posterior satisfiability check with false positive calculation

For efficiency reasons it is more sensible that the *posterior statisfiability* test does not compute the whole set of extractions. Instead iteratively enumerating and testing the extractions according to their membership to VS decreases the learning and runtime behavior. Basically, the *posterior satisfiability* test can be reasonably reduced, if the new rule is not tested on all documents contained in the training set, but only on those documents from which the examples have been taken that were used for the current rule construction. This can delay the detection of a bad rule, but it minimizes the number of query computations. The test runs presented in Chapter 10 were done with this setting.

7.2. Properties of basic-BFOIL

Obviously *basic-BFOIL* always terminates, because in the outer while loop the *pool*, i.e. set of examples for which a rule has to be learned, is always reduced by one element. The inner loop tries to reduce the set of example clauses, but never increases the number of examples. Therefore the *pool* set in the succeeding outer loop contains at least one element less than *pool* in the previous round. From this follows that *basic-BFOIL* always returns a hypothesis.

In the worst case *basic-BFOIL* learns by rote, if every *clause lgg* for any two examples results in a clause violating the *posterior satisfiability* or *range restrictedness.* The other extreme is, that the learned hypothesis contains solely one single clause. In this case *basic-BFOIL* behaves just like *basic-OSL*, but with significant higher computational effort. The following theorems summarize consistency properties of the *basic-BFOIL* algorithm.

Theorem 7.2.1 basic-BFOIL *learned AV,CD and RTD-wrappers are consistent under a floundering free calculus.*

Proof 7.2.2 (Theorem 7.2.1)

Proof Assumption: Let P be an appropriate clause set definition of the hypothesis language for a wrapper model M (e.g. AV,CD, or RTD). Let E be an exhaustive set of examples such that for all $e \in E$ an example description clause $cd_P^C(e, I)$ exists. Let $E^+ \subset E$ and $VS = E_M(E)$ be the validation set. Further let D be the set of appropriate document representations of D' occurring in E^+ with respect to the wrapper model M (e.g. $AV(D')$ and $TD(D')$).

Proof Assertion: Let H be a set of basic-BFOIL *learned clauses. For each $e \in E_M(E^+)$ there exists a $R \in H$ and a grounding answer substitution θ such that $P \wedge D \wedge R \vdash e'\theta$ and $e'\theta = e$.*

Proof by contradiction

Assumption: $\exists e \in E_M(E^+) \forall c \in H : P \wedge D \wedge c \not\vdash e'\theta$ *with* $e'\theta = e$.

Proof: Let $e \in E_M(E^+)$ *then there must be a* $E_i \subseteq E^+$ *such that* $E_M(E_i) \subseteq E_m(E^+)$ *and* $e \in E_M(E_i)$.

Let $c = clause\_set\_lgg(CD_M(E_i))$ *and* $c \in H$. *Further let* $c' = cd_P^C(x)$ *be the* example description clause *regarding example* x *for* e *with* $e = E_M(\{x\})$. *Because* $e \in E_i$ *it follows from the proof assumption and the Definition 2.1.18 that* $c \preceq c'$ *and that there exists a substitution* θ *such that* $c\theta \subseteq c'$. *By Definition 5.5.1 and 5.5.2 respectively it holds that* $e \in c'$. *Since all clause templates contain the literal of the target predicate, namely extract, there also must be an* $e' \in c$ *with* $e'\theta = e$, *because* $c \preceq c'$. *From* basic-BFOIL*'s range restricted test it follows that only extraction rules are added to the set of learned rules yielding ground answers*[1]. *Hence,* c *is range restricted and under the discussed restrictions for hypothesis languages (Section 5.2.1) it follows that only ground answers are computed for the query* $e'\theta$. *Then it holds that* $P \wedge D \wedge c \vdash e'\theta$ *with* $e'\theta = e$, *because* $e'\theta$ *is positive in* c *and from* $c \preceq c'$ *follows* $c \models c'$ *by Theorem 2.1.1 there must be an answer substitution* θ *under SLD/NF-refutation such that* $e'\theta = e$. *This contradicts the contradiction assumption, hence* basic-BFOIL *is consistent under a floundering free calculus.* *q.e.d.*

In fact the use of the range restrictedness is an improvement or refinement of the very basic *BFOIL* version. If this requirement is dropped, *basic-BFOIL* is not guaranteed to learn consistent wrappers anymore. If a generalized clause is not range restricted and a non ground answer substitution is computed, it remains to the implementation how to interpret not grounded slot filler arguments. Either they are interpreted as empty strings; in this case they will eventually cover a subset of the presented examples, or they are kept not grounded. In this case a mere set membership test for testing consistency fails. In this case the *posterior satisfiability* is violated and *BFOIL* remains consistent. In the first case, it still can be the case that *BFOIL* covers some of the positive examples (e.g. if the example set contained only two different types regarding empty and missing slots). But as soon as the example set becomes too diverse, the learned wrappers will probably produce only false positives. This again would be detected by the *posterior satisfiability* check. Nevertheless, there may be inconsistent wrappers learned. This can be prevented by the range restrictedness test.

The *BFOIL* wrapper learning algorithm was first mentioned in [Thomas, 2003]. There a consistency refinement is mentioned explicitly checking if the set of computed answers (Function 5 *Extractions*) is a subset of the set of already presented learning examples. This test always requires the computation of answers by an SLD/NF-resolution procedure. The method proposed in this thesis is significantly less complex, because merely testing for range restrictedness of a learned clause guarantees a consistent wrapper. A range restrictedness test can be done in linear time complexity in the number of variables occurring in a clause.

Corollary 7.2.3 *RTD-wrappers learned by* basic-BFOIL *without range restrictedness test are not necessarily consistent.*

[1] This obviously is only true if the subgoals concerning the hypothesis predicates do not yield non grounded answer substitutions. Therefore only ground literals are allowed to be derivable from the chosen hypothesis language.

Corollary 7.2.3 follows directly from Theorem 6.4.5. Its proof is analogously to Proof 6.4.6.

Although consistency is an affordable property for a learned wrapper in some contexts it might be also of great importance to be sure that all performed extractions are correct. At least for extractions resulting from documents belonging to the training set this can be assured for wrappers learned by the *basic-BFOIL* algorithm.

Proposition 7.2.4 basic-BFOIL *learns* 100% *precise AV,CD and RTD-wrappers regarding the set of training documents.*

The Proposition 7.2.4 is a trivial consequence from the function *check* used in the *basic-BFOIL* algorithm. Because no clause c is added to the set of learned rules that produces any false extractions from the presented training documents only clauses not violating the *posterior satisfiability* are contained in the set of learned rules. Consequently applying a learned wrapper on documents from which training examples were taken results in correct extractions only.

The *basic-BFOIL* algorithm has a significant weak point in that it does not guarantee to output the best hypothesis regarding its possible predictive strength obtainable under clause set lgg. In other terms, the resulting generalized clause set is not always the best complete and sound clause set, with respect to the training examples and the target wrapper.

This means, although the learned hypothesis is consistent it can be suboptimal such that the resulting wrapper yields less extractions (has lower recall rates) than a wrapper learned from other possible partitions. The reason for this is that the quality of the learned hypothesis is not independent of the order in which the algorithm selects example description clauses for generalization. From a theoretical point of view *basic-BFOIL* can be easily modified to output the best possible hypothesis but this results in a much higher computational complexity.

In fact this would require to compute all subsets of the example description clauses for which the *clause set lgg* yields a consistent rule. From the so obtained subsets a partition of the example description clauses has to be built such that the lggs of the partitions compute the maximal possible number of examples.

Note that in general we can not say in advance without a query test, which *clause set lgg* is consistent and provides the best extraction results. Only those subsets can be excluded containing clause pairs for which it already has been shown that their *lgg* yields inconsistent extraction rules. Hence, the number of required query computations is proportional to the number of subsets under these restrictions.

Secondly, in general there is not a unique partition with the necessary properties. For instance, given three clauses c_1, c_2 and c_3. Assume the *lgg* of c_1 with c_2 and c_1 with c_3 provide consistent rules and no other combination. Depending on the clause *basic-BFOIL* selects first, it either outputs $lgg(c_1, c_2) \vee c_3$ or $lgg(c_1, c_3) \vee c_2$. Now assume the rule obtained from $lgg(c_1, c_2)$ covers more examples of the validation set than $lgg(c_1, c_3)$. In this case a *best BFOIL* algorithm should output $lgg(c_1, c_3) \vee c_2$. But if both lggs cover the same number of examples it is unclear what generalization to prefer, since we do not know which rule will perform better on future documents.

The crucial performance lack of *basic-BFOIL*'s hypothesis construction is the time needed to evaluate a hypothesis by logical answer computation. For practical reasons, the conscious decision is made to use the proposed suboptimal algorithm to minimize the number of queries. Clearly this can result in lower recall rates but still preserves high precision rates. From this observations Proposition 7.2.5 follows.

Proposition 7.2.5 *The* basic-BFOIL *algorithm is not guaranteed to learn the best complete wrapper with respect to the validation set.*

One important result of this section is Theorem 7.2.1. We can expect that any learned *AV*, *CD* and *RTD-wrapper* is capable of extracting the text tuples it has be trained on. Since this result depends on the *range restrictedness* criterion, the *range restrictedness* test can be used to force *BFOIL* to learn consistently or not.

This gives no hint how strong its predictive properties are, or in other terms how good it will perform on future documents. Especially, the observation that in the worst case the *basic-BFOIL* algorithm demotes to simply storing all example description clauses shows that for very differing examples low recall rates are expectable.

Nevertheless, by Theorem 7.2.1 and Corollary 7.2.3 it becomes clear that the *range restrictedness* test of generalized example description clauses is sufficient in this context to evaluate wether the learned rule is consistent or not. Following strictly the *IE-ILP setting* and testing after every *clause lgg* computation if the resulting clause still covers the *posterior satisfiability* obviously yields wrappers providing only correct extractions (Proposition 7.2.4) from the training documents.

It is important to point out that *basic-BFOIL* does not try to find those subsets which lggs cover the most examples of the validation set. The reason is that this would require a much higher number of query computations. Instead the presented heuristic approach partitions the example set with the aim to provide consistent wrappers with as high recall rates as possible, but using as less query computations as possible to keep the required learning time low.

In comparison to *basic-OSL* and ϵ-*OSL* the *basic-BFOIL* algorithm is always consistent. As expected it requires longer learning time due to the complexity of answer derivations. It always yields correct extractions from the training documents, but will probably have lower recall rates on future documents, because it tends to learn less general wrappers due to the *posterior satisfiability check*.

7.3. Refinements of BFOIL

The *basic-BFOIL* algorithm offers two basic options to be modified. Either the quality regarding precision and recall rates of learned wrappers can be improved, or the runtime and learning time behavior can be advanced. The following sections present modifications and refinements.

7.3.1. Rule Quality Threshold

One basic characteristic almost all information extraction and information retrieval systems have in common is the anti-proportional behavior of precision and recall. Modifying a learning algorithm to produce hypothesis that yield higher recall rates in almost all cases reduces the precision of the learned hypothesis, and vice versa. From the previous discussion about *basic-BFOIL*'s qualities a reasonable improvement would be to raise the recall rate, since it is quite low for some of the experimental test cases. In general, it is always desirable for a learning algorithm to be adjustable regarding specific problem cases. This includes the possibility to bias the learning into a certain direction, as for example constructing wrappers with higher recall or precision rates.

In the *basic-BFOIL* algorithm this can be easily achieved by modification of the *posterior satisfiability* check. Until now, only those *lggs* are used for further generalization steps, that do

function *check*(R, P, D, VS)

input : R =rule ; P = logic program ; VS = validation set ;
D = set of sets of logical document representation wrt. VS

$PosteriorSatisfiability = true$
while $PosteriorSatisfiability \wedge D \neq \emptyset$ **do**
  $d \in D$
  $D = D \setminus d$
  while $P \cup d \cup R \vdash R_{head}\theta$ *and* $PosteriorSatisfiability$ **do**
    if $R_{head}\theta \notin VS$ **then**
      $PosteriorSatisfiability \leftarrow false$

Function 6: posterior satisfiability check with *one false* stopping criterion

not yield any false extraction. Weakening this strict criterion obviously leads to a wrapper with lower precision rates. On the other hand, this will probably increase the recall rate, because a larger number of extractions can be obtained. If the extraction rule does not become overly general, the percentage of correct extractions will likely increase.

For the basic algorithm there is obviously no need to compute the whole set of false extractions of the current rule to evaluate its quality. Because as soon as one false extraction is computed, the *posterior satisfiability* test can be negatively terminated. Consequently, estimating a quality measure for the current hypothesis based on the number of false extractions requires a different computation of the *posterior satisfiability* criterion. Apparently, calculating the complete number of false extractions conflicts with the previous idea of a faster satisfiability test. The question remains, if there is a heuristic which does not need to compute the complete set of false extractions and nonetheless defines a reasonable measure to estimate the rules quality according its coverage of false extractions. Systems like *Whisk* [Soderland, 1999] use a Laplacian expected error ($\frac{p+1}{n+1}$ with n the false extractions and p the correct extractions). The rule value used in the *RAPIER* system [Califf, 1998] is based on the informativity of a rule (i.e. the degree a rule separates positive and negative examples) a modified Laplacian estimation plus a bias for shorter rules ($-log_2(\frac{p+1}{p+n+2}) + \frac{ruleSize}{p}$). But since both metrics require the number of false extractions (negative examples covered by a rule) we have chosen a different metrics.

As a first step in finding a suitable modification such that the *basic BFOIL* algorithm accepts clauses with a small degree of false extractions, consider Function 6. This is an implementation of the *posterior satisfiability* check with the *one false* stopping criterion. The algorithm assumes that the used theorem prover and logic programming system respectively is capable of computing and enumerating answers for a query iteratively. As for example via backtracking like PROLOG systems do.

In this case the inner loop calculates all extractions from one specific document or it stops when a false extraction from the document is detected. If we stick to this method, a weaker method to measure the false behavior of the current rule is to count on how many training documents it fails. Therefore solely the outer while condition has to be modified and the failures have to be counted as shown in Function 8.

Several points should be noted. Instead of passing the complete validation set to the function

input : $E \subset \mathcal{W}$ an exhaustive example set E for a wrapper $\mathcal{W}$
$E^+ \subset E$ training examples
M the wrapper model to be learned
P a logic program implementing the hypothesis language of M
$\mathcal{D}$ set of sets of logical document representation regarding E
T a rule quality threshold

```
Pool ← CD_M(E+)
LearnedRules ← ∅
while Pool ≠ ∅ do
    Rule ∈ Pool
    Pool ← Pool \ {Rule}
    VS ← E_M(E).Rule
    D ← 𝒟.Rule
    Remains ← ∅
    while Pool ≠ ∅ do
        X ∈ Pool
        VS ← VS ∪ E_M(E).X
        D ← D ∪ {𝒟.X}
        R ← clause_lgg(Rule, X)
        if range_restricted(R) ∧ check(R, P, D, VS, T).PosteriorSatisfiability then
            Rule ← R
        else
            Remains ← Remains ∪ {X}
            VS ← VS \ E_M(E).X
            D ← D \ {𝒟.X}
        Pool ← Pool \ {X}
    LearnedRules ← LearnedRules ∪ {Rule}
    Pool ← Remains
return LearnedRules
```

Algorithm 7: *threshold-based BFOIL*

function $check(R, P, D, VS, T)$

input : R =rule
P = logic program
VS = validation set
T = failure threshold
D = set of sets of logical document representation wrt. VS

$PosteriorSatisfiability \leftarrow true$
$failures \leftarrow 0$
$pos \leftarrow 0$
while $D \neq \emptyset$ **do**
  $d \in D$
  $D \leftarrow D \setminus \{d\}$
  $fp \leftarrow false$
  while $P \cup d \cup R \vdash R_{head}\theta$ and $\neg fp$ **do**
    if $R_{head}\theta \notin VS$ **then**
      $fp \leftarrow true$
      $failures \leftarrow failures + 1$
    else
      $pos \leftarrow pos + 1$
  $NegRate \leftarrow \frac{failures}{|D|}$
  $PosRate \leftarrow \frac{pos}{|VS|}$
  $failure\_quality \leftarrow \frac{1+(NegRate-PosRate)}{2} * NegRate$
if $failure\_quality \geq T$ **then**
  $PosteriorSatisfiability \leftarrow false$

Function 8: posterior satisfiability check with quality estimation *check* as in the *basic-BFOIL* version, now the validation set consists solely of the exhaustive example set regarding the documents and examples from which the current rule has been generalized. This requires a small modification of the main algorithm as shown in Algorithm 7. For representational issues $VS \leftarrow E_M(E).Rule$ denotes the exhaustive example set regarding the example description rule *Rule* with respect to E, i.e. subset of E regarding *Rule*. The associated logical document representation of an example description rule X is denoted by $\mathcal{D}.\mathcal{X}$. Only documents given by D, consisting of sets of logical document representations, are used for measuring the quality of the current rule. Depending on these two sets and the number of correct and wrong extractions the quality of the wrapper ($failure\_quality$ in Function 8) is calculated as a weighted ratio of failure documents to the total number of documents in D. If both correct and wrong extractions have maximum values, the quality of the rule is higher as if only wrong extractions are obtained by the current rule. It has to be pointed out that *PosRate* does not represent the real number of correct extractions, since the estimation stops when the first wrong extraction is computed.

In fact, this is just a heuristics, because in the worst case for every document the first computed extraction is a wrong one and all succeeding computations would be correct ones. But since the inner loop terminates after the first wrong extraction, the positive ones are not computed.

Nevertheless, assuming that there are no conditional probabilities for wrong and correct extractions regarding the order in which they are computed, it can be assumed that those which build the majority are computed first with higher probability. For the actual computed rule quality this means that a rule covering all examples in the validation set and also extracts one wrong extraction from each document in D still has a quality of 0.5. A rule performing correct extractions only yields a $failure\_quality$ value of 0. Hence, the value computed by the quality function is a measure for the failure behavior of the wrapper. The number of positive extractions are used as a factor to reduce the failure rate (*NegRate*). Finally, a threshold value T determines if the rule is accepted for further generalizations or if it is rejected. It depends on T and the $failure\_quality$ if the rule fulfills the *posterior satisfiability*. This does not conform to the original *posterior satisfiability* condition defined in the *IE-ILP setting*. Therefore we call it *weak* or *threshold-based posterior satisfiability*.

A few remarks about the efficiency and complexity of the proposed refinement. In the best case threshold-based posterior satisfiability requires $|\mathcal{D}| - 1$ more query computations than the *one false stopping* version. In the worst case first all correct extractions are computed from each document and after that one false. For the threshold-based posterior satisfiability this results in $|VS| + |\mathcal{D}|$ query computations. If from the "last" document a wrong extraction is derived, the worst case for the *one false stopping* check is the number of query computations $|VS| + 1$.

If we think of practical applications where a wrapper is learned from 5 to 20 examples in average taken from 10 documents or less, the difference for the best and worst case seems to be ignorable small.

However, it might be crucial for practical applications, because each query computation can vary a lot in its computation time depending on the generalization degree of the extraction clause. Figure 7.3 shows two wrappers one learned with *basic-BFOIL* and one with *threshold* refinement with $T = 0.25$.

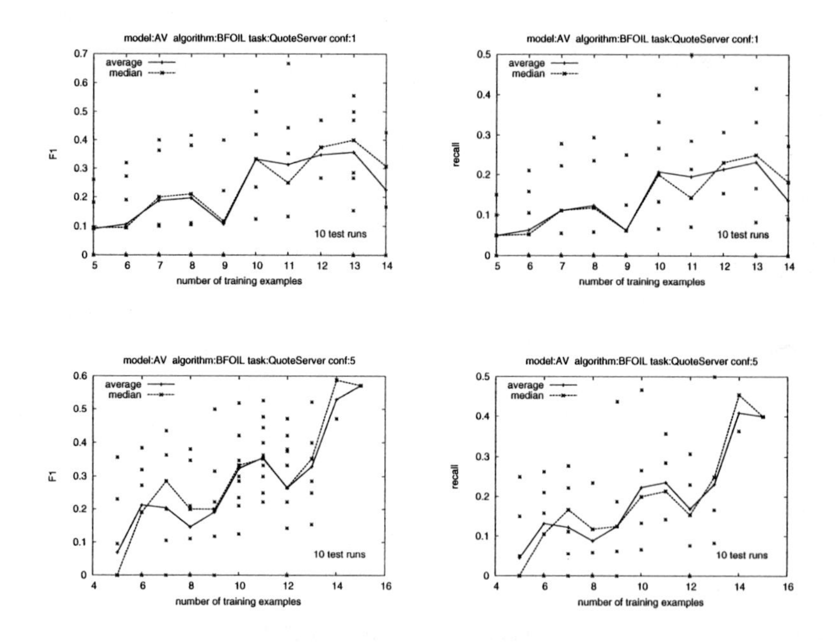

Figure 7.3.: F1 and recall graphs of *AV linear-x-*ϵ^+*- wrapper* learned with *basic-BFOIL* (c1) and *threshold-based BFOIL* (c5) with $T = 0.25$

7.3.2. Semantic LGG Operators

So far, we assumed that the *lgg* operation is a reasonable operation for generalizing clause sets. However, depending on the clauses to be generalized, in some cases even for single atoms, the use of the least general generalization operator results in too general atoms. For instance, assume the *lgg* of the two *TDOM predicates* $xspan(1, ([0, 2, 1], 4, 8), t_1)$ and $xspan(1, ([0, 2, 1], 2, 9), t_2)$, given by $xspan(1, ([0, 2, 1], X, Y), t')$. Focusing on the generalization of the left and right span borders it is not very reasonable that the *lgg* of these two atoms allows arbitrary values. For this special case it would be much more reasonable to define a constraint allowing X only to take values of $\{2, \ldots, 4\}$ and Y either 8 or 9. As with the $xspan$ predicate there may be many other predicates in the chosen hypothesis language, for which a finer generalization operation is desirable.

So, instead of using only the *lgg* operator, a reasonable step to guide the learning process is a modification that adds specific constraining predicates for certain hypothesis literals. Such a modification can be understood as a sort of program transformation or inference rule with respect to least general generalization. For instance, the semantic lgg rule shown in Example 7.3.1 states "Whenever the *lgg* of two atoms **xspan** occurring in clause C_1 and C_2 has to be computed, the resulting semantically modified *lgg* of these two atoms is given by a clause containing the *lgg* of the two atoms and two additional atoms `member`, constraining the range of the left and right borders of the `xspan` *lgg* wrt. to their borders". The predicate `member(L,List)` tests if there is a term L' contained in list *List* such that L unifies with L'. Consequently, literals added by the modified *lgg* operation have to be defined in the background knowledge, i.e. the logic program.

Example 7.3.1 (Semantic LGG inference rule) *Let* C_1 *and* C_2 *be* RTD *example description clauses. The following inference rule defines the* semantic lgg operation *for the literal* $xspan$.

$$\frac{C_1\backslash\{xspan(D_1,(N_1,L_1,R_1),TL_1)\} \quad C_2\backslash\{xspan(D_2,(N_2,L_2,R_2),TL_2)\}}{\{member(L,[L_1,...,L_2]),member(R,[R_1,...,R_2])\} \cup CL}$$

$$\text{with } CL = clause\_lgg(C_1, C_2) \text{ and } xspan(D,(N,L,R),TL) \in CL.$$

Additionally, the *clause lgg* computation has to be modified such that previously semantically generalized clauses have to be treated in a special way. This means literals contained in a clause that are not contained in the hypothesis language, have to be detected and incorporated into succeeding *lgg* computations according to the semantics of the inference rule used.

This method opens up a wide choice of possibilities to guide the learning process with additional domain knowledge. One can think of even more sophisticated extensions following the proposed idea. A further idea could be to assume that an additional background theory is given which is use to compute logical consequences regarding some literals occuring in the clause. For this, designated literals contained in C_1 and C_2 are added to the theory. Then the computation of logical consequences might yield helpful literals to replace or to add to the semantically generalized clause. A somewhat similar idea is known under the notion of *relative least general generalization (rlgg)* and is used in the ILP-system *GOLEM* [Muggleton and Feng, 1992]. It computes the *lgg* of a clause relative to a given background knowledge.

This type of refinement is not investigated further in this thesis, because the original motivation for the introduction of the *semantic lgg operations* were by practical observations about very bad learning and runtime behavior of the implemented methods in the system LIPX[2]. In almost all cases *xspan* literals tended to become too general regarding the left and right borders and node identifier terms. This resulted in a dramatically large number of possible instantiations, because almost all possible combinations of subtrees of a *TDOM-tree* are covered by the predicate. This also increases the number of backtracking points in a standard SLD/NF-resolution proof, which obviously results in far too long answer derivations during the hypothesis evaluation and the later application of a learned wrapper.

Reconsidering one of the basic motivations for using a pure bottom-up *lgg* based learning technique for learning wrappers was it to minimize the search space for adequate rule literals and the therewith associated large search space of possible instantiations and combination of hypothesis literals. But as the query computation with overly general clauses show, this problem is an eminent hindrance. Either a top-down approach has to use intelligent techniques to search and select reasonable literals from the huge hypothesis space, or the bottom-up approach has to detect at an early stage if the current learned rule is tractable for evaluation. If such a decision procedure would exist then a sufficiently good learning algorithm must exist as well. Such a procedure would be capable to decide if the current rule is a good one or

[2] *LIPX* is a wrapper learning system using wrapper models and algorithms presented in this thesis.

not, hence it would decide if the rule allows to derive only correct extractions. Obviously, this procedure must contain a more efficient logical calculus. It follows, that the use of a *better* calculus than the one used in standard PROLOG might be a solution.

Nevertheless the proposed method of a *semantic lgg operator* can significantly decrease the evaluation and also the runtime behavior of the *basic* and *threshold-based BFOIL* technique. The learning time is increased by a linear factor in the number of literals added to a clause by the *semantic lgg operator*. Hence, succeeding *clause lgg* computations take longer time, since the number of literals per clause is larger. Besides these issues this enhancement can be easily integrated into the *clause lgg* computation and shows improved wrapper application times. This refinement approach has been tested with *RTD-wrapper models* with the presented *semantic lgg rule* for *xspan* predicates. A theoretical framework and more elaborated tests are necessary to evaluate this idea more detailed. Nevertheless, experiments showed that this refinement improves the quality of learned wrappers.

7.4. BFOIL Observations

OSL based learning methods showed already promising results for simple wrapper tasks like learning *linear-$\wedge$ wrapper classes*. For the more sophisticated extraction problems involving the learning of *non-linear*, ϵ, ϵ^+, or *MVclasses* the application of incremental learning techniques as proposed by the *BFOIL* method seems to be a reasonable step to improve the quality. Three basic ideas motivated the development of *basic-BFOIL*: 1) Increase the precision rate of wrappers by incremental construction and evaluation to sort out clauses causing false extractions when generalized. 2) By learning sets of rules, avoid over-generalizations (i.e. number of false positives), which may appear in *OSL* if it learns from too many and too different examples. Central idea is that this can model exceptional cases among the examples and yields better $F1$ rates, without to lower recall rates.

The mentioned problem of expectable too high query computation times turned out to be the crucial problem of the proposed *BFOIL* approach. The learning results presented in this thesis are all obtained under a time limit for the learning phase of the wrapper induction system *LIPX*. This limit was set to 30 minutes. For several problem classes, this resulted in untimely terminated processes. Especially for the multi-slot extraction problems this resulted in some poor scores (Chapter 10). Figure 7.4 illustrates the dramatic decrease of median and average F1 scores against increasing number of training examples due to exceeded learning times. However, for single slot extraction tasks the *BFOIL* based approaches showed better $F1$ scores than *OSL* based learning methods for almost all investigated problems.

Comparing the untimely terminated *BFOIL* results with *OSL* learned results does not help to gain reasonable information about the quality of *BFOIL* learned wrappers, except for the fact that *BFOIL* needs far too long time for learning. Nevertheless from the single slot results and the observations taken from a combined method using clustering techniques in combination with *BFOIL* this is a promising approach worth be studied and refined.

One significant observation concerning the crucial learning time behavior of *BFOIL* is that up to a certain threshold regarding the number of training examples, the algorithm showed acceptable results. This and the circumstance that not for all problem tasks the learning time limit was exceeded will be taken into consideration when discussing and comparing some of the basic observations in the following.

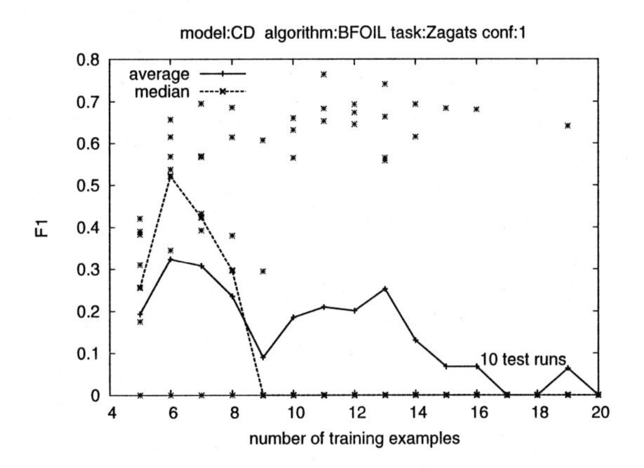

Figure 7.4.: reduced F1 scores due to exceeded learning times

Improving Precision

AV-wrappers are the weakest models among the three wrapper models regarding expectable precision rates. This is explained by the fact, that they exclusively use left and right delimiter patterns for extractions. They do not use any additional information about the text structure of constraining features, like *CD* or *RTD-wrappers* do. They tend to become too general compared to the other two models.

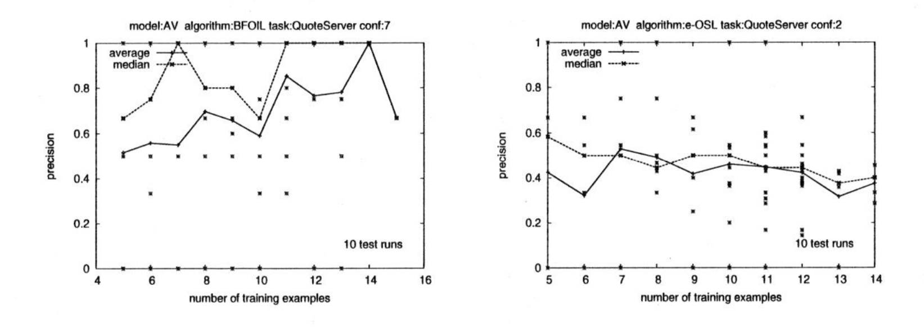

Figure 7.5.: increased precision of *BFOIL* learned *AV-wrapper* compared to *OSL*

BFOIL should yield *AV-wrappers* with improved precision results in comparison to *OSL* learned *AV-wrappers* in almost all of the problem cases investigated (Chapter 10), because it already detects overly general rules with respect to the training documents. Figure 7.5 shows the behavior explicitly for the case of learning a *linear*-x-$\vee$-ϵ^{+} wrapper.[3]

Avoiding Overly General Rules While Keeping High Recall Rates

Reducing the number of false positive extractions is one of the major goals of *basic-BFOIL* and its refinements. Basically, this is done by tuning a retrieval algorithm so that it yields higher precision rates, because of the reciprocal relationship between precision and recall this results in general in lower recall rates.

Since *BFOIL* learns a set of rules in contrary to *OSL*, the effect of decreasing recall rates can be kept quite small, as long as the set of examples is sufficiently heterogeneous. In this case each rule models specific properties of some subsets of the training example. For cases where specific properties differ strongly *BFOIL* can be expected to overcome the problem of over-generalization.

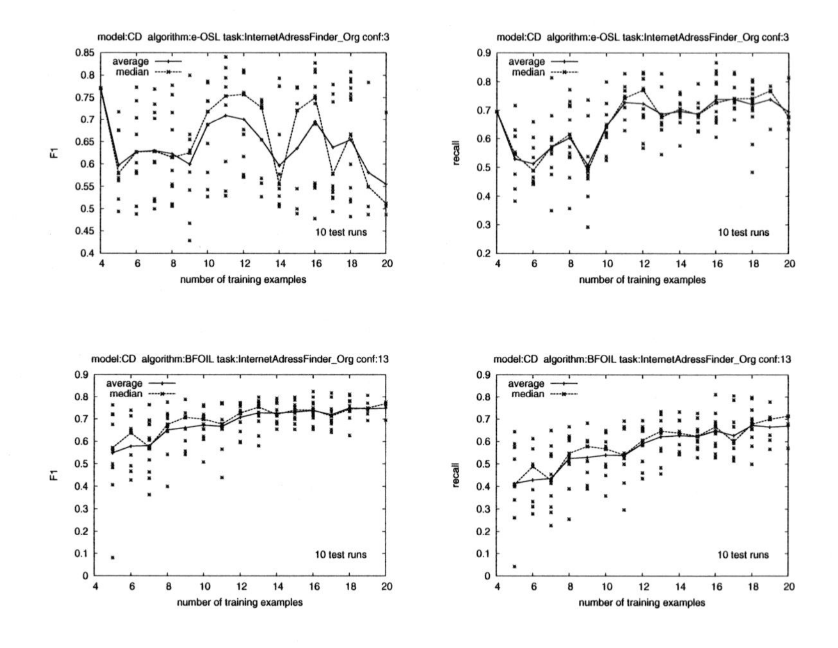

Figure 7.6.: *BFOIL* improved F1 scores with slightly decreased recall rates

[3]The BFOIL graph shows decreased precision scores for 15 examples due to missing examples based on the random example selection used for testing.

Figure 7.6 illustrates for a single slot extraction task, how precision and F1 scores are increased by *BFOIL* with only slightly decreasing recall rates. Both test runs were run with 5 to 20 training examples. Average results calculated for 12 to 20 examples yield following scores for *BFOIL* under the indicated refinement configuration c_i (see Chapter 10): positives: 20.3, false positives: 3.09, recall: 64.22%, recall: 87.09%, F1: 73.54% and for ϵ-*OSL*: positives: 22.84, false positives: 19.1, recall: 71.57%, precision: 61.92%, F1: 64.04%. The graphs show the typical growth of precision rates compared with the *basic-OSL* learned wrappers. Note that this does not imply that the quality in terms of F1 score is increased, as well.

Context Length Observations

Observations on *OSL* learned wrappers with increasing *mdl* values showed increased precision rates for the cost of lower recall rates. So, using longer delimiter lengths yields more specific wrappers with respect to a fixed number of training examples. We observe the same for *BFOIL* based learning. Figure 7.7 shows precision scores for learning a single slot *RTD-wrapper* with increasing context lengths.

Greater delimiter lengths and context lengths do not always guarantee better precision rates. Figure 7.8 shows decreasing average values for longer delimiter lengths, because for some of the evaluation documents the learned wrappers become too specific. Some wrappers provide no extraction from the testing documents, therefore the precision value is 0 (Definiton 3.2.4). This example demonstrates, that too long delimiters can decrease the quality of wrappers. In this case, the F1 score is decreased from 52.63% to 44.44% and 37.50%.

Apparently, choosing the right delimiter and context length has a significant impact on the quality of the learned wrapper. From a mere theoretical point of view a reasonable value can be determined incrementally by computing the peak of the precision rate with respect to increasing context lengths. Repeated learning of wrappers and evaluation, starting with the shortest context length and incrementing the length for every next learning-evalution cycle finds the precision peak and the optimal delimiter and context length.

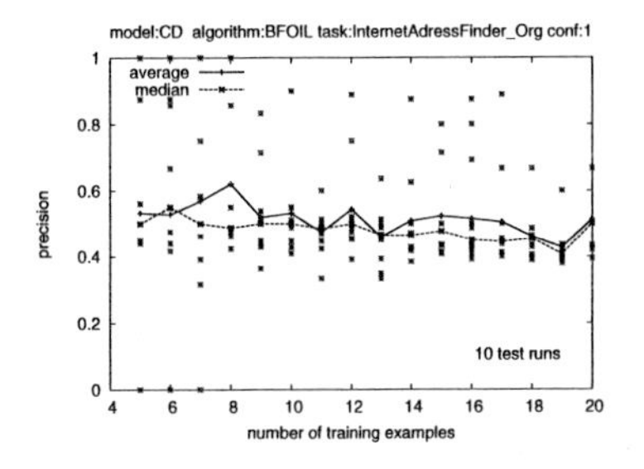

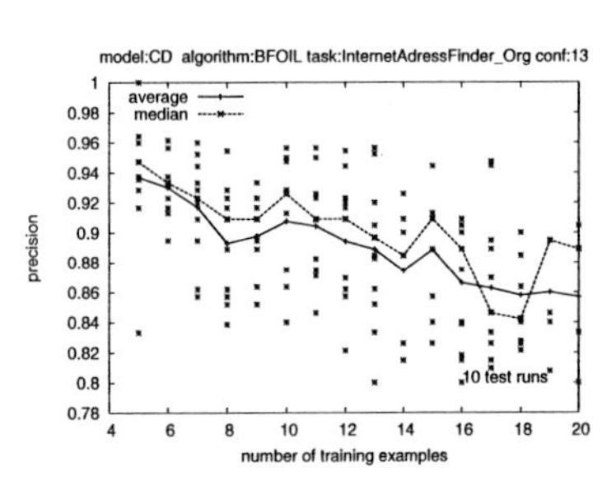

Figure 7.7.: *BFOIL* learned single slot *RTD-wrappers*, context lengths 3 (c1) and 7 (c13)

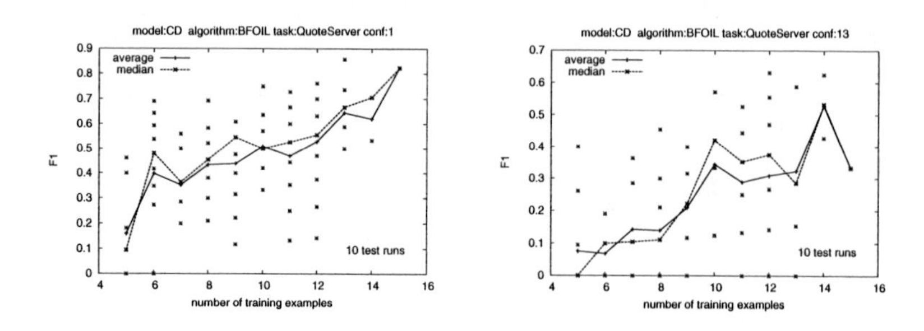

Figure 7.8.: *BFOIL* learned multi-slot *CD-wrappers*, delimiter length 3 (c1) and 7 (c13)

Learning Time Observations

Long learning times are the crucial shortcoming of the *BFOIL* learning approach. Figure 7.9 shows a typical learning time graph. Depending on a certain number of training examples (here 15) the learning time exceeds the predefined acceptable duration of 30 minutes (1800 cpu seconds) which results in an unexpected zero F1 score.

To vindicate the proposed method it has to be remarked that very few effort has been spent on optimizing the implementation of the presented algorithms. The *LIPX* system is almost a one-to-one implementation of the algorithms presented. The results obtained show, that the query computation time for huge instance spaces is a serious problem, not only in this application domain but also for the field of inductive logic programming in general.

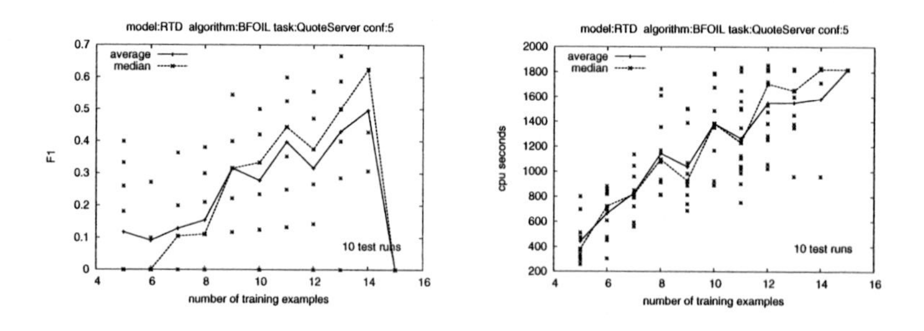

Figure 7.9.: F1 and learning time graph of *BFOIL* learned *RTD-wrapper*

Just like the learning times also the time for information extraction was limited for all experiments. Independently of the used learning algorithm and wrapper model the available extraction time was limited to 7 minutes per document. Especially for not so structured documents, where slot fillers are not nested in *HTML* environments or where delimiters consist of varying natural text, multi-slot learned wrappers run into difficulties. For some slots the related delimiter patterns and property describing predicates become too general. As a consequence the number of possible instantiations is dramatically increased, which results in very long extraction times.

Noisy Data

If the training data for learning wrappers with *basic-BFOIL* contains noisy data we have to distinguish the following cases : 1) The training data contains only true positive examples but the validation set contains false positive examples. In this case *basic-BFOIL* might reject certain generalized rules providing only correct extractions, because they yield extractions which are due to the noise not contained in the validation set. On the other hand generalized rules yielding false extractions might be accepted because they cover exactly the false positive data contained in the validation set. Obviously the same observations can be made if both training and validation set contain false positive examples with the additional case that a rule learned from false positive data is accepted by the *posterior satisfiability* check because it covers the false positive examples in the validation set. 2) The training set contains false positive examples and the validation set contains no noisy data. In this case *basic-BFOIL* will provide a hypothesis almost identical to one learned from none noisy data, except that the hypothesis covers the false positive training examples. This is explained by the fact that no matter if a rule was learned from false positive data or true positive data it is rejected if it does not meet the *posterior satisfiability* check. So depending on the degree of noise probably each false positive will be covered by one ground extraction rule.

The *threshold-based BFOIL* approach might help to reduce the introduced error by noisy data. Therefore assume the previously discussed case 1) and *threshold-based BFOIL*: probably some of the correct rules that are rejected by *basic-BFOIL* are now accepted due to the threshold based *posterior satisfiability* check. But on the other hand the positive effect of case 2) is reduced or eliminated, because evidently false rules learned from false positive data are accepted as long as they cover some of the true positive examples contained in the validation set. In contrast to *OSL* based approaches the *BFOIL* based approaches seem to be more roboust against noisy data, as long as the noise contained in the validation set is kept very small. But since learning from noisy data is not in the focus of this thesis, a more elaborated investigation is omitted.

Complexity Issues

In the best case *basic-BFOIL* outputs one single clause. In the worst case when all the corresponding generalizations do not suffice the *range restrictedness* and *posterior satisfiability* criteria it simply stores all example description clauses.

As discussed in Section 6.5 in the worst case for *basic-OSL* and ϵ-*OSL* the number of *clause lgg* computations is linear to the number of examples. In the best case *basic-BFOIL* computes $n-1$ clause lggs. In the worst case it computes $\sum_{i=1}^{n-1}(n-i) = \frac{n(n-1)}{2}$ times a *clause lgg*. Thus for the worst case its complexity is quadratic in the number of examples. Beside the *lgg* computations a second component of the *basic-BFOIL* algorithm increases the learning time significantly in contrast to *OSL* based approaches. It is the check if the current rule produces false positives, the *posterior satisfiability* test. Here the full time and space complexity of SLD/NF-resolution proofs with respect to the type of background knowledge and clause templates take place. Since each computed *clause lgg* is evaluated, the number of queries is identical to the number of *clause lgg* computations.

8. Combining Clustering Methods and BFOIL

In Chapter 7 the *BFOIL* algorithm was presented. One of the basic properties of this algorithm is characterized by the partition of the set of example description clauses. *BFOIL* finds a partition such that each *clause set lgg* of it results in an error free extraction rule with respect to the evaluation set. As discussed this approach is computationally expensive, since after every calculation of an *lgg* of two clauses the resulting clause has to be evaluated by a query computation. A similar partition approach is used with the ϵ-*OSL* algorithm. But there the partition process was solely based on certain assumptions regarding the occurrence of empty slot fillers. In contrast to *basic-BFOIL* the ϵ-*OSL* algorithm offers much shorter learning times, since no hypothesis evaluation is carried out during the partition process. This leads to less precise wrappers and higher extraction times for some extractions tasks.

Hence, combining both approaches, a pre-partition or clustering of the example set and if needed a *basic-BFOIL* learning applied to each of the obtained clusters, may improve the learning time and retains the wrapper's quality regarding its precision. Figure 8.1 depicts the underlying idea of merging an iterated agglomerative clustering technique with changing vector representations and distance metrics with the *basic-BFOIL* algorithm.

8.1. Clustering of Example Description Clauses

The question arises, how does a clustering method finds reasonable subsets such that each lgg of a cluster fulfills the *IE-ILP posterior satisfiability* condition. Following the depicted step-wise clustering idea in Figure 8.1 the ideal case is given if the algorithm terminates with good clusters after the application of the first clustering. In this case for each cluster only one *clause set lgg* computation and rule evaluation is necessary to complete the wrapper construction. Thus it follows that cluster techniques that minimize the number of clusters but still fulfill the requested quality are preferred. For a basic understanding of clustering techniques, we briefly illustrate four general needed components by means of partition example description clauses.

Reconsider the clustering used in the partition based *one step learning* algorithm ϵ-*OSL*. Choosing a slightly more abstract view on this problem, an example description rule for means of clustering can be represented by a binary vector of length n with n the number of slots. Such a vector representation indicates if a slot filler is empty or not. For instance, `<1,1,0,1>` and `<1,0,0,1>` represent two clauses (i.e. examples) simply stating that the third slot of the first example is empty and in the second example slots 2 and 3 are empty.

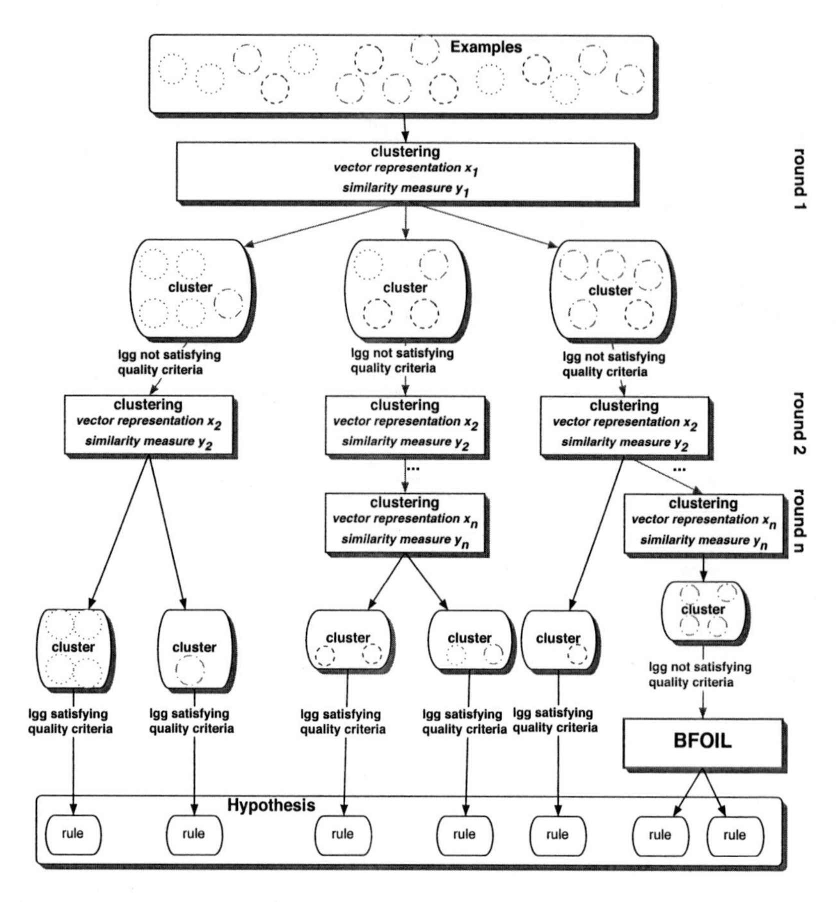

Figure 8.1.: idea of *cluster-BFOIL*

The second step in this clustering approach requires a comparison of two vectors to estimate how similar the associated examples are. The easiest comparison or similarity measure is simply to check if both vectors have the same component values. If this is true, then the related example description clauses belong into one cluster. Otherwise they have to be put into different sets. Obviously, *ϵ-BFOIL* uses exactly this *vector representation* and *similarity measure* to cluster the set of example description clauses. And not surprisingly in the research area of *Clustering* [Aldenderfer and Blashfield, 1984; van Rijsbergen, 1980] the *vector representation* and *similarity or distance metrics* are central topics of investigation to improve clustering methods.

Besides these two issues there is a third very important component, that decides if a new element is to be added to a cluster or not. A comparison policy is needed stating exactly with which element or calculated value of the elements the new potential candidate has to be compared. Shall the element in the cluster representing the average of the cluster be used (*centroid* method), or the element nearest to the candidate (*single link*) or the one most different to it (*complete link*) or is it sufficient if the average of all distances of the cluster elements is beneath a certain threshold (*group average*)? It has to be noted that the notion of similarity and distance is coupled with the idea of a predefined threshold. That means, as soon as there is a more elaborated similarity function as the illustrated binary one for the special case of empty slots a pre-defined threshold is used to decide if an element is added to a cluster or not. For further readings on these topics see [Hatzivassiloglou *et al.*, 2000; Miyamoto, 1990; Rasmussen, 1992; Steinbach *et al.*, 2000].

```
function cluster(M, D, T)
input : M = set of tuples of clause vector representation
        D = a distance function
        T = a threshold

C ← ∅
i ← 1
while i ≤ (|M| − 1) do
    x ← nth(i, M)
    j ← i + 1
    while j < |M| do
        y ← nth(j, M)
        if D(x.vec, y.vec) < T then
            if ∃c ∈ C : x ∈ c ∧ not conflict(y, c) then
                c ← c ∪ {y}
            else
                if ∃c ∈ C : y ∈ c ∧ not conflict(x, c) then
                    c ← c ∪ {x}
                else
                    C ← C ∪ {{x, y}}
        j ← j + 1
for e ∈ M do
    if ¬∃c ∈ C : e ∈ c then
        C ← C ∪ {{e}}
return C
```

Function 9: agglomerative, complete-link clustering

A fourth component of a clustering algorithm is the method how to build clusters. The general methods are distinguished into those starting with one large cluster (*partitive*) or methods starting with a set of clusters, where each datum from the corpus is set as a cluster (*agglomerative*). An iterated agglomerative clustering method is used by *cluster-BFOIL*. The start configuration consists of an empty set of clusters C and the set of example description clauses and their vector representations M. Then an element e_i is drawn from M and compared once with every other element $e_j \in M$. If the distance or similarity value of e_i and e_j meet a given threshold value, the following cases are considered: 1) if there is a cluster $c \in C$ containing either e_i or e_j the other is added to c if it does not conflict with any element in c. 2) if there is no cluster $c \in C$ containing either e_i or e_j a new cluster is added to C containing both element. After all elements from M have been compared with each other, there may remain elements that have not been added to any cluster in C. These elements build one element clusters. The second step can lead to overlapping clusters, which means one element can occur in more than one cluster. Function 9 summarizes this complete-link agglomerative clustering method used by the *cluster-BFOIL* implementation.

All of the mentioned methods have pro's and con's depending on the application domain, vector representation, distance measures and used clustering methods. For an exhaustive discussion on these methods the reader is referred to [Cutting *et al.*, 1992; Lewis, 1992; Miyamoto, 1990; Rasmussen, 1992]. In the remainder of this chapter we illustrate how some of the very basic clustering techniques can be used to improve the *basic-BFOIL* algorithm. The following sections are by no means intended to give an overview or introduction into the active research field of *Clustering*. On the contrary, they demonstrate the manifold to enhance and mix the pure ILP-based learning algorithm with other successful techniques.

8.2. Rule Vector Representation and Similarity Measures

The first step of a standard clustering technique is to find a suitable numerical vector representation of the data [Salton, 1989; Sable and Church, 2001]. Commonly it is derived from certain features of each object in the data set. For example, one common method to represent documents for document clustering tasks is to use a vector of term and word frequency based measures, as for example the *TF/IDF* (term frequency and inverse document frequency) measure [Salton and Buckley, 1988]. The difference between standard clustering application areas and the proposed idea to use these techniques to partition logical clauses is the significantly smaller size of the data corpus and the problem to find a suitable vector representation for clauses.

For example, in document clustering there is in general a large term space and each document contains hundreds of different words. In comparison to this term space and data corpus, the occurrence of different literals within one clause is significantly smaller.

From these observations it turns out that the whole idea of clustering example description clauses might only be a promising approach for *RTD-wrapper models*, because they consist of non rigid clauses with varying literals. Nevertheless, also for *AV* and *CD-wrapper models* the clustering approach is a reasonable method, because it incorporates the idea of ϵ-*OSL* as long as a reasonable vector representation is chosen.

The central idea of the used clustering method to use changing vector representation and similarity respectively distance measures is briefly discussed in the following. In principle, the best clustering method is the one that is yielding the largest clusters and therefore smaller number of clusters fulfilling the *posterior satisfiability* requirement regarding the *clause set lgg* of the cluster, because this would minimize the number of evaluation tests (i.e. query computations). Hence, a reasonable idea is to assume that a vector representation and well chosen similarity measure that roughly partitions the set of example clauses complies with this intuition. This idea is assured by the observations taken from the ϵ-*OSL* approach. Consequently, using a hierarchy of rough to more elaborated vector representations and distance measures results in more and fine granular clusters. But as a consequence of creating more clusters also more evaluation tests are necessary and probably wrappers with higher precision but lower recall rates are learned. So, a larger number of clusters with less examples means that the generalization degree regarding the *clause set lgg* of each cluster is lower and therefore more specific rules are constructed.

The basic intention of *cluster-BFOIL* is to find (maximal) subsets of the example set in reference to their *clause set lgg* and compliance of the *posterior satisfiability* criterion. The *cluster-BFOIL* algorithm applies a step-by-step clustering starting with very rough clustering parameters such that resulting clusters are formed according to concise (e.g. empty slots) features. With each next clustering round more subtle ones (e.g. number of occurrences of a literal and distribution among all clauses) are built. Accordingly, we investigated this conception with three different vector representations and two different similarity resp. distance measures.

Basically two different representation approaches for clustering clauses can be chosen, one that represents each clause as either a binary, integer or real valued vector [Salton, 1989; Sable and Church, 2001], or clauses are kept unchanged. If clauses are used unchanged the standard distance metrics can not be applied and specific distance metrics for clauses [Nienhuys-Cheng, 1997; Hutchinson, 1997; Markov and Marinchev, 2000; Ramon and Bruynooghe, 1998; Ramon *et al.*, 1998; Ramon and Raedt, 1999] have to be used.

	binary literal $< a,b,c,d >$	TF/IDF $< a,b,c,d >$	mod-TF/IDF $< a,b,c,d >$
$r_1 : e \leftarrow a,c,a,d$	$< 1,0,1,1 >$	$< 0,0,0,0.103 >$	$< 0,0,0,0.415 >$
$r_2 : e \leftarrow a,b,c,a,d$	$< 1,1,1,1 >$	$< 0,0.083,0,0.083 >$	$< 0,0.415,0,0.415 >$
$r_3 : e \leftarrow a,b,a,b,c$	$< 1,1,1,0 >$	$< 0,0.138,0,0 >$	$< 0,0.830,0,0 >$
$r_4 : e \leftarrow a,b,c,d,a,b,c$	$< 1,1,1,1 >$	$< 0,0.118,0,0.059 >$	$< 0,0.830,0,0.415 >$

Table 8.1.: binary literal occurence, TF/IDF and mod-TF/IDF clause vectors

cluster-BFOIL uses a vector representation of clauses for clustering the example description set, because this offers the possibility to conveniently evaluate standard clustering techniques in this domain. Except for the binary *empty slot vectors* (Definition 8.2.1) a clause is represented as a vector with n components, where each component corresponds to one predicate symbol and n is the total number of different predicates occurring in the set of example description clauses. Depending on the chosen representation each vector component is assigned a specific value estimating the relevance of each literal under a certain theory of relevance. Table 8.1 illustrates this technique. Since we only use predicate symbols for the different vector representations and similarity metrics introduced in the sequel, the literals of a rule are represented in the following examples only by their functor.

Definition 8.2.1 (Empty Slot Vector) *Given a set of example description clauses C. The* empty slot vector *representation for each $c \in C$ is given by the vector $\vec{v} =< c_1, c_2, \ldots, c_n >$ with $c_i \in \{0,1\}$ indicating if slot i of c is empty by $v.i = 0$ or if it is not empty by $v.i = 1$.* □

The *binary literal occurence vector* (Definition 8.2.2) simply indicates which of the hypothesis language literals occur in a given example description clause. It neither counts the number of occurrences nor does it estimate and respect any distribution or frequency ratios of literals among the set of example description clauses. Obviously, this representation yields more and more identical vectors with increasing number of slots. In Addition, this is only a reasonable representation for *RTD-wrapper models.*

Definition 8.2.2 (Binary Literal Occurrence Vector) *Given a set of non prefix protected example description clauses C. The* binary literal occurrence vector *representation for $c \in C$ is defined as follows: L is the set of all predicate symbols occurring in C and v is an n-ary vector with $|L| = n$ such that there is a bijective mapping ϕ from components i of v to $l \in L$ and $v.i = x$ with $x \in \{0, 1\}$ indicating if literal $l = \phi(i)$ occurs in c.* □

Using the presented clustering Algorithm 9 on the clause set and binary literal occurrence representation from Table 8.1 yields the clusters shown in Table 8.2.

	r_2			r_3			r_4		
	d	g	l	d	g	l	d	g	l
r_1	1.00	0.25	0.50	2.00	0.50	1.00	1.00	0.25	0.50
r_2				1.00	0.25	0.50	0.00	0.00	0.00
r_3							1.00	0.25	0.50

$$Clusters_{threshold=0.25}\{\{r_2, r_4\}, \{r_1\}, \{r_3\}\}$$

d := Euclidean distance
g := global scaling factor: $0.25 = \frac{1}{n}$ with $n = 4$ arity of vectors
l := local scaling factor: 1/2

Table 8.2.: Euclidean distance matrix of binary literal occurrence vectors and clusters

The third vector representation used in this thesis is a modified version of the *TF/IDF* representation. Usually, a selected set of n words from a document corpus is chosen. Then for every document an n-ary document vector is constructed, such that each component is assigned a numerical value that represents the importance of the word w_i regarding the whole corpus. This value is estimated by the *TF/IDF* measure, which weighs the term frequency of term w_i in D_j with a factor that reduces the importance of w_i if it appears in very many documents. Or in other words, *TF* measures the term density within a record and *IDF* measures the informativeness or rarity of a term across the whole corpus. *TF/IDF* is one of the most widespread vector representations for document clustering. Its standard definition is as follows:

$$TF/IDF(w, D) = \frac{\text{occurences of w in D}}{\text{total number of words in D}} \times log_2 \left(\frac{\text{number of documents}}{\text{no. documents w occures in}} \right)$$

At first sight the *TF/IDF* measure also appears to be a reasonable representation method to rate clauses regarding the literals they contain. Focusing on *RTD-wrapper models* and sets of example description clauses the standard *TF/IDF* measure has a shortcoming. Basically, in the investigated context it is less important how many literals a clause has. But it is much more interesting if some literals occur or not. Because in the resulting *lgg* of two clauses those literals not contained in both clauses are eliminated. Therefore it is more reasonable to build clusters containing clauses having as many literals in common as possible. If clusters are built according to this rule, clauses with important property describing literals are grouped together and as few literals as possible will be eliminated by the *clause lgg* operator. Table 8.3 shows clustering results using standard *TF/IDF* based vector repesentation for clauses from Table 8.1.

	r_2			r_3			r_4		
	d	g	l	d	g	l	d	g	l
r_1	0.007	0.00	0.25	0.030	0.00	1.00	0.016	0.00	0.53
r_2				0.010	0.00	0.33	0.002	0.00	0.06
r_3							0.004	0.00	0.13

$$Clusters_{threshold=0.25} : \{\{r_2, r_4\}, \{r_3, r_4\}, \{r_1\}\}$$

d := Euclidean distance
g := global scaling factor: $0.125 = 1/(2 * log_2(|r|/1)^2)$ $n = 2$ max l. occ.
l := local scaling factor: $33.3 = 1/0.030$

Table 8.3.: Euclidean distance matrix of TF/IDF vectors and clusters

It appears that the total number of literals in a clause is less relevant under the *lgg* based wrapper learning scenario and thus should not be used to estimate the relevance of a single literal. Nevertheless, it is of interest how often a literal occurs within one clause, because under *prefix protection* each of its occurrences is equivalent to a different literal (i.e. linked with a slot, representing certain properties of the slot). The term frequency of the *TF/IDF* measure is not perfectly modeling this idea, since only the number of occurrences and not its frequency shall be taken into account. Actually it suffices to simply drop the denominator of the term frequency function, which obviously correlates to the idea to focus only on the number of appearances and not on the literal's frequency within a clause. Keeping the inverse document frequency has proven to be still a good function to measure the relevance of a literal. Roughly speaking, *IDF* covers the idea that a literal contained in an example description clause describes an extraordinary property of the text example, if this literal occurs less often within other clauses. Definition 8.2.3 summarizes this modification. Table 8.4 displays the resulting clusters using *mod-TF/IDF*[1] and Algorithm 9.

Definition 8.2.3 (Modified TF/IDF Vector) *Given a set of non prefix protected example description clauses C. A* mod-TF/IDF vector *representation for $c \in C$ is defined as follows: L is the set of all predicate symbols occurring in C and v is an n-ary vector with $|L| = n$ such that there is a bijective mapping ϕ from components i of v to $l \in L$ and $v.i = mod - TF/IDF(l, c)$ with:*

$$mod - TF/IDF(l, R) = (occurrences\ of\ l\ in\ R) \times log_2 \left(\frac{number\ of\ rules}{no.\ rules\ l\ occurs\ in} \right)$$

□

[1] A more moderate weighing of the number of occurrences for a specific literal is conceivable. Therefore instead of the linear occurrence weight a logarithmic one like log_2(occurrences of l in R) can be used.

	r_2			r_3			r_4		
	d	g	l	d	g	l		g	l
r_1	0.172	0.00	0.20	0.861	0.01	1.00	0.689	0.01	0.80
r_2				0.344	0.01	0.40	0.172	0.00	0.20
r_3							0.172	0.00	0.20

$$Clusters_{threshold=0.25} : \{\{r_1, r_2\}, \{r_2, r_3\}, \{r_3, r_4\}\}$$

d := Euclidean distance
g := global scaling factor: $1/64 = 1/\sum_1^4 (n * log_2(4/1))^2$ $n = 2$ max l.occ.
l := local scaling factor: $1/0.861$

Table 8.4.: Euclidean distance matrix of modified TF/IDF vectors and clusters

As similarity measure for *empty slot vectors* the already discussed identical vector measure is used. It simply compares pairs of all components $v_1.i$ and $v_2.i$ of vectors v_1 and v_2 for identical values. If all components are identical the similarity of v_1 and v_2 is 1 otherwise 0. For *binary literal occurrence* and *mod-TF/IDF* vectors a slightly modified version of the Euclidean distance metrics is used to measure the similarity of two vectors. Again, this is one of the standard similarity measures used within common clustering systems. We also experimented with other similarity and distance metrics as for example *dice* and *cosinus coefficient* [Rasmussen, 1992; Miyamoto, 1990]. But for the proposed representations and *RTD-wrapper models* resp. example description clauses a slightly modified Euclidean distance measure showed the most promising results. The modification simply is to omit the square root to yield a stronger quadratic separation of literal rule vectors. The slightly modified Euclidean distance is given by $d_{modEuclid}(\vec{v_1}, \vec{v_2}) = \sum_k (v_1.k - v_2.k)^2$.

Defining Thresholds

Some remarks concerning the use of a common threshold for all three methods have to be given. Basically a mapping of distances on the interval [0..1] is desirable, because then one common threshold T can be used, independently of the chosen vector representation and distance metrics. This can be easily achieved by using a suitable scaling factor as long as the distances are measured in the Euclidean space. Therefore in theory only the maximum expectable distance between two vectors has to be determined.

For *binary literal occurence vectors* the theoretical maximum distance is given by n the number of vector components. Assuming a vector $\vec{v_1}$ represents a rule containing all literals, and a vector $\vec{v_2}$ represents a rule with empty body, then $\sum_{i=1}^{n}(v_1.i - v_2.i)^2 = n$. Hence a reasonable scaling factor is $\frac{1}{n}$ for the computed distance values. In the special case of example description clauses the theoretical maximum does not hold, because example description clauses have per definition at least one literal describing at least one non empty slot filler. So, the theoretical maximum will always over-estimate the practical maximum distance. This is acceptable if we only want the calculated distances to fall into the interval [0..1], but if one common threshold shall be used a more precise scaling factor is needed. *AV* and *CD-wrapper* clauses have a fixed number of literals and for *RTD-wrapper* example description clauses, there exists a certain set of always occurring literals, as for instance the predicates `xspan`, `xpath`, and `star_and_end_nodes`. Depending on the chosen hypothesis language m such predicates can be identified, thus the maximum is given by $n - m$.

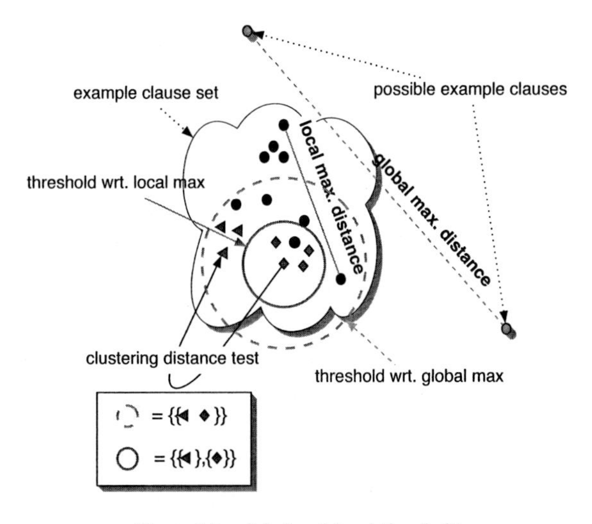

Figure 8.2.: global and local threshold

For *mod-TF/IDF vectors* the crucial point in estimating the maximum distance consists in determining the maximum number of occurrences of a literal within one clause. In fact this number is limited by the definition of *RTD-wrapper clause templates* (Section 5.4), stating that the length of a clause is limited by at most $n \times$ (|delimiter predicates + content predicates + structural predicates|) + $\frac{n(n-1)}{2} \times$ |relational span predicates| literals taken from $\mathcal{L}_{RTD}$ with n the number of slots. A relational span literal can occur at most $\frac{n(n-1)}{2}$ times within a clause. The maximum *mod-TF/IDF* value for a relational span literal is given by $max_r = \frac{n(n-1)}{2} \times log_2(\frac{\text{no. of rules}}{1})$ for all other literals $max_o = n \times log_2(\frac{\text{no. of rules}}{1})$. If m is the number of vector components and r the number of relational span literals in $\mathcal{L}_{RTD}$, then the maximum distance of two *mod-TF/IDF* vectors of *RTD-wrapper* clauses under modified Euclidean distance is given by $max = \sum_{i=1}^{m-r} max_r^2 + \sum_{i=m-r}^{r} max_o^2$ and the scaling factor by $\frac{1}{max}$.

Although these are theoretically correct scaling factors to map distance results onto the interval [0..1] their application leads to poor clustering results. Instead of comparing the distances between the given example description clauses relative to the maximal possible distance it is more reasonable in this setting to compare the distances with a local maximum. Using the maximal distance between vectors of the given example description clause set yields much better results as Table 8.4 and 8.3 and Figure 8.2 illustrate.

Cluster results and resulting generalized clauses for the running example are summarized in Table 8.5. To illustrate the difference between generalizing protected clauses and non protected ones both resulting *lgg* sets are presented and literals are marked with indices regarding the associated slot. So, each multiple occurrence of one literal describes a property of a different slot. As in the previous examples only the literal functors are used, because only those are relevant for the proposed vector representations and similarity metrics.

rules		
$r_1 : e \leftarrow a_1, c_1, a_2, d_2$ $r_2 : e \leftarrow a_1, b_1, c_1, a_2, d_2$	$r_3 : e \leftarrow a_1, b_1, a_2, b_2, c_2$ $r_4 : e \leftarrow a_1, b_1, c_1, d_1, a_2, b_2, c_2$	
binary literal	TF/IDF	mod-TF/IDF
	clusters	
$c_1 = \{r_2, r_4\}$ $c_2 = \{r_3\}$ $c_3 = \{r_1\}$	$c_1 = \{r_2, r_4\}$ $c_2 = \{r_3, r_4\}$ $c_3 = \{r_1\}$	$c_1 = \{r_1, r_2\}$ $c_2 = \{r_2, r_4\}$ $c_3 = \{r_3, r_4\}$
	lgg	
$c_1 : (e \leftarrow a, b, c, d)\Theta$ $c_2 : r_3$ $c_3 : r_1$	$c_1 : (e \leftarrow a, b, c, d)\Theta$ $c_2 : (e \leftarrow a, b, c)\Theta$ $c_3 : r_1$	$c_1 : (e \leftarrow a, c, d)\Theta$ $c_2 : (e \leftarrow a, b, c, d)\Theta$ $c_3 : (e \leftarrow a, b, c)\Theta$
	prefix protected lgg	
$c_1 : (e \leftarrow a_1, b_1, c_1, a_2)\Theta$ $c_2 : r_3$ $c_3 : r_1$	$c_1 : (e \leftarrow a_1, b_1, c_1, a_2)\Theta$ $c_2 : (e \leftarrow a_1, b_1, a_2, b_2, c_2)\Theta$ $c_3 : r_1$	$c_1 : (e \leftarrow a_1, c_1, a_2, d_2)\Theta$ $c_2 : (e \leftarrow a_1, b_1, c_1, a_2)\Theta$ $c_3 : (e \leftarrow a_1, b_1, a_2, b_2, c_2)\Theta$
	with $\Theta = \theta^{-1}$ if $c \preceq c' \equiv c\theta \subseteq c'$ and $c'' \subseteq c'$ and $c''\theta^{-1} = c$	

Table 8.5.: example cluster results

8.3. The cluster-BFOIL Algorithm

The central idea of the algorithm *cluster-BFOIL* is to find subsets of the given examples by means of clustering techniques, such that the *clause set lgg* of those subsets fulfill the IE-ILP *posterior satisfiability* condition. For this purpose the initial example set is step-wise partitioned by applications of the previously discussed clustering methods.

A order of clustering methods is defined stating which vector representation and which distance metrics to use. If a resulting cluster resp. its *clause set lgg* does not fulfill the desired quality criteria (i.e. *range restrictedness*, *posterior satisfiability* or *weak satisfiability*) and all clustering methods have been already applied to this subset, the *basic-BFOIL* or *threshold-based BFOIL* is applied to it. Algorithm 10 shows the *cluster-BFOIL* algorithm.

Some remarks concerning the notation and functions used in Algorithm 10. By $C(i)$ the i-th element from list C is denoted and by $C(i).v$ it is referred to the vector representation of the i-th tuple of C. Similar to this notation $Cluster.Examples$ refers to the original set of examples of the example description clauses contained in the cluster $Cluster$. The function $cluster(M, D, T)$ is shown in Function 9 and function $vector\_representation(CD_M(E^+), C(i).v)$ computes the vector representation given by $C(i).v$ of a set of example description clauses.

The *cluster-BFOIL* algorithm recursively tries to partition a given set of extraction examples

---

input : $E \subset \mathcal{W}$ an exhaustive example set E for a wrapper $\mathcal{W}$
$E^+ \subset E$ training examples
M the wrapper model to be learned
P a logic program implementing the hypothesis language of M
D logical document representation of documents in E
C ordered list of tuples (v, d) with v a vector representation and d a distance metrics.
T_R a rule quality threshold ; T_C a cluster threshold ; i a index, initial value 0

$Vecs \leftarrow vector\_representation(CD_M(E^+), C(i).v)$
$Clusters \leftarrow cluster(Vecs, C(i).d, T_C)$
$VS \leftarrow E_M(E)$
$LearnedRules \leftarrow \emptyset$
while $Clusters \neq \emptyset$ **do**
  $Cluster \in Clusters$
  $Clusters \leftarrow Clusters \setminus \{Cluster\}$
  $E^+ \leftarrow Cluster.Example$
  $R \leftarrow clause\_set\_lgg(Cluster)$
  if $range\_restricted(R) \wedge check(R, P, D, VS, T_R).PosteriorSatisfiability$ **then**
    $NewRules \leftarrow \{Rule\}$
  else
    if $i < length(C)$ **then**
      $NewRules \leftarrow cluster\_bfoil(E, E^+, M, P, D, C, T_R, T_C, i + 1)$
    else
      $NewRules \leftarrow bfoil(E, E^+, M, P, D, T_C)$
  $LearnedRules \leftarrow LearnedRules \cup NewRules$
return LearnedRules

Algorithm 10: *cluster-BFOIL*

---

according to specific clustering parameters (i.e. $C(i)$), whereas the recursion depth is limited by the number of clustering configurations (i.e. the length of the list of clustering parameters in C). If *cluster-BFOIL* is called with $i = 0$ it is guaranteed to terminate for three reasons:

1. There is a fixed upper bound on the recursion depth, $i < length(C)$.
2. The evaluation loop of clusters terminates, because for every iteration one element is removed from the set of clusters, i.e. $while\ Clusters \neq \emptyset$.
3. *basic* and *threshold-based BFOIL* terminate and therefore also function *check* terminates.

8.4. Properties of cluster-BFOIL

Properties of *basic-BFOIL* as the consistency results (Theorem 7.2.1, Corollary 7.2.3) also hold for the *cluster-BFOIL* algorithm using a rule quality threshold of $T_R = 1.0$. This is obvious, since *cluster-BFOIL* either learns rules by calling *basic-BFOIL* or it uses identical conditions as *basic-BFOIL* to decide if a rule is acceptable or not.

The clustering algorithm used by *cluster-BFOIL* allows to construct overlapping clusters, i.e. one clause can be contained in more than one cluster. There are two advantages to use overlapping clusters in this context: 1) the clustering is biased to produce slightly more general rules, because on average more clauses are contained in a cluster and thus the *clause set lgg* yields a more general clause than generalizing sets with less clauses. 2) the order of selecting clauses to be compared for being added to a cluster is irrelevant.

If the overlapping property is disallowed, then the order in which description clauses are edited can yield different clustering results. This is a general problem of clustering algorithms. So, the independence of the order of examples is one criterion for *theoretical soundness* [van Rijsbergen, 1980].

One serious decision influences the behavior of *cluster-BFOIL*, namely the order in which the different vector representations and distance metrics are to be applied. Since the central idea is to become more and more fine grained (i.e. larger number of clusters) with each clustering round, it makes sense to order vector representations and distance metrics according to their type, length and range. For instance, the *empty slot vector* is a binary vector, with the shortest length of all three discussed representations and the distance metrics (*identical vector components*) maps onto the set $\{0, 1\}$. Roughly speaking, the *empty slot vector* representation depends on far less properties and uses a much rougher metrics than the *mod-TF/IDF* vector representation with an Euclidean distance metrics. Thus, a clustering with the latter parameters should be applied last.

8.5. cluster-BFOIL Observations

Two basic reasons motivated the development of *cluster-BFOIL*. Firstly, in some cases *BFOIL* uses too long learning times, therefore *cluster-BFOIL* should be faster. Secondly, basic ideas observed from ϵ-*OSL*, using information about the literal occurrences in example description clauses are to be used on a more general and not handcrafted level as done by ϵ-*OSL*. And thirdly, the two proposed methods shall provide comparable *basic-BFOIL* quality results.

From the experimental results discussed in Section 10 it is observable, that *cluster-BFOIL* fulfills all of the three purposes. It shows faster learning times for all investigated problems with a speed-up factors ranging from 1.7 to 8.8. The F1 quality score of *cluster-BFOIL* learned wrappers are in general ±4 % different from *threshold-based BFOIL*'s results[2]. Figure 8.3 illustrates a typical comparison of *threshold-based BFOIL* and *cluster-BFOIL* results[3].

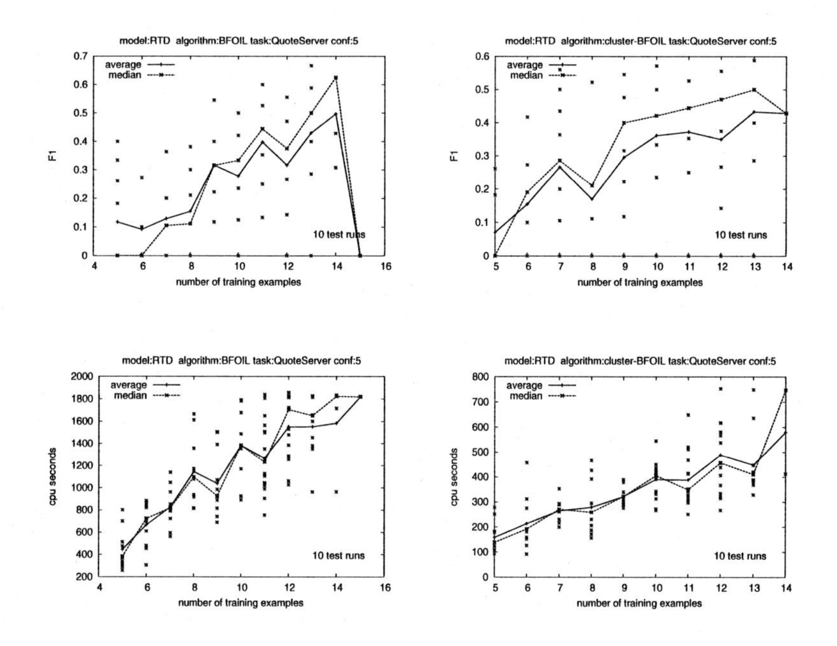

Figure 8.3.: *RTD-wrapper* learned by *threshold-based BFOIL* and *cluster-BFOIL*

Another evidence that the proposed *cluster-BFOIL* method and the discussed clustering techniques work well is the observation, that almost the same number of clauses for a wrapper have been learned by *cluster-BFOIL* as by *threshold-based BFOIL*. There is only a small variance of ±1 rule among the presented median results. This fact together with the negligible loss in extraction quality with respect to $F1$ scores underlines that *cluster-BFOIL* is indeed an improvement in efficiency of *threshold-based BFOIL*.

Complexity Issues

As with the other presented learning algorithms the two crucial points, the number of *clause lgg* and query computations, are briefly investigated. The smallest number of *clause lgg* com-

[2]Only for two of the best median results presented in Section 10 there are two fugitives of 22% and 32%.

[3]The BFOIL graph shows zero and decreased F1 scores for 15 examples due to exceeded learning time (1800 cpu seconds, see learning time graph).

putations is given if the algorithm terminates after the first clustering round with n clusters each one containing exactly one example description rule. Apparently, then learning demotes to memorizing. Assuming the case that at least each cluster yields two clauses, such that a reasonable learning can take place, then *cluster-BFOIL* has to compute $\frac{n}{2}$ clause lggs and has to compute $\frac{n}{2}$ queries. In contrast, in the best case *basic-BFOIL* needs $n-1$ *clause lgg* computations. Though in theory both have a linear complexity in practice it can make a great difference to reduce the number of *lgg* computations and queries by a factor of 2.

In the naive worst case *cluster-BFOIL* attempts to cluster m-times the initial example set, where each clustering attempt results in the given starting set. In this case it computes $m*(n-1)$ times a *clause lgg* with n the size of the example set. Finally after these m clustering attempts it applies the *basic-BFOIL* algorithm to the initial example set, with worst case effort of $\sum_{i=1}^{n-1}(n-i) = \frac{n(n-1)}{2}$ computations. Thus, a naive implementation of *cluster-BFOIL* never needs more than $m*(n-1)$ *clause lgg* computations than *basic-BFOIL*.

best case	OSL	ϵ-OSL	basic-BFOIL	cluster-BFOIL
clause lgg	$n-1$	$\frac{n}{2}$	$n-1$	$\frac{n}{2}$
	$O(n)$			
query	0	0	$n-1$	$\frac{n}{2}$
			$O(n)$	

$n =$ number of examples

worst case	OSL	ϵ-OSL	basic-BFOIL	cluster-BFOIL
clause lgg	$n-1$	$n-1$	$\frac{n(n-1)}{2}$	$m*(n-1)-\frac{m(m+1)}{2}$
	$O(n)$		$O(n^2)$	$O(mn)$
query	0	0	$\frac{n(n-1)}{2}$	$\frac{n}{2}+\sum_{i=1}^{m}\frac{n}{2^i}$
			$O(n^2)$	$O(nm)$

$n =$ number of examples; $m =$ number of cluster rounds

Table 8.6.: best and worst case effort for *clause lgg* and query computation

For estimating the number of query computations in the worst case assume that *cluster-BFOIL* caches edited clusters and uses them to avoid redundant computations. Then a maximal number of queries is computed if in every cluster round only new clusters are computed such that the number of clusters in each round is maximal. The maximal number of clusters for which lggs and queries have to be computed is $\frac{n}{2}$ with n the number of examples. This can only occur under caching in the last clustering round m. Otherwise one of the clusters would be cached in the next round and thus would minimize the number of *lgg* computations. Note, for worst case investigation, we have to assume that no *lgg* of a cluster yields a satisfying clause. So, the maximal number of clusters is given if each cluster after the last clustering round contains exactly two clauses. Consequently the clusters can be pairwise grouped such that they descend from one cluster from the previous round. Thus, there can be at most $\frac{n}{4}$ clusters in round $m-1$. Finally, in the worst case the proposed clustering method with caching can at most create $\sum_{i=1}^{m}\frac{n}{2^i}$ different clusters and therefore the same number of query computations is needed. In comparison, *basic-BFOIL* always needs the same number of query and *clause lgg* computations which in the worst case is $\sum_{i=1}^{n-1}(n-i)$.

In addition to the effort for *lgg* and query computation the complexity of clustering has to be estimated. Clustering a set of n clauses according to the presented Function 9 takes $\sum_{i=1}^{n-1}(n-i)$ comparisons of vector distances, whereas vector distances are normally calculated only once and stored, which also has a time complexity of $\sum_{i=1}^{n-1}(n-i) = \frac{n(n-1)}{2}$. In the best case (i.e. only one cluster round) $\frac{n(n-1)}{2}$ comparisons are needed with n the number of elements in the set to be clustered. In the worst case without caching $m * \frac{n(n-1)}{2}$ comparisons are needed, because every round yields the same cluster. With caching in every round the resulting cluster size is decreased by one. Actually, in practice the time needed for cluster computations is a minor matter, because once the distance matrix is computed the clustering itself consists of simple lookup and comparison operations. Table 8.6 summarizes the best and worst case complexities of the four presented learning algorithms of this thesis. For the best case estimations only the non-trivial best cases are considered.

9. Summary Part Two

This second part of the thesis started with a brief review of the standard task of inductive logic programming [Muggleton, 1991; Muggleton and Raedt, 1994] followed by presenting the most commonly associated semantics in Section 5.1. Starting from these different semantics we investigated in Section 5.2 which one is suited best to be used in the application domain of automatic wrapper learning. This discussion led to the insight, that a modified version of the *example setting* [Muggleton and Raedt, 1994] is suited best. Consequently, we introduced a new ILP setting named *IE-ILP semantics* tailored for wrapper learning tasks. Especially the fact that wrappers are to be learned from positive examples only lead to this setting.

Problems how to derive implicitly negative examples from certain types of knowledge and under varying semantics have been exemplified in Section 5.2.1. Furthermore similarities between the *IE-ILP setting* and existing semantics have been pointed out. Additionally their intersections with techniques used in other ILP systems (e.g. *FOIDL* [Mooney and Califf, 1995]) have been exposed.

Section 5.3 gave a short retrospect on fundamental ILP methods and techniques [Bergadano and Gunetti, 1996; de Raedt, 1996; Lavrac and Dzeroski, 1994; Lavrac and Dzeroski, 2001; Wrobel, 1996]. In Section 5.3.2 standard *ILP operators* were discussed. A detailed explanation on the *least general generalization (lgg)* operator was given in Section 5.3.2 concluding with the outlook to automatically construct extraction rules by means of *lgg based bottom-up algorithms.*

A framework for representing wrapper models and extraction examples as introduced in Part I in the context of *ILP* and especially in the *IE-ILP setting* was presented in Section 5.4. The notion of *wrapper clause templates* was introduced which is somewhat related to work on *rule models* and *schemata* by [Bergadano *et al.*, 1989; Kietz and Wrobel, 1991]. According to this, an example representation was introduced under the notion of *example description clauses*, which is based on instantiated clause templates. Thus, a set of extraction examples is transformed into a set of instantiated wrapper clauses and considered to be a set of most specific wrapper clauses (i.e. extraction rules), which is used as starting point for the introduced *lgg* based learning algorithms.

The Chapters 6, 7 and 8 presented step-wise refined *lgg based bottom-up learning algorithms.* In each chapter the proposed algorithm respectively refinements were motivated by weak spots of the previous learning algorithms. Some of the basic properties of each learning algorithm were identified and formally proven. Furthermore advantages and shortcomings were discussed by means of empirical and theoretical observations.

Thesis Contributions of Part Two

The following list summarizes the contributions of Part Two of the thesis:

- a general ILP setting for learning logic information extraction programs from positive examples only (Section 5.2)

- a clause template based language bias for learning logical wrapper models (Section 5.4)
- a one step bottom-up learning algorithm for inducing wrappers solely based on *clause lgg* operation and clause set partition refinement (Chapter 6)
- an incremental *clause lgg* based bottom-up learning algorithm for inducing wrappers with rule quality threshold and semantic lgg refinements (Chapter 7)
- an ILP algorithm combining standard clustering techniques with the incremental bottom-up learning algorithm (Chapter 8)

Part III.

Results & Discussion

10. Comparisons and Experiments

This chapter presents an overview about existing *Machine Learning based Information Extraction systems* in Section 10.1 and compares methods and experimental results of those systems with the presented approaches of this thesis. The emphasis for the method comparison is set on existing IE-systems using machine learning and especially ILP based methods, this is discussed in Section 10.1.1. In Section 10.2 the test cases, extraction tasks and test configurations used for experiments are introduced and a classification of the *RISE* [Muslea, 1998] extraction tasks according to the wrapper classes introduced in Section 3.1 is presented. Section 10.3 summarizes the best median learning results regarding the three different wrapper models *AV*, *CD* and *RTD* and the learning methods *ϵ-OSL*, *threshold-based BFOIL* and *cluster-BFOIL* on the extraction tasks. In Section 10.4 the results are compared to experimental results of existing state-of-the-art systems. Additionally a critical discussion on the results of *OSL*- and *BFOIL*-based wrapper learning techniques is given. The chapter concludes with Section 10.5 containing an empirical learnability analysis.

10.1. ML-based IE-Systems

A vast number of *IE-systems* and also *ML based IE-systems* have been developed over the last decade. In the following we enumerate the most significant features of existing systems to provid a basis for comparing the relevant systems with the *LIPX* system, which is the implementation of the methods presented in this thesis.

Although one associates with the concept of informtation extraction a procedure that provides text fragments from a document there are also approaches that use a slightly different ingress to the field of *IE*. For instance, marking relevant text parts by annotating a text with tags is especially since the widespread use of XML documents a reasonable technique. Insofar, the *traditional* methods differ from those *tagging* approaches in that the techniques for *tagging* learn rules for the insertion of tags. By repeatedly processing a document such tags can then be inserted in a document. This offers a possibility to easily *tag* nested or multi valued extractions by identifying or learning single slot rules. One system of this class is *PINOCCHIO* [Ciravegna, 2000] using a rule learning algorithm called LP^2 for tagging which is not closeley related to *LP* or *ILP*.

Existing systems differ in how they make use of implicit given information about the text properties and therefore what a wrapper is learned to detect. The *delimiter approach* simply tries to detect left and right text anchors, ignoring any structural information as paragraphs, chapters or semi-structured annotated information given by tags. The *hierarchical approach* uses the additional environmental information given either by natural language text formatting (e.g. chapters, section, paragraphs) or by semi-structured information (e.g. *SGML*, *XML*, or *HTML* tags) or type-setting and formatting commands (e.g. LaTeX). A third less widespread idea used by the *ROADRUNNER* system [Crescenzi *et al.*, 2001] is to learn wrappers for

HTML documents based on the comparison of web pages and the observable similarities and differences.

Further differences exist in the way *ML based IE systems* construct hypotheses (i.e. wrappers). They either serially process example by example and thus learn wrappers in an *incremental* manner or they compute wrappers in *one step* from the whole set of examples. *One step learning* is also called *Batch Learning*. *Batch Learning* is always an *off-line learning* process, whereas *incremental learning* approaches offer the possibility to be embedded into running systems such that they improve over time by examples given to them. Besides the majority of automatic wrapper construction systems there are also some systems that rely on user interaction like *AUTOSLOG* [Riloff, 1994]. The assumption that always a sufficiently large or good training corpus is present is not realistic in some problem domains. The research field of *active learning* investigates methods to learn from very small data sets and how to use unlabeled data to improve the learner's quality. Such an approach for wrapper learning is presented by [Muslea *et al.*, 2000]. Another very promising approach presented by [Freitag and Kushmerick, 2000] uses *Boosting* [Shapire and Singer, 1998] to induce wrappers. *Boosting* improves the quality of weak learners by repeatedly applying the weak algorithms to a changing training set. The training set is changed in that training examples are weighted according the previous failures of the weak learner.

As there are different extraction tasks existing systems are also differently capable of solving them. Basically, it is distinguished between *single* and *multi slot* extraction tasks. Additionally, some assumptions about the number of extractions per document can be made either there is only *one per document* or there are *many per document* extractions to be performed. The task becomes more complex depending on the arity of the extraction construction task coupled with aspcects like the occurrence order of slot fillers in the text. For instance, under a *one per document* assumption *multi slot* wrapper learning can be eventually reduced to constructing n single slot wrappers. Consequently, there are also approaches that investigate how to group extracted fields under the *many per document* assumption [Jensen and Cohen, 2001].

With the emerging flood of semi-structured and structured documents triggered by the spreading of the World Wide Web and also the XML movement, the interest in these document types among the *IE* community also increased. Especially, because for semi-structured documents available from the web commonly the pattern and document structure based extraction approaches show better results than semantical or computational linguistic based ones. Hence, some systems are exclusively tuned to work with tagged documents whereas others only process natural language text (*free text*). Since the notions of semi-structured and structured documents are differently used in some publications, we use both notions for *SGML-based* documents (e.g. web pages, XML documents). This is a somewhat rougher classification of text types, since some researchers classify newsgroup postings, or emails to be semi-structured text. They argue that such texts are in general grammatically messy and contain some significant text fields indicating text to be extracted. Nevertheless, such fields can also exist in free text and since those texts do not contain structuring elements in the sense of hierarchical tagging environments we classify them as free text. Also connected with these representational issues is the system's internal representation of documents. Several representations are used, documents and thus wrapper learn on a *character level*, *word level*: character sequences are grouped as they appear in the text, *token level*: words from a document are tokenized or classified (e.g. int, alpha), *feature structures (fs)*: words are transformed and eventually extended into lists of attribute value pairs, or the document is internally represented as a *tree structure* based on structural information present in the document (e.g. paragraphs, sections, *DOM*, *TDOM*). Commonly, the used feature structures are only *flat*, but some systems like the *WHISK* system [Soderland, 1999] also use more complex feature structures, where the value of an feature can be a feature structure. For the following we do not separate between flat and non flat structures.

Some systems enrich the preprocessing of documents, which is needed for the construction of the used internal document representation, by incorporating computational linguistic (CL) analyzation tools. Such tools as *part of speech* taggers can especially aid the wrapper learning process for natural language texts, because they provide additional semantic knowledge. For instance generalizing the symbol sequences `six` and `five` under mere *character level* aspects is not promising, whereas with additional *CL* knowledge a reasonable generalization is a pattern like `NUMERAL`. Linguistic pre-processing is one method other techniques to add domain specific knowledge either by special token types or extended feature structures are best described as *semantic tagging*. For instance, adding information of the length of a word, if it starts with capital letters, or the information if a sequence of numbers represents a telephone number, date or price belong to such preprocessing techniques. Instead of using only one fixed document representation and therefore pre-processing method the *WhizBang Labs Wrapper Learner* (WL^2) [Cohen *et al.*, 2002] follows a strategy to incorporate several freely chosen document representations. Therefore they define the concept of *builders* that can be integrated into the basic wrapper learning algorithm as long as a *builder* provides a predefined set of functions, namely *generalize* and *refine*.

Widely discussed in previous chapters of this thesis, the most desirable way to learn a wrapper is to present only a very small number of positive examples. Because of theoretical reasons shown by Gold [Gold, 1967] or by the PAC-learnability [Valiant, 1984; Kushmerick, 1997; Dzeroski *et al.*, 1992] theory, there are certain boundaries towards the expectable quality of learned wrappers. Hence, most of the systems make different assumptions concerning the number of positive examples required and the requirement of negative examples. Some systems use heuristic methods to build negative examples [Freitag, 1998]; rely on the explicit preparation of them; use strong assumption on the positive examples to implicitly derive negative ones (*LIPX*) and others work without them and use other heuristic techniques to evaluate the quality of wrappers during learning. Related to the problem of learning from sparse examples is the research work on using selective sampling techniques [Muslea *et al.*, 2001] and boostraping methods [Jones *et al.*, 1999] to overcome the need for tedious labeling of extraction examples.

Similar to extending documents with additional knowledge during the pre-processing phase some systems make use of additional knowledge sources during the learning phase of wrappers. Methods vary from exclusively using linguistic knowledge about syntactic phrases (e.g. nominal phrases, prepositional phrase) to the use of ontologies. Basically, by means of additional knowledge the learning of syntactical properties can be improved and a semantical content analysis and generalization can be provided. For instance, the *RAPIER* system [Califf, 1998] operates in this manner.

Surly, the *ILP* based way to learn wrappers proposed in this thesis is only one out of many techniques researched. There are a manifold of other techniques used by the existing systems. Commonly, these techniques can be grouped into *Finite State Automata* based on *Grammatical Inference* [Murphy, 1996; Parekh and Honavar, 1998] and *Relational Rule Learner* based on *ILP* or ideas derived from the *ILP* area. *Grammatical Inference (GI)* can be best understood as the learning of finite state automata such that the target domain is a formal language and the hypothesis space consists of a family of grammars. Given a set of text examples the task is then to find a minimal automat accepting the learning examples.

At the present time, learning of wrappers include techniques like the learning of *prefix trees*, *Finite State Automata* [Chidlovskii *et al.*, 2000], *Naive Bayes based frequency tables* [Freitag, 1998], *Hidden Markov Models* [Freitag and McCallum, 2000], *Regular Expressions* [Soderland, 1999], *tree automata* [Kosala *et al.*, 2003], *path expression* [Taniguchi *et al.*, 2001] or *logic programs* [Junker *et al.*, 1999; Thomas, 2003]. Finally, there are also approaches to combine several different type of learners to automatically build wrappers as investigated in [Freitag, 1998]. Table 10.1 gives an overview about the most important systems and works compared under some of the mentioned properties of *ML based IE-systems*.

List of Brief System Descriptions

To complete the overview about existing systems and ML based approaches to automatic wrapper construction a list with very brief system descriptions concerning the used technique, wrapper representation respectively implementation, and the closely related machine learning field is given next. Other overviews can be found in [Muslea, 1999; Kushmerick and Thomas, 2003]

System: **AUTOSLOG** (*linguistic frame patterns*, -)
A one step learning system, learning from positive examles, involving no induction steps that uses a linguistic taxonomy for learning, and user interaction.

System: **CRYSTAL** (*linguistic frame patterns*, -)
Uses a *bottom-up sequential covering algorithm*, learns from positive and negative examples, generates linguistic patterns by means of syntactical and semantical generalizations based on linguistic knowledge and a given taxonomy.

System: **LIEP** (*linguistic frame patterns*, -)
Uses a taxonomy of syntactic relationships among linguistic sentence constituents to construct generalized pattern templates from positive and negative examples obtained by user interaction.

System: **PALKA** (*linguistic frame patterns*, *version space*)
Uses a *candidate elimination algorithm* [Mitchell, 1977; Mitchell, 1982] to learn linguistic phrase patterns. Patterns are constructed by generalizing and specializing terms with terms derived from a given taxonomy. The system learns from positive and negative examples.

System: **HASTEN** (*linguistic frames*, *instance based learning*)
Learns ranked extraction templates. Weights for pattern templates from positive and negative examples by means of *k-nearest neighbor learning* [Atkeson *et al.*, 1997] are learned.

System: **BWI** (*simple reg. exp.*, *boosting & GI*)
Given only positive examples simple best prefix and suffix pattern (delimiters) are learned which are improved by *Boosting* [Shapire and Singer, 1998] methods.

System: **WIEN** (*constrained reg. exp.*,)
Learns delimiter strings from positive examples. Several wrapper classes are investigated, stating

system	wrapper		document		
	slot	idea	type	rep.	pre-proc.
AUTOSLOG [Riloff, 1994]	single	linguistic			
RAPIER [Califf, 1998]		phrase			
HASTEN [Krupka, 1995]		patterns	free	fs	tokenize, CL
LIEP [Huffman, 1996]	multi	or			
PALKA [Kim and Moldovan, 1995]		templates			
Bayes [Freitag, 1998]					
Bayes+GI [Freitag, 1998]					
GI [Freitag, 1998]	single		free	fs	tokenize
SRV [Freitag, 1998]		left and			
PINOCCHIO (LP^2)[Ciravegna, 2000]		and right		fs	tokenize, CL
HMM [Freitag and McCallum, 2000]		char, word,		word	tokenize
[Junker *et al.*, 1999]		token, or fs		word	tokenize
BWI [Freitag and Kushmerick, 2000]		delimiter		word, token	tokenize
WIEN [Kushmerick, 1997]				char	-
SoftMealy [Hsu and Chang, 1999]				word	tokenize
CRYSTAL [Soderland, 1997]	multi		semi, free	char, word,	tokenize, CL
WHISK [Soderland, 1999]		(reg. exp.)		token, fs	
[Chidlovskii *et al.*, 2000]				token	tokenize
k-testable [Kosala *et al.*, 2003]	single	tree, path		tree, word	mod. DOM
STALKER [Muslea *et al.*, 1999]		delimiter		*EC*-tree, token	*EC*, tokenize
ROADRUNNER [Crescenzi *et al.*, 2001]		pattern and	semi	tree, token	DOM
LIPX [Thomas, 2003]	multi	hierarchical		tree, fs	mod. DOM, AV
WL^2 [Cohen *et al.*, 2002]		properties		tree, (+other)	mod. DOM, (+other)

Table 10.1.: overview ML based IE-systems: wrapper & document features

different constraint resp. wrapper classes.

System: **Chidlovskii et al.** (*version of FSA*, *GI*)
By means of *edit distances* [Aho, 1990] and positive examples only *linear finite state automata* are learned for detecting text delimiters.

System: **SoftMealy** (*version of FSA*, *GI*)
From positive and negative examples the graph structure of *Soft Mealy Machines* [Hopcroft and Ullman, 1979; Hsu and Dung, 1998] are learned by means of a *set covering algorithm* [Haussler, 1988].

System: **Bayes, GI, Bayes+GI** (*version of FSA*, *GI*)
Given positive examples only, Bayes learns *delimiter texts frequency tables* using *Naive Bayes* based methods. GI induces grammars (finite state automata) for detecting delimiters by means of standard *Grammatical Inference* algorithms (Alergia [Carraso and Oncina, 1994], ECGI [Rulot and Vidal, 1988]). Finally Bayes+GI combines *Naive Bayes* methods with these standard algorithms.

System: **STALKER** (*version of FSA*, *GI*)
Learns disjunctions of linear finite state automata (landmark automata) by a *sequential covering* method. Learned landmark automata are used to identify relevant extractions in a hierarchical, tree structure document representation, called *embedded catalog EC*.

System: **HMM** (*version of FSA*, *GI*)
Hidden Markov Models are learned from positive examples only. To find the optimal automata structures *shrinkage* [McCallum, 1999] and *stochastic methods* are used.

System: **k-testable** (*tree automata*, *GI*)
Given positive examples *k-testable tree automata* [Rico-Juan *et al.*, 2000] are induced for detection of paths in tree structured documents.

System: **ROADRUNNER** (*html tree pattern*, *unlabelled data*)
Instead of learning extraction patterns or rules from a set of example extractions the system compares documents to find a common pattern that serves as extraction rule. Therefore a method called *ACME, align, collapse under mismatch, and extract* is presented that learn from unlabelled data.

System: **LP²** (*symbolic rules*, -)
A bottom-up inductive approach to learn symbolic tagging rules and correction rules, to repair erroneous tagginng rules. The algorithm learns from positive examples, assuming exhaustive labeling and uses implicitly negative examples.

System: **WHISK** (*regular exppressions*, -)
A top-down induction method for learning regular expressions from positive examples only. It incorporates the use of syntactic and semantic class knowledge for the preprocessing step and uses heuristics for rule selection whilst it can use semantic classes for learning.

System: **LIPX** (*FOL rules*, *ILP*)
A bottom-up ILP approach learning sets of extraction rules (Horn clauses). Each learned wrapper rule follows a predefined wrapper model, i.e. clause template. The wrapper language is freely interchangeable. Uses *one step* or *incremental* learning, standard clustering techniques, specific generalization operators, works on modified DOM trees and feature structure sequences. Only positive examples are given assumed to be exhaustively labeled.

System: **Junker et al.** (*FOL rules*, *ILP*)
A pure logical, ILP *separate and conquer* algorithm [Fürnkranz, 1999] using specific clause refinement operators to learn *Prolog* rules from positive and intensionally defined negative examples.

System: **RAPIER** (*FOL rules*, *ILP*)
Learns from positive and negative examples. It uses a combined bottom-up & top-down based ILP algorithm similar to *CHILLIN* [Mooney *et al.*, 1994]. Semantic generalization is based on the semantic hierarchy of *WordNet* [Fellbaum, 1998].

System: **SRV** (*FOL rules*, *ILP*)
A *FOIL* [Quinlan, 1990] based approach to learn logical delimiter rules based on predicates describing token features and word occurrence orders. It assumes exhaustively enumerated positive examples and uses implicitly given negative ones.

System: **WL^2** (*FOL rules*, *ILP*)
A *FOIL* based approach using *lgg* concepts to learn sets of rules. It learns from exhaustively enumerated positive examples and negative examples are derived implicitly. Different wrapper languages and hence document representations can be integrated by so called *builders*.

10.1.1. Closely related IE-Systems

In proportion to the number of existing systems only a few, namely [Junker *et al.*, 1999], *RAPIER* system [Califf, 1998], SRV [Freitag, 1998], WL^2 [Cohen *et al.*, 2002] are ILP systems or at least make intensive use of common *ILP* techniques. The other class of related systems are those that also try to learn or incorporate hierarchical information obtainable from a tree-view on documents, *STALKER* [Muslea *et al.*, 1999], *ROADRUNNER* [Crescenzi *et al.*, 2001] *k-testable* [Kosala *et al.*, 2003], and [Taniguchi *et al.*, 2001]. These systems are not really closely related to the presented methods in this thesis in that they only share some wrapper ideas and views on document representational issues and do not use related learning techniques. *STALKER* for example has a more or less closeness, because it also represents documents as tree structures and comparable to *TDOM* path predicates like `xpath` it also tries to identify certain path properties by finite state automata. This also holds for the other two systems *ROADRUNNER*, *k-testable* and approaches like that by [Taniguchi *et al.*, 2001]. They all try to induce path or tree properties either represented directly as automata or by patterns which can be easily transfered into *FSA's*. Hence, in the following we focus in more detail on related *ILP* based *IE-systems*.

The proposed *separate-and-conquer* strategy in [Junker *et al.*, 1999] learns standard logic programs for extraction. For refinement and generalization they use handcrafted operators tailored for a set of predicates describing word and word occurrence properties. Though an explicit learning algorithm is not presented the standard *separate-and-conquer* approach used is a top-down approach starting with a unit clause, which is then by application of the operators refined. Hence, this approach differs from *LIPX* bottom-up methodology. Similar to the use of *semantic lgg operators* as introduced in 7.3.2 [Junker *et al.*, 1999] also use special operators for generalizing clauses. Therefore they define generalization operators almost identical to the presented ones except that they do not incorporate *lgg* operations. Another distinguishable property is the requirement and treatment of negative examples. [Junker *et al.*, 1999] defines negative examples by providing a set of fragments from a document which are guaranteed to not contain any information to be extracted plus an intensional definition (i.e. program rule) based on the given positive examples. In contrary, *LIPX* does not need any provided negative examples, but assumes exhaustively enumerated positive examples. [Junker *et al.*, 1999] does not discuss any methods for multi-slot extractions nor are any experimental results presented. Although, they aim at learning standard logic programs the proposed refinement operators are bound to the underlying hypothesis language, whereas the learning methods used in *LIPX* do not depend on a special hypothesis language (except for the use of *semantic lgg operators*).

The systems SRV [Freitag, 1998] and WL^2 [Cohen *et al.*, 2002] also use *separate-and-conquer* strategies, which are very closely related to *FOIL* [Quinlan, 1990]. Similar to [Junker *et al.*, 1999] and the method presented in this thesis *SRV* uses a set of so-called *token features* for which several test predicates are introduced. For instance a predicate like `some(?A [previous token] in_title true)` is true if the previously parsed token also occurs in the *HTML* `title` environment of a document. Similar concepts like the *maximal delimiter length* or *context distance* are also covered by certain predicates. Although, *SRV* has a strong logical background it does not use a standard logic programming syntax like [Junker *et al.*, 1999] or *LIPX* do. Nevertheless the hypothesis language used by the *SRV* system possesses strong similarities to the ideas used in the *AV* and *RTD-wrapper languages.*

One of the major differences of *SRV* to other *ILP* related *IE* approaches discussed here, is the fact that it does not learn rules that extract information but it learns rules that decide if a given text fragment belongs to the set of intended extractions or not. This covers almost exactly the discussed way to interpret wrapper learning as concept learning in Section 2.2. In addition to this only single slot wrappers are learned. Since *SRV* is based on *FOIL* and uses the same basic algorithm it also uses the almost identical *information gain* based metric to select literals for refining the current rule. So, compared to [Junker *et al.*, 1999] and *LIPX* the rule construction and evaluation is based on information theory measures, whereas [Junker *et al.*, 1999] uses handmade refinement operators, and *LIPX* guides the generalization process by a quite strong logical consequence (answer derivation) test. Since, *SRV's* learning algorithm is more or less a general purpose rule induction algorithm it is widely independent of the chosen wrapper (hypothesis) language, which also holds for *LIPX*. As mentioned in the previous chapters this allows easily to incorporate domain knowledge by extending the hypothesis language by additional predicates. In [Freitag, 1998] he reports a representational and predicative extension of *SRV* with *WordNet* [Fellbaum, 1998], surprisingly the chosen semantic extension does not yield much improved extraction results on the tested free text experiments.

One step further regarding the flexible use of different wrapper languages is the approach of the *WhizBang Labs Wrapper Learner (WL^2)* [Cohen *et al.*, 2002]. WL^2 focuses on the extraction from *HTML* document advocating that restricting a wrapper learning system to solely one document representation and wrapper language is not adequate to achieve perfect extraction procedures. Therefore they developed a technique incorporating several different extraction languages into their wrapper learning algorithm. In [Cohen *et al.*, 2002] they assume that a set of different extraction languages is given such that each language provides some predicates for detecting relevant properties. Additionally for each language there must be a generalization and refinement function provided, which allows to compute the *lgg* of a training set regarding the language and for refinement reasons a set of concepts covering some of the training examples. These extraction languages are called *builders*. *LIPX* adopted the idea used by [Cohen *et al.*, 2002] to represent documents as *DOM* trees and examples as *spans*. Similar as *LIPX* builds *RTD-example description rules* from span representations WL^2 uses slightly different span based example representations to instantiate *builder* predicates describing example properties. Obviously, the basic motivation by [Cohen *et al.*, 2002] to incorporate many different wrapper languages and therefore eventually different views on documents can be adopted to the *LIPX* system by appropriate extension of the used hypothesis language. Nevertheless, the advantage of WL^2 system is that it uses some sort of *encapsulated* wrapper language dependent generalization and refinement operations, which allows to integrate sophisticated language and domain dependent techniques. In contrary, *LIPX* uses more or less a brute force *lgg* application on clauses and predicates regardlessly of the semantics of the predicates. Except, *semantic lgg operators* are intensively used. WL^2 general learning algorithm also differs from the presented *lgg* based bottom-up learning algorithms. WL^2, similar to *FOIL*, consists of an outer, set-covering loop, which repeatedly learns a single rule covering some examples. The inner loop, uses the *builder's lgg* computed predicates to refine the current rule. Negative examples are induced whilst learning. As reported in [Cohen *et al.*, 2002] another major difference is that WL^2 is in fact, not a multi-slot rule learner like *LIPX* is, because extraction tasks are restricted to *binary extraction tasks*. Actually, rules are learned (similar to *SRV*), that decide wether or not a presented substring from the document is to be extracted or not.

Basically, *RAPIER* is like *LIPX* a bottom-up learner also based on the idea to use *least general generalization* based techniques to induce wrappers. But there are several points in which these two systems differ. Firstly, *RAPIER* makes extensive use of computational linguistics and additional semantic knowledge. It uses *part-of-speech* tagging information for document representation and learned wrapper rules. Rules consist of *pre-filler*, *filler*, and *post-filler* patterns mainly using linguistic features. Secondly, to overcome certain computational complexity issues, which are caused by using a modified *lgg* operator it incorporates a top-down component. Thirdly, the *lgg* operator is modified such that for differing constraints (i.e. features) a disjunction is computed consisting of an empty set of constraints and the union of them. Furthermore, the *lgg* is tailored computing the generalization of semantic constraints based on linguistic and semantical knowledge obtained from a taxonomy (WordNet [Fellbaum, 1998]). Although an extraction rule consists of the three components *pre,filller,post* pattern, *RAPIER* starts to generalize patterns for the *filler pattern* and keeps empty *pre* and *post* pattern. The *pre* and *post filler* patterns are specialized in the inner loop of *RAPIER's* algorithm by adding generalized (lgg) constraints (features) to them. Basically, this is done by a beam-search, whereas the current rules are hold in a priority queue regarding the rules quality. But instead of processing all constructed rules only the k best are kept and older ones are eliminated if better new ones are added to the queue. For estimating a rule's quality *RAPIER* uses a measurement of informativity, based on the number of correct and false extractions biased by favoring simpler rules over complex ones. *RAPIER* derives negative examples implicitly under identical assumptions as *LIPX* does. It is assumed that all documents are completely labeled (i.e. positive examples are exhaustively enumerated).

Summarizing the observations leads to the following cognition: from an algorithmic point of view *RAPIER* is closest to *LIPX*, because it uses a *bottom-up lgg* based rule induction algorithm. With the work of [Junker *et al.*, 1999] the same motivation regarding a pure logic programming and ILP framework is shared. Ideas to provide as much flexibility regarding the use of different hypothesis (extraction) languages and document representations is closely related to the approach by [Cohen *et al.*, 2002].

10.2. Test Settings

Each of the presented learning algorithms and wrapper models offers a wide variety to adjust parameters like the *maximum delimiter length*, *rule quality threshold* or *distance threshold* criteria. Therefore, several trials with a fixed set of configurations have been conducted. The specific test configurations for *ϵ-OSL*, *threshold-based BFOIL*, and *cluster-BFOIL* are given in Section 10.2.2.

Testing was done in the following way. For every task each wrapper model has been learned by each of the three learning methods *ϵ-OSL*, *threshold-based BFOIL* and *cluster-BFOIL.* A test run for a wrapper model and learning method consisted of 10 repetitions for each number of randomly drawn training examples. Except for some extractions tasks after 10 iterations the number of training examples was increased by one from 5 upto 20. For the selection of positive examples and the required exhaustively enumerated example sets the available set of documents is first randomly split into half.

Then from one half of the document corpus n positive examples were selected assuming equal distribution probability of examples among the sample documents. These n positive examples randomly selected from m randomly chosen documents where used for learning and the exhaustive example enumeration set was build from the m documents (training set). This assures that the exhaustive enumeration set does not consists of the complete available data set. Learned wrappers were evaluated on the complete document set, but not on the training examples. This process was repeated ten times for each number of training examples and every test configuration. Figure 10.1 shows a typical graph with F1 scores for learning a *RTD-wrapper* with the *threshold-based BFOIL* algorithm.

As mentioned in Chapter 7 and 8 additional time limits had been setup for all extraction tasks. The upper time limit for learning a wrapper was 30 cpu minutes and the maximum time a wrapper was allowed to take for extracting information from one document was set to 7 minutes. If the processes exceeded these time limits they were terminated and in case of exceeded extraction time the maximum time was taken and the so far yielded extractions were counted.

Most of the wrappers were learned, given 5 up to 20 examples for each extraction task (Table 10.3). The *QuoteServer* test runs were limited to 5 up to 15 examples because of its small number of examples. Because of the same reason the single slot tasks *Internet Address Finder, Altname* and *QuoteServer Date* and *Vol* were limited to 2 up to 6, 5 up to 10 and 5 up to 12 examples. Although by its nature the *Seminar Announcement* test set is a multi-slot extraction tasks, no multi-slot wrapper results are presented, because none of it terminated under the given time constraints (see Section 10.4 for an explanation). Because it turned out that this extraction class needed extremely long learning and extraction time the single slot test runs were carried out with a reduced document set (120 instead of 400).

attribute	meaning
algorithm	the used learning algorithm (*e-OSL* $\hat{=}$ *ϵ-OSL*, *c-BFOIL* $\hat{=}$ *cluster-BFOIL*, *BFOIL*$\hat{=}$ *threshold-based BFOIL*)
conf	used configuration (e.g. mdl, threshold)
ld	number of training documents from which examples are taken
p	total number of correct extractions
fp	total number of false extractions
rec	the recall rate in percent
pre	the precision rate in percent
f1	the F1 rate in percent
lt	time to learn a wrapper in cpu seconds
tpd	the median time a wrapper needed to process one document
r	the number of learned extraction rules

Table 10.2.: result table legend

For learning *RTD-wrappers* the hypothesis language as introduced in Section 3.4.2 was chosen. But the three structural level description predicates `xnright_brother/5`, `xfather/3` and `xchild/4` were omitted for two reasons: a) tests showed that they increase the search space without yielding better results which leads to longer learning and extraction times b) some of the information they represent becomes redundant if other more informative predicates are used (e.g. the information about a `father` node is implicit given by the `xpath` predicate).

In the following sections only accumulated results are presented. The median score of a test run is computed by building the median of each trial regarding the number of examples for which the best results were obtained. We consciously choose to represent the median results of the trials, because due to restricting the test runs by time limits the average estimate counts those test runs as zero extractions. The median estimate represents a more optimistic view on those premature terminated wrappers. Generally, it is observable that almost only *threshold-based BFOIL* needs too long learning times. But it is also observable that all *BFOIL* based techniques show improved F1 scores in proportion to the number of examples. Hence, it is more reasonable to use the median estimate. Each presented result table contains the attributes described in Table 10.2.

A short remark concerning the number of training examples. If a *BFOIL* based technique is said to learn from n examples, then this means that the final wrapper was constructed from n example description clauses. Since *BFOIL* based techniques require a validation set sometimes it is also of interest to know the size of the validation set. The size can be estimated by computing the average number of occurrences of extractions per document multiplied by the number of training documents, which is the number of documents from which the learning examples are taken. The approximate size of the validation set under the assumption of equally distributed extraction examples among the documents is given by: $\frac{\text{total number of examples}}{\text{total number of documents}} \times$ number of training documents.
For instance, if a wrapper for *BigBook* was learned from 15 examples drawn from 5 training documents, the validation set contains approximately $\frac{204}{11} * 5 \approx 93$ examples and the learned wrapper was tested on $204 - 15 = 189$ examples.

To keep the size of result presentations small but still comprehensive for each wrapper model only the best wrapper obtained by each of the three learning algorithms is presented. The F1 scores are used for comparison and all trials were made with the wrapper induction system *LIPX*. The *LIPX* system is the implementation of methods and techniques presented in this thesis. The tests were executed on Intel Pentium 4, 2.4 GHz, 512 MB main memory machines, running LINUX Suse 8.0. The *LIPX* wrapper induction system is implemented using *ECLIPSE* a *Prolog* and constraint programming system [ECLiPSe, 2004].

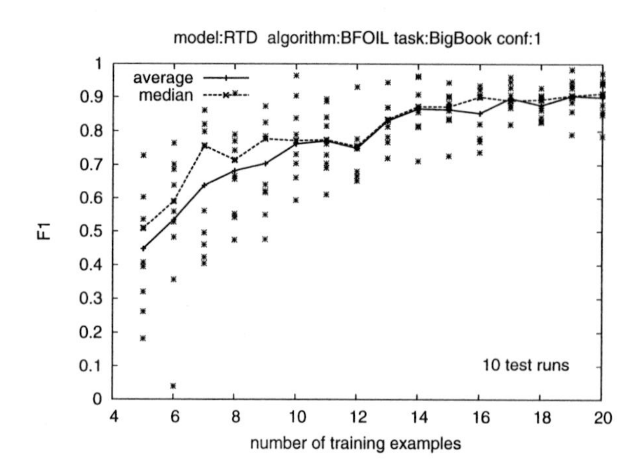

Figure 10.1.: graph of 10 test runs with increased number of examples from 5 up to 20

10.2.1. Extraction Tasks

The most common test set used by recent IE approaches is the *RISE repository* [Muslea, 1998]. *RISE* is a collection of semi-structured and non semi-structured electronical document sets like newsgroup postings and web pages. These documents have been used for evaluation reasons in almost all relevant research publications regarding ML based wrapper learning.

RISE provides for every extraction problem an extraction specification. Though many of the existing approaches only investigate on a subset of these specifications, namely on single-slot extraction tasks, we evaluated the presented wrapper models and learning algorithms on the original multi-slot extraction task specifications and on single slot variants mentioned in related publications[1]. Although the main aim of this thesis is to develop ILP based techniques for multi slot wrapper learning the single slot extraction tasks have been investigated, because they provide a basis for comparison to state-of-the-art IE systems.

[1]For *LA-Weekly* and *Zagats* containing a multiple value slot (MV-class) with n values a learning example is represented as n examples where each variant contains one of the n slot fillers from the multiple value slot. For *Zagat* a reduced variant was investigated consisting only of *[name,food,address,telephone]*

RISE problem	class	documents	examples	slots
multi-slot extraction tasks				
BigBook	linear, $\wedge$	11	204	6
Internet Address Finder	non-linear, ϵ^+,$\vee$	10	84	6
LA-Weekly	linear, MV, ϵ^+, $\vee$	28	547	4
Okra	linear, $\wedge$	252	3335	4
QuoteServer	linear, ϵ^+, x, $\vee$	10	25	18
Zagats	linear, MV, $\wedge$	91	140	4
single-slot extraction tasks				
CS, name	$\vee$	30	1149	1
IAF, alt. name	$\wedge$	5	11	1
IAF, organization	$\wedge$	10	52	1
LA-Weekly, credit cards	$\wedge$	28	144	1
QuoteServer, date	$\vee$	10	24	1
QuoteServer, vol	$\vee$	10	25	1
Seminar Announcements speaker	$\vee$	120	124	1
Seminar Announcements stime	$\vee$	120	242	1
Seminar Announcements etime	$\vee$	120	119	1
Seminar Announcements location	$\vee$	120	133	1
Zagats, address	$\wedge$	91	140	1

Table 10.3.: investigated extraction tasks

In total 6 different multi slot extraction tasks and 11 single slot extraction tasks have been investigated (Figures 10.2 - 10.9). All but one problem class the *Seminar Announcements*, which are newsgroup postings, consists of extraction tasks from *HTML* documents. The reason is that the presented wrapper languages are developed to be used with *HTML* documents and not with natural text. Anyhow, it is of interest to see how wrapper models using a quite weak attribute value representation in comparison to linguistic pre-processing methods perform on one of the most difficult extraction tasks.

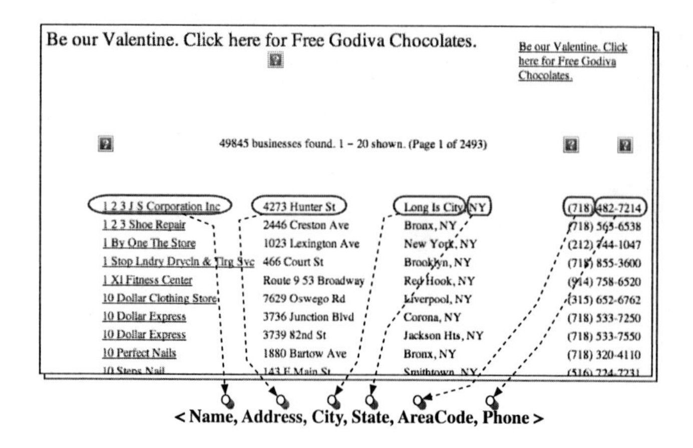

Figure 10.2.: BigBook extraction task

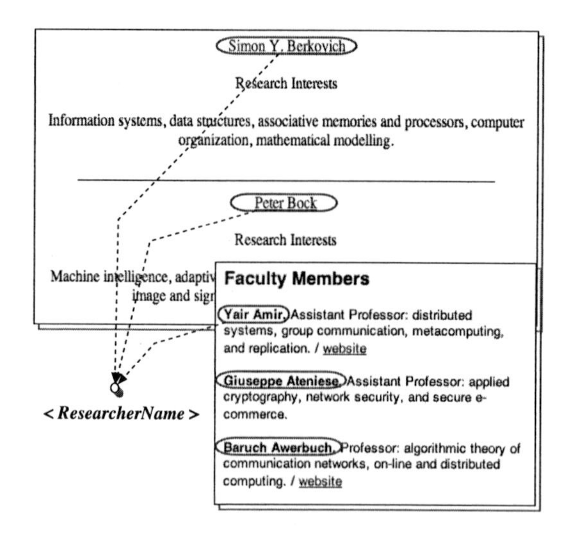

Figure 10.3.: CS faculty member extraction task

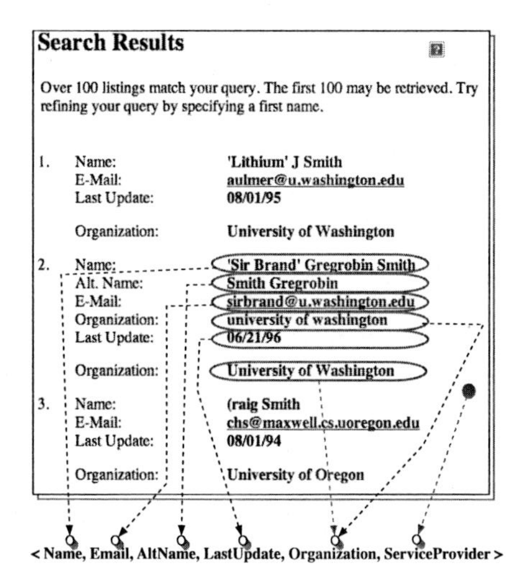

Figure 10.4.: Internet Address Finder extraction task

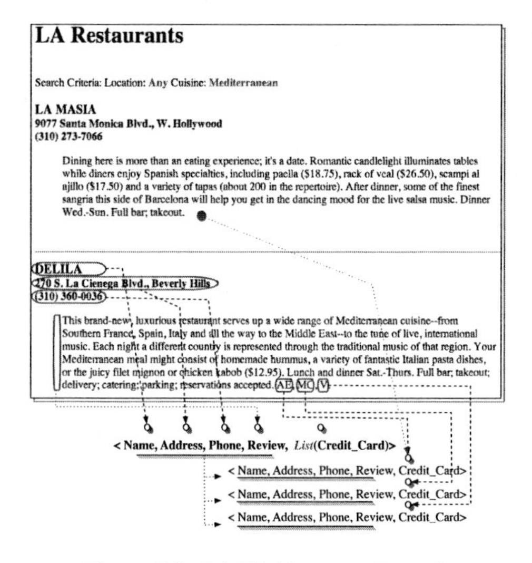

Figure 10.5.: LA-Weekly extraction task

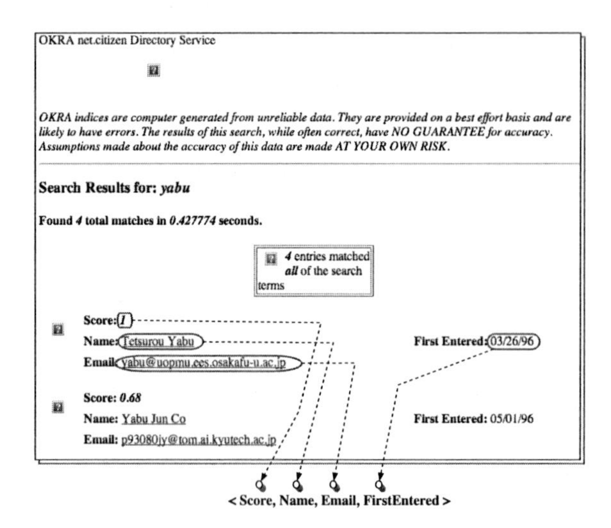

Figure 10.6.: OKRA extraction task

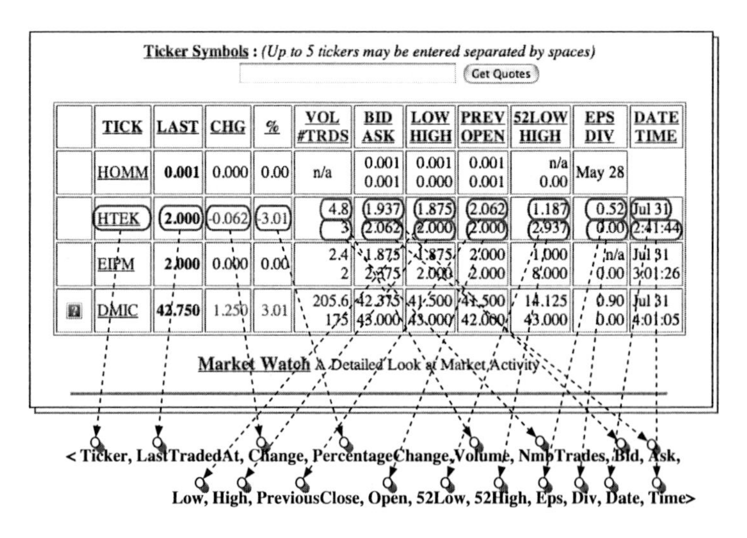

Figure 10.7.: QuoteServer extraction task

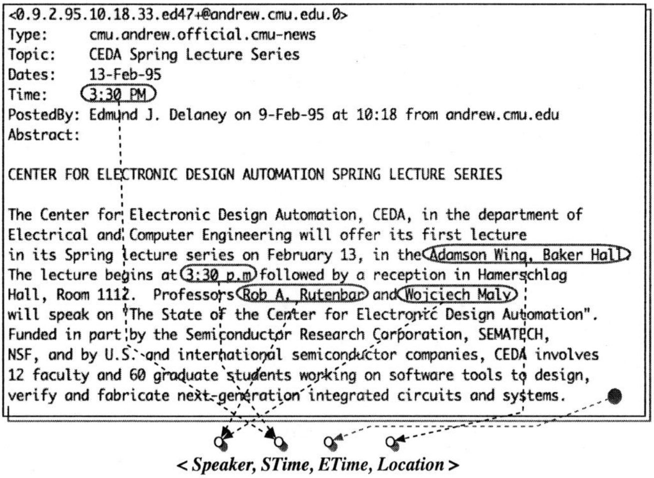

```
<0.9.2.95.10.18.33.ed47+@andrew.cmu.edu.0>
Type:      cmu.andrew.official.cmu-news
Topic:     CEDA Spring Lecture Series
Dates:     13-Feb-95
Time:      3:30 PM
PostedBy: Edmund J. Delaney on 9-Feb-95 at 10:18 from andrew.cmu.edu
Abstract:

CENTER FOR ELECTRONIC DESIGN AUTOMATION SPRING LECTURE SERIES

The Center for Electronic Design Automation, CEDA, in the department of
Electrical and Computer Engineering will offer its first lecture
in its Spring lecture series on February 13, in the Adamson Wing, Baker Hall.
The lecture begins at 3:30 p.m followed by a reception in Hamerschlag
Hall, Room 1112.  Professors Rob A. Rutenbar and Wojciech Maly
will speak on "The State of the Center for Electronic Design Automation".
Funded in part by the Semiconductor Research Corporation, SEMATECH,
NSF, and by U.S. and international semiconductor companies, CEDA involves
12 faculty and 60 graduate students working on software tools to design,
verify and fabricate next-generation integrated circuits and systems.
```

< Speaker, STime, ETime, Location >

Figure 10.8.: Seminar Announcement extraction task

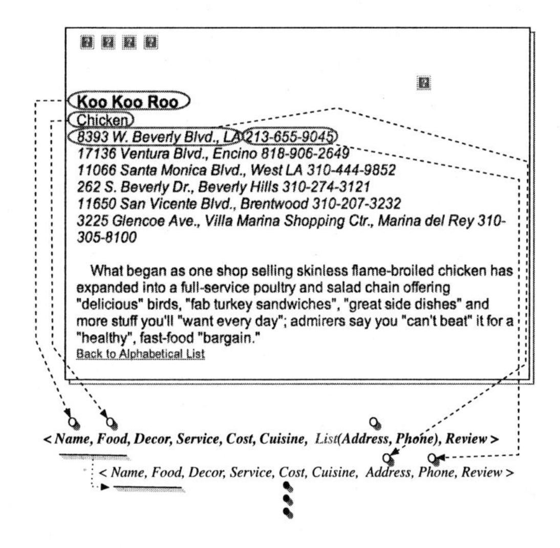

Figure 10.9.: Zagats extraction task

10.2.2. Test Configurations

ϵ-OSL Configurations

One step learning of *AV* and *CD-models* and *RTD-models* has been tested under four configurations (*c1* to *c4*). For each configuration the *maximum delimiter length* and *context length* was set to: c1 = 3, c2 = 5, c3 = 7, and c4 = 9.

threshold-based BFOIL Configurations

threshold-based BFOIL was evaluated on all three wrapper models according to the test configurations listed in Table 10.4. In all test runs *false positive evaluation* was performed solely on already seen documents. Additionally, some of the refinements from Section 7.3 have been used as shown in the table. The flag *rule consistency* indicates if a rule is tested for covering all so far presented learning examples or not (i.e. application of the range restrictedness test).

configuration			mdl cl			rule threshold	rule consistency
*	⋆	∘	*	⋆	∘		
c1	c7	c13	3	5	7	0	true
c2	c8	c14	3	5	7	0	false
c3	c9	c15	3	5	7	0.18	true
c4	c10	c16	3	5	7	0.18	false
c5	c11	c17	3	5	7	0.25	true
c6	c12	c18	3	5	7	0.25	false
mdl = maximal delimiter length, cl = context length							

Table 10.4.: test configurations *threshold-based BFOIL*

cluster-BFOIL Configurations

cluster-BFOIL has been tested using three clustering rounds. The first clustering uses the *empty slot vector* representation and *identical vector* similarity metrics. The second level uses a *binary literal occurence* representation and in a third cluster attempt the *modified TF/IDF* representation is used. Both use as distance metrics the slightly modified *Euclidean distance. cluster-BFOIL* is implemented using caching, thus it skips already processed clusters. It has only been evaluated on learning *RTD-wrappers*, because *AV*and *CD-models* have almost an identical number and set of literals. In the case of different number of empty slots ϵ-*OSL* is sufficient enough to cluster *AV* and *CD-models.* The used configurations for evaluating *cluster-BFOIL* are listed in Table 10.5. The notion *cluster threshold* denotes the maximal acceptable similarity respectively distance between clauses belonging to one cluster.

configuration			mdl cl			cluster threshold	rule threshold
*	⋆	∘	*	⋆	∘		
c1	c7	c13	3	5	7	0.18	0
c2	c8	c14	3	5	7	0.18	0.18
c3	c9	c15	3	5	7	0.18	0.25
c4	c10	c16	3	5	7	0.25	0
c5	c11	c17	3	5	7	0.25	0.18
c6	c12	c18	3	5	7	0.25	0.25
mdl = maximal delimiter length, cd = context length							

Table 10.5.: test configurations *cluster-BFOIL*

10.3. Results: Multi and Single Slot Wrappers

BigBook

BigBook is one of the most highly structured document sets and the easiest extraction task among the investigated ones. The best learned wrapper of all trials is a *CD-bfoil* learned wrapper using configuration *c6*, 10 examples taken from 5 documets with scores for *precision*: 100%, *recall*: 99.48%, and *F1*: 99.74%. Because of the regular document structure there is almost no observable difference in quality between results yielded by ϵ-*OSL* and the more sophisticated *cluster-BFOIL* algorithms. Only the learning time is higher for *BFOIL* learning and in average more rules have been learned by *BFOIL* for *AV-wrapper models.*

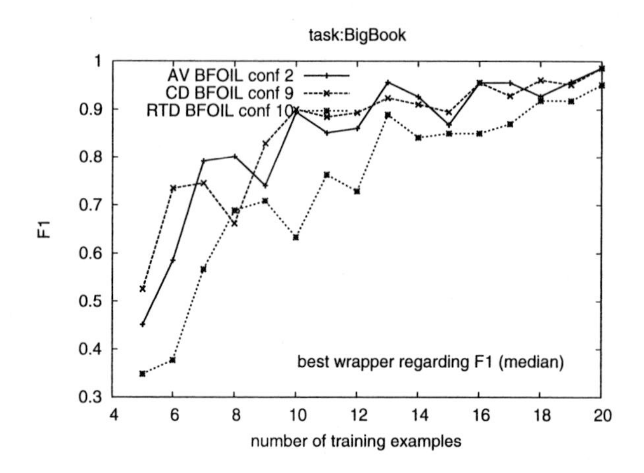

Figure 10.10.: best median f1 wrapper models for BigBook

algorithm	conf	ld	p	fp	rec	pre	f1	lt	tpd	r
BFOIL	c2	5.00	179.00	0.00	97.28	100.00	98.62	208.59	1.24	2.00
ϵ-OSL	c1	5.00	168.00	0.00	91.30	100.00	95.45	13.70	1.45	1.00

Table 10.6.: best AV-Wrappers for BigBook, median f1, 20 examples

algorithm	conf	ld	p	fp	rec	pre	f1	lt	tpd	r
BFOIL	c9	5.00	179.00	0.00	97.28	100.00	98.62	295.81	1.44	1.00
ε-OSL	c1	5.00	169.00	0.00	92.35	100.00	96.02	16.56	1.38	1.00

Table 10.7.: best CD-Wrappers for BigBook, median f1, 20 examples

algorithm	conf	ld	p	fp	rec	pre	f1	lt	tpd	r
BFOIL	c10	5.00	166.00	0.00	90.71	100.00	95.13	441.16	3.10	1.00
ε-OSL	c3	5.00	163.00	0.00	88.59	100.00	93.95	36.59	3.76	1.00
c-BFOIL	c5	5.00	160.00	0.00	87.43	100.00	93.29	209.59	9.13	1.00

Table 10.8.: best RTD-Wrappers for BigBook, median f1, 20 examples

CS - Name

Different from the other investigated tasks the *CS-Name* task contains web pages collected from different web sites. This is one reason why the documents differ strongly in their structure. The extractions task is to detect staff member names on these web pages. The difficulty is that staff member names are in almost all cases not surrounded by identical environment tags or other keywords on the different web pages. Hence, from the test runs it was observable that the quality of the wrapper was extremely influenced by the distribution of randomly selected learning examples. The best learned wrapper for this task is a *AV-wrapper* learned by *BFOIL* with 18 examples from 10 documents with *precision*: 86.3%, *recall*: 66.1% and *F1*: 74.9% using test configuration *c*3.

algorithm	conf	ld	p	fp	rec	pre	f1	lt	tpd	r
BFOIL	c6	8.00	441.00	125.00	38.68	77.83	50.66	245.81	7.86	4.00
ε-OSL	c2	8.00	741.00	2466.00	65.00	25.02	36.13	30.62	15.59	1.00

Table 10.9.: best AV-Wrappers for CS_Name, median f1, 11 examples

algorithm	conf	ld	p	fp	rec	pre	f1	lt	tpd	r
ε-OSL	c2	8.00	716.00	342.00	62.81	64.05	56.27	44.60	15.94	1.00
BFOIL	c3	9.00	468.00	71.00	41.05	88.16	55.95	186.64	9.02	5.00

Table 10.10.: best CD-Wrappers for CS_Name, median f1, 11 examples

algorithm	conf	ld	p	fp	rec	pre	f1	lt	tpd	r
ε-OSL	c1	8.00	404.00	333.00	35.44	59.59	43.17	152.45	63.66	1.00
c-BFOIL	c3	8.00	133.00	13.00	11.67	71.89	20.68	1845.80	15.83	2.00
BFOIL	c3	8.00	103.00	20.00	9.04	81.99	16.31	579.90	24.76	2.00

Table 10.11.: best RTD-Wrappers for CS_Name, median f1, 11 examples

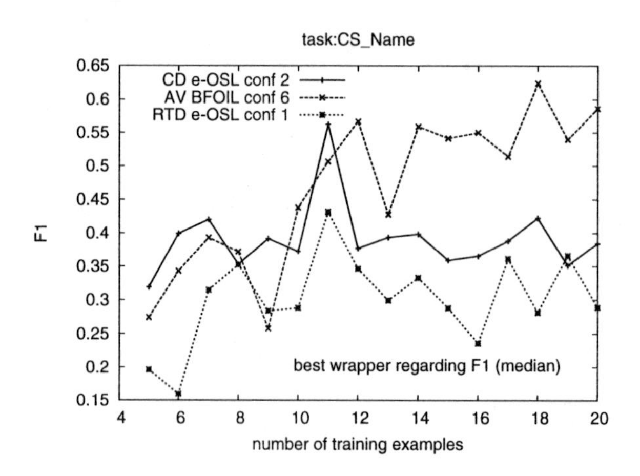

Figure 10.11.: best median f1 wrapper models for CS_Name

Internet Address Finder

The *Internet Address Finder* extraction task belongs to the more difficult ones among the investigated problems. To wrap successfully the relevant data a learned wrapper must be capable of handling missing slots (ϵ^+ class) and varying order of slot fillers (*non linear-wrapper class*). Additionally, some attribute fields are occurring more than once, as shown in Figure 10.4 The best learned multi-slot wrapper for this task is a *CD-model* learned by ϵ-*OSL* given 20 examples from 5 documents with *precision*: 90.9%, *recall*: 62.5% and *F1*: 74.1% using test configuration *c*3. The best learned single-slot wrapper for *Altname* is a *CD-wrapper* learned by *threshold-based BFOIL* given 4 examples from 2 documents with *precision*: 87.5%, *recall*: 100% and *F1*: 93.3% using test configuration *c*13. The best learned single-slot wrapper for *Org* is a *CD-wrapper* learned by *threshold-based BFOIL* given 14 examples from 5 documents with *precision*: 84.8%, *recall*: 93.3% and *F1*: 88.9% using test configuration *c*18.

Multi-Slot Results

algorithm	conf	ld	p	fp	rec	pre	f1	lt	tpd	r
ϵ-OSL	c3	5.00	28.00	5.00	40.00	88.57	56.57	7.29	266.41	4.00
BFOIL	c3	5.00	14.00	5.00	20.00	77.78	32.18	1850.25	23.46	4.00

Table 10.12.: best AV-Wrappers for InternetAdressFinder, median f1, 14 examples

algorithm	conf	ld	p	fp	rec	pre	f1	lt	tpd	r
ϵ-OSL	c3	5.00	26.00	4.00	37.14	88.89	51.49	7.79	191.93	4.00
BFOIL	c3	5.00	10.00	3.00	14.29	76.92	24.69	1548.55	14.82	5.00

Table 10.13.: best CD-Wrappers for InternetAdressFinder, median f1, 14 examples

algorithm	conf	ld	p	fp	rec	pre	f1	lt	tpd	r
c-BFOIL	c12	5.00	8.00	0.00	11.43	100.00	20.51	97.61	1.61	7.00
ϵ-OSL	c4	5.00	17.00	99.00	24.29	13.74	17.28	11.68	4.90	4.00
BFOIL	c16	5.00	6.00	0.00	8.57	100.00	15.79	171.71	1.32	8.00

Table 10.14.: best RTD-Wrappers for InternetAdressFinder, median f1, 14 examples

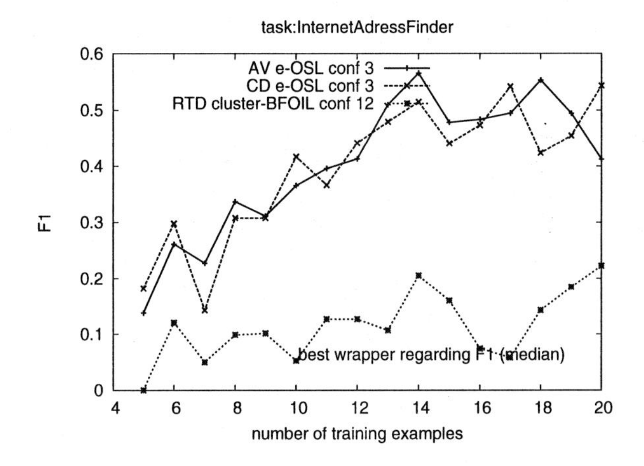

Figure 10.12.: best median f1 wrapper models for InternetAdressFinder

Single-Slot Results: Altname

algorithm	conf	ld	p	fp	rec	pre	f1	lt	tpd	r
BFOIL	c13	2.00	5.00	1.00	71.43	87.50	83.33	5.18	0.62	3.00
ϵ-OSL	c4	2.00	5.00	104.00	71.43	4.59	8.62	1.60	2.17	1.00

Table 10.15.: best AV-Wrappers for InternetAdressFinder_Altname, median f1, 4 examples

algorithm	conf	ld	p	fp	rec	pre	f1	lt	tpd	r
BFOIL	c15	2.00	5.00	1.00	71.43	87.50	83.33	6.04	0.85	3.00
ϵ-OSL	c3	2.00	5.00	112.00	71.43	4.27	8.06	1.52	1.73	1.00

Table 10.16.: best CD-Wrappers for InternetAdressFinder_Altname, median f1, 4 examples

algorithm	conf	ld	p	fp	rec	pre	f1	lt	tpd	r
BFOIL	c16	2.00	6.00	0.00	85.71	100.00	92.31	8.05	1.75	2.00
c-BFOIL	c8	2.00	3.00	0.00	42.86	100.00	60.00	2.35	0.56	3.00
ϵ-OSL	c3	2.00	5.00	34.00	71.43	15.00	25.53	1.61	1.91	1.00

Table 10.17.: best RTD-Wrappers for InternetAdressFinder_Altname, median f1, 4 examples

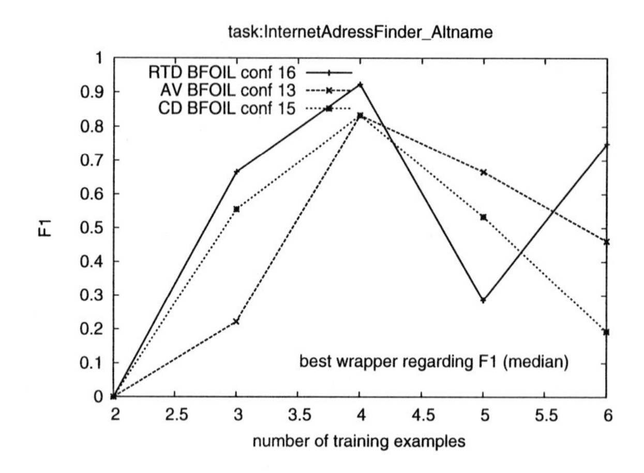

Figure 10.13.: best median f1 wrapper models for InternetAdressFinder_Altname

In Figure 10.13 we observe a dramatic decrease of $F1$ scores for more than 4 training examples. This is explained by the small number of available training documents and the chosen policy for select training examples. For the *altname* extraction task only 5 documents are available. Randomly splitting this document set in training and testing sets results in 2 learning documents from which in most of the cases not more than 4 different training examples could be drawn. For instance, in the 10 test-trials for *RTD BFOIL test configuration 16* only two times 5 examples could be selected. Hence this configuration was tested only twice on 5 training examples instead of ten times. Unfortunately in both cases these 5 training examples resulted in learned wrappers with reduced $F1$ scores. A more elaborated training example selection would probably eliminate this behavior. The same effect appears for the *IAF Organization* extraction task for more than 18 examples as shown in Figure 10.14.

Single-Slot Results: Org

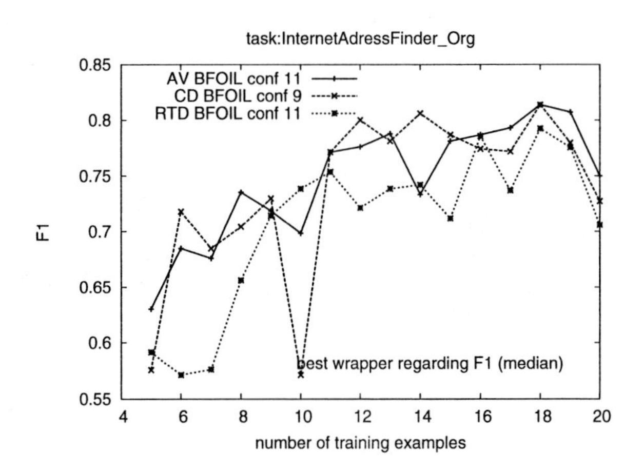

Figure 10.14.: best median f1 wrapper models for InternetAdressFinder_Org

algorithm	conf	ld	p	fp	rec	pre	f1	lt	tpd	r
BFOIL	c11	5.00	23.00	5.00	80.00	82.14	81.36	25.63	0.69	3.00
ϵ-OSL	c2	5.00	22.00	5.00	77.42	80.77	75.86	5.32	0.66	1.00

Table 10.18.: best AV-Wrappers for InternetAdressFinder_Org, median f1, 18 examples

algorithm	conf	ld	p	fp	rec	pre	f1	lt	tpd	r
BFOIL	c9	5.00	24.00	5.00	80.00	82.76	81.36	28.83	0.74	5.00
ϵ-OSL	c3	5.00	22.00	5.00	74.19	78.95	66.67	4.83	0.69	1.00

Table 10.19.: best CD-Wrappers for InternetAdressFinder_Org, median f1, 18 examples

algorithm	conf	ld	p	fp	rec	pre	f1	lt	tpd	r
BFOIL	c11	5.00	21.00	5.00	77.78	81.82	79.25	95.41	1.31	2.00
c-BFOIL	c12	5.00	20.00	4.00	72.41	82.61	77.19	79.40	1.27	3.00
ϵ-OSL	c3	5.00	21.00	5.00	68.00	77.27	72.34	6.01	0.90	1.00

Table 10.20.: best RTD-Wrappers for InternetAdressFinder_Org, median f1, 18 examples

LA-Weekly

LA-Weekly demands wrappers belonging to the class of *linear*, ϵ^+, and *MV*. In this task a usual address information of a restaurant contains additional descriptions in the form of a list of accepted credit cards at the end of each description text. The items of this list have to be individually made available by the wrapper, hence each accepted credit card must be extracted separately. This makes the task particular difficult, since the list and the single credit card information are not embedded in significant structural environments, the credit card names occur in changing order and for some restaurants there is no list given. The best learned multi-slot wrapper for this task is a *RTD-wrapper* learned by *cluster-BFOIL* given 17 examples taken from 9 documents with *precision*: 94.5%, *recall*: 64.7% and *F1*: 76.8% using configuration *c*5. The best learned single-slot wrapper for *CreditCards* is a *AV-wrapper* learned by *BFOIL* from 19 examples taken from 10 documents with *precision*: 86.2%, *recall*: 63.6% and *F1*: 73.2% using configuration *c*5.

Multi-Slot Results

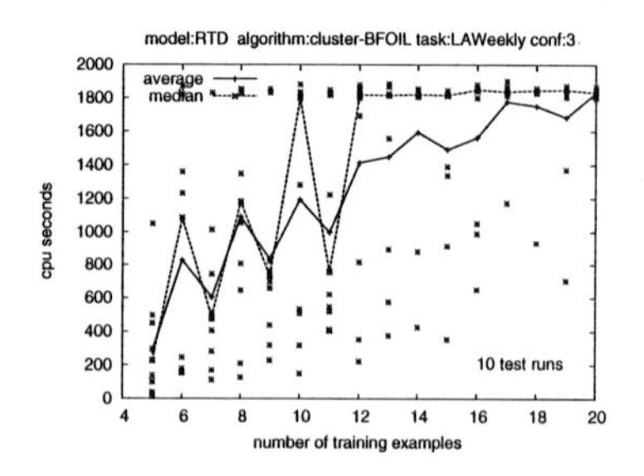

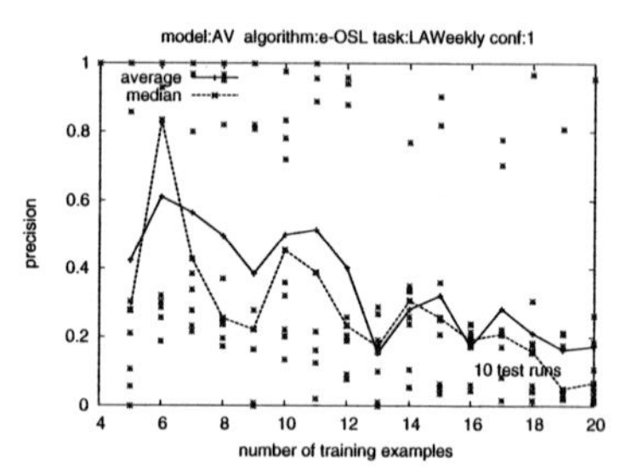

Figure 10.15.: precision AV ϵ-*OSL* conf. 1 and ltime RTD cluster-BFOIL conf. 3

Figure 10.16 displays two typical behaviors already discussed in Section 6.5 and Section 7.4: exceeded learning time and over-generalization due to increasing number of examples (Figure 10.15).

algorithm	conf	ld	p	fp	rec	pre	f1	lt	tpd	r
ϵ-OSL	c1	6.00	21.00	50.00	3.90	22.22	7.45	17.67	185.64	2.00
BFOIL	c1	7.00	0.00	0.00	0.00	0.00	0.00	1842.80	0.75	0.00

Table 10.21.: best AV-Wrappers for LAWeekly, median f1, 9 examples

algorithm	conf	ld	p	fp	rec	pre	f1	lt	tpd	r
BFOIL	c1	6.00	0.00	0.00	0.00	0.00	0.00	1855.58	0.80	0.00
ϵ-OSL	c1	7.00	0.00	0.00	0.00	0.00	0.00	46.11	251.05	1.00

Table 10.22.: best CD-Wrappers for LAWeekly, median f1, 9 examples

algorithm	conf	ld	p	fp	rec	pre	f1	lt	tpd	r
c-BFOIL	c3	6.00	175.00	10.00	32.53	94.87	48.75	734.47	40.19	2.00
ϵ-OSL	c1	7.00	77.00	27.00	14.31	39.57	24.72	83.40	307.53	2.00
BFOIL	c1	7.00	0.00	0.00	0.00	0.00	0.00	1914.18	0.66	0.00

Table 10.23.: best RTD-Wrappers for LAWeekly, median f1, 9 examples

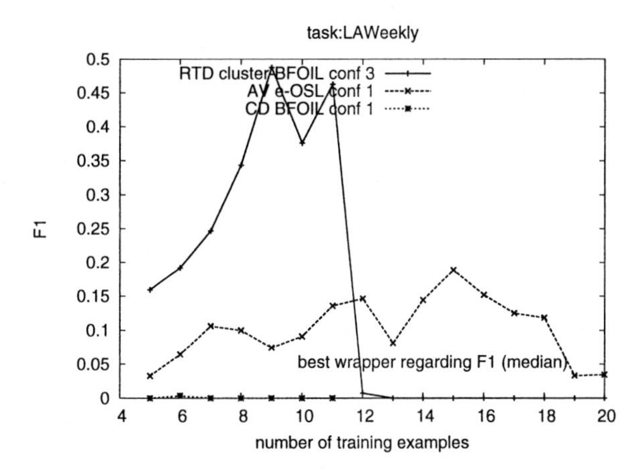

Figure 10.16.: best median f1 wrapper models for LAWeekly

Single-Slot Results: Credit Cards

algorithm	conf	ld	p	fp	rec	pre	f1	lt	tpd	r
BFOIL	c6	6.00	55.00	6.00	42.64	96.36	57.61	48.79	1.41	1.00
ϵ-OSL	c2	8.00	54.00	48.00	42.52	48.94	46.15	11.60	2.18	1.00

Table 10.24.: best AV-Wrappers for LAWeekly_CC, median f1, 7 examples

algorithm	conf	ld	p	fp	rec	pre	f1	lt	tpd	r
BFOIL	c1	6.00	52.00	2.00	41.67	96.30	58.43	31.29	1.09	2.00
ϵ-OSL	c4	8.00	53.00	18.00	44.00	79.69	55.56	11.27	3.30	1.00

Table 10.25.: best CD-Wrappers for LAWeekly_CC, median f1, 7 examples

algorithm	conf	ld	p	fp	rec	pre	f1	lt	tpd	r
ϵ-OSL	c1	6.00	24.00	52.00	19.35	51.06	23.81	9.51	340.39	1.00
BFOIL	c1	6.00	0.00	0.00	0.00	0.00	0.00	1831.89	0.85	0.00
c-BFOIL	c1	6.00	0.00	0.00	0.00	0.00	0.00	1849.25	0.53	0.00

Table 10.26.: best RTD-Wrappers for LAWeekly_CC, median f1, 7 examples

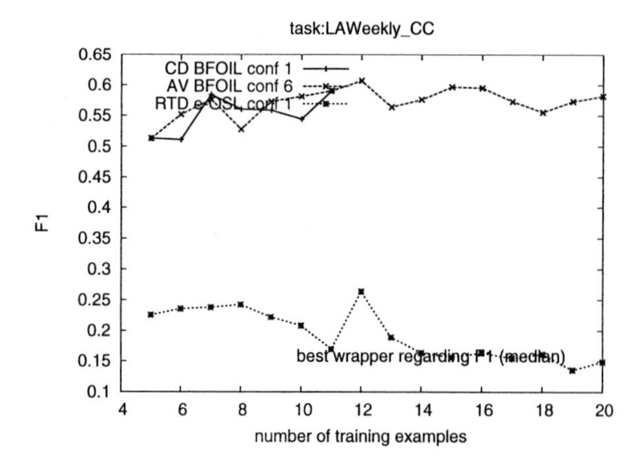

Figure 10.17.: best median f1 wrapper models for LAWeekly_CC

Okra

The *Okra* extraction task belongs to the same wrapper class as *BigBook*. It requires at least *linear-$\wedge$-wrappers* to be wrapped perfectly. *threshold-based BFOIL* shows distracting results for *AV* and *CD-wrappers*, although a suitable wrapper must be easily learnable, because of the highly structured content. The reason for this behavior is that the learning time given 18 examples exceeds the time limit. Because learned *AV* and *CD-wrappers* have for this task higher match complexities than *RTD-wrappers* they need longer evaluation (i.e. query computation) time. The best wrapper is a ϵ-*OSL* learned *AV-wrapper* given 19 examples taken from 17 documents with *precision*: 100 %, *recall*: 99.5%, and *F1*: 99.8% with configuration *c*1.

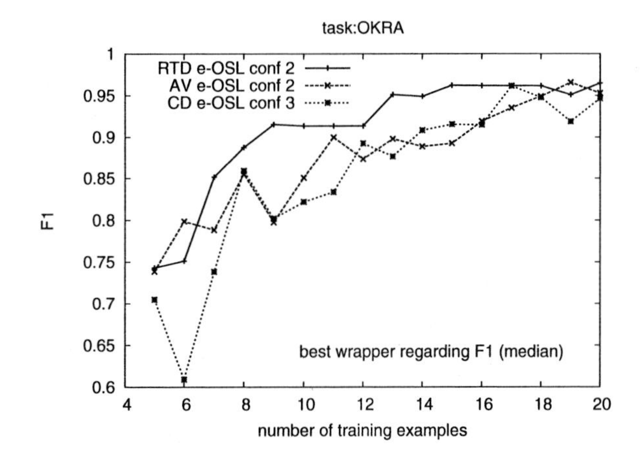

Figure 10.18.: best median f1 wrapper models for OKRA

algorithm	conf	ld	p	fp	rec	pre	f1	lt	tpd	r
ϵ-OSL	c2	18.00	2996.00	0.00	90.32	100.00	94.92	39.76	11.40	1.00
BFOIL	c1	16.00	0.00	0.00	0.00	0.00	0.00	1917.56	1.26	0.00

Table 10.27.: best AV-Wrappers for OKRA, median f1, 18 examples

algorithm	conf	ld	p	fp	rec	pre	f1	lt	tpd	r
ϵ-OSL	c3	17.00	2987.00	0.00	90.05	100.00	94.77	45.71	5.71	1.00
BFOIL	c4	18.00	2416.00	0.00	72.86	100.00	84.30	1565.17	14.61	1.00

Table 10.28.: best CD-Wrappers for OKRA, median f1, 18 examples

algorithm	conf	ld	p	fp	rec	pre	f1	lt	tpd	r
ϵ-OSL	c2	17.00	3073.00	0.00	92.64	100.00	96.18	54.23	3.10	1.00
c-BFOIL	c12	17.00	3041.00	0.00	91.68	100.00	95.66	383.86	3.16	1.00
BFOIL	c10	18.00	2939.00	0.00	88.60	100.00	93.96	1411.85	3.04	1.00

Table 10.29.: best RTD-Wrappers for OKRA, median f1, 18 examples

QuoteServer

Although the *QuoteServer* task seems to be a simple extraction exercise, because of its tabular structure, it contains some nasty pitfalls. First, some of the column entries plus its delimiters can be missing (ϵ^+-class). Second, the string "n/a" denotes either one or two missing items (x-class). Third, some of the tables contain wrong date entries, in that they appear in the wrong column of the table. Another obstacle is the quite low amount of learning data, only 25 examples. Nevertheless, these documents are highly structured and are reasonably good wrapped by a *linear-x-ϵ^+-wrapper.* Flashy is the difference of precision scores for *ϵ-OSL* learned *AV* and *CD-wrappers.* This is due to missing constraints where *AV-wrappers* have difficulties to detect the correct delimiters, which are almost identical for all slots. The best learned wrapper for this task is a *CD-wrapper* learned by *threshold-based BFOIL* from 13 examples taken from 5 documents with *precision*: 100%, *recall*: 75% and *F1*: 85.7% using configuration *c*1. For the single-slot extraction task *Date* for all models and learning algorithms a 100% F1 score wrapper was learned. For the single-slot extraction task *Vol threshold-based BFOIL* yielded for all model a wrapper with 100% F1 score. Whereas 6 examples from 5 documents were sufficient for *threshold-based BFOIL* to learn a 100% F1 score *RTD-wrapper* using configuration *c*16.

Multi-Slot Results

algorithm	conf	ld	p	fp	rec	pre	f1	lt	tpd	r
BFOIL	c15	5.00	5.00	2.00	50.00	71.43	58.82	179.22	1.48	6.00
ϵ-OSL	c1	5.00	3.00	7.00	30.00	30.00	30.00	22.92	1.04	7.00

Table 10.30.: best AV-Wrappers for QuoteServer, median f1, 15 examples

algorithm	conf	ld	p	fp	rec	pre	f1	lt	tpd	r
BFOIL	c1	5.00	7.00	0.00	70.00	100.00	82.35	191.48	0.90	5.00
ϵ-OSL	c2	5.00	6.00	0.00	60.00	100.00	75.00	67.91	1.56	6.00

Table 10.31.: best CD-Wrappers for QuoteServer, median f1, 15 examples

algorithm	conf	ld	p	fp	rec	pre	f1	lt	tpd	r
c-BFOIL	c1	5.00	5.00	0.00	50.00	100.00	66.67	657.31	0.96	2.00
ϵ-OSL	c1	5.00	5.00	0.00	50.00	100.00	66.67	604.50	1.80	6.00
BFOIL	c11	5.00	4.00	0.00	40.00	100.00	57.14	1805.73	1.02	2.00

Table 10.32.: best RTD-Wrappers for QuoteServer, median f1, 15 examples

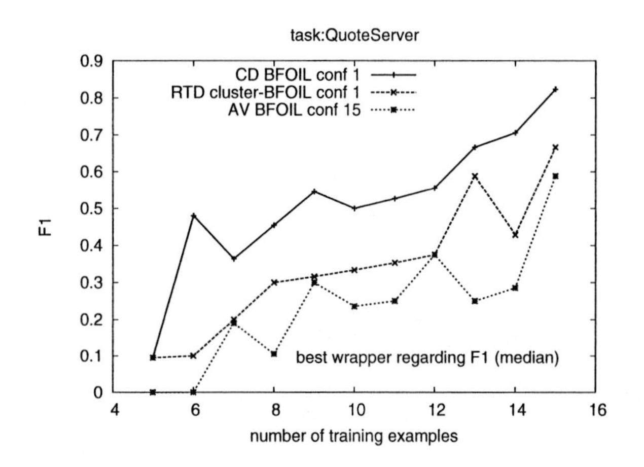

Figure 10.19.: best median f1 wrapper models for QuoteServer

Single-Slot Results: Date

algorithm	conf	ld	p	fp	rec	pre	f1	lt	tpd	r
BFOIL	c10	4.00	19.00	0.00	100.00	100.00	100.00	3.68	0.43	1.00
ϵ-OSL	c2	4.00	19.00	0.00	100.00	100.00	100.00	0.87	0.27	1.00

Table 10.33.: best AV-Wrappers for QuoteServer_Date, median f1, 5 examples

algorithm	conf	ld	p	fp	rec	pre	f1	lt	tpd	r
BFOIL	c10	4.00	16.00	0.00	100.00	100.00	100.00	2.76	0.26	1.00
ϵ-OSL	c2	4.00	19.00	0.00	100.00	100.00	100.00	1.04	0.32	1.00

Table 10.34.: best CD-Wrappers for QuoteServer_Date, median f1, 5 examples

algorithm	conf	ld	p	fp	rec	pre	f1	lt	tpd	r
BFOIL	c10	4.00	16.00	0.00	100.00	100.00	100.00	1.46	0.29	1.00
c-BFOIL	c10	4.00	16.00	0.00	100.00	100.00	100.00	1.28	0.30	1.00
ϵ-OSL	c3	4.00	16.00	0.00	100.00	100.00	100.00	1.02	0.30	1.00

Table 10.35.: best RTD-Wrappers for QuoteServer_Date, median f1, 5 examples

Single-Slot Results: Vol

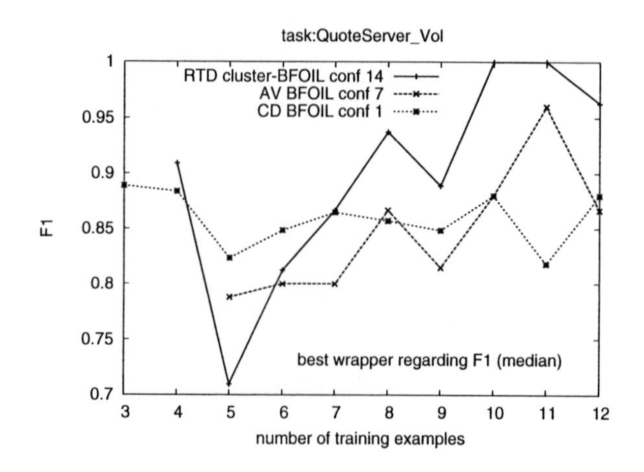

Figure 10.20.: best median f1 wrapper models for QuoteServer_Vol

algorithm	conf	ld	p	fp	rec	pre	f1	lt	tpd	r
BFOIL	c7	5.00	11.00	0.00	84.62	100.00	88.00	18.50	0.36	2.00
ϵ-OSL	c4	5.00	13.00	12.00	100.00	52.00	65.00	2.01	0.47	1.00

Table 10.36.: best AV-Wrappers for QuoteServer_Vol, median f1, 10 examples

algorithm	conf	ld	p	fp	rec	pre	f1	lt	tpd	r
BFOIL	c1	5.00	11.00	0.00	84.62	100.00	88.00	7.28	0.34	2.00
ϵ-OSL	c4	5.00	13.00	12.00	92.86	52.00	59.26	2.25	0.49	1.00

Table 10.37.: best CD-Wrappers for QuoteServer_Vol, median f1, 10 examples

algorithm	conf	ld	p	fp	rec	pre	f1	lt	tpd	r
c-BFOIL	c14	5.00	12.00	0.00	100.00	100.00	100.00	4.68	0.23	2.00
BFOIL	c11	5.00	13.00	0.00	100.00	100.00	96.30	5.02	0.31	2.00
ϵ-OSL	c4	5.00	13.00	5.00	100.00	72.22	83.87	2.34	0.64	1.00

Table 10.38.: best RTD-Wrappers for QuoteServer_Vol, median f1, 10 examples

Seminar Announcements

The *Seminar Announcement* problem is a collection of several newsgroup postings announcing seminars. The postings have been collected from university newsgroups. This is the only *free text* document set the *LIPX* system is tested on. The *Seminar Announcement* extraction tasks are a challenge for most of the existing *IE systems*. The single-slot wrappers for *stime* and *etime* are easier to learn than *speaker* and *location*, because there are more or less fixed key phrases occurring before and after the start and end time information. Learning multi-slot wrappers as shown in Figure 10.8 is beyond *LIPX* capabilities. Reasons for this are given in Section 10.4. This task was only investigated using *cluster-BFOIL*, because *threshold-based BFOIL*'s learning times exceeded the time limits. Because of the strongly differing delimiters ϵ-*OSL* was not tested.

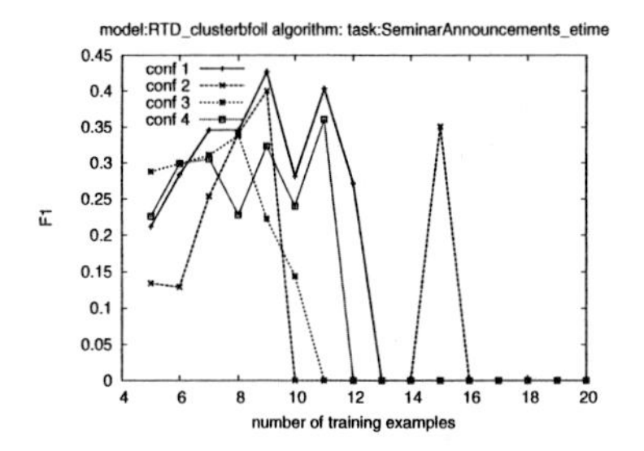

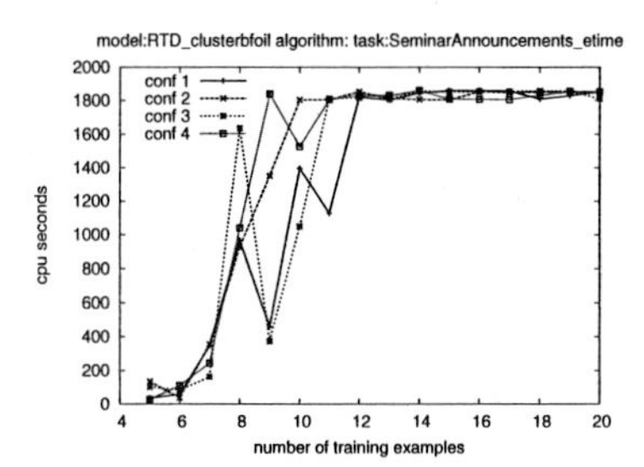

Figure 10.21.: SA-etime F1 and ltime RTD cluster-BFOIL

The best learned single-slot wrapper for *etime* is a *RTD-wrapper* learned by *cluster-BFOIL* from 10 examples taken from 9 documents with *precision*: 100%, *recall*: 41.7 % and *F1*: 58.8% using configuration *c6*. The best learned single-slot wrapper for *stime* is a *RTD-wrapper* learned by *cluster-BFOIL* from 13 examples taken from 13 documents with *precision*: 72.0%, *recall*: 45.8% and *F1*: 56.0% using configuration *c3*. The best learned single-slot wrapper for *speaker* is a *RTD-wrapper* learned by *cluster-BFOIL* from 7 examples taken from 7 documents with *precision*: 23.0%, *recall*: 16.2% and *F1*: 19.0% using configuration *c2*. The best learned single-slot wrapper for *location* is a *RTD-wrapper* learned by *cluster-BFOIL* from 12 examples taken from 12 documents with *precision*: 47.8%, *recall*: 9.4% and *F1*: 15.7% using configuration *c4*.

Single-Slot Results: etime, location, stime, speaker

algorithm	conf	ld	p	fp	rec	pre	f1	lt	tpd	r
c-BFOIL	c1	9.00	29.00	2.00	27.62	78.57	42.65	459.05	2.29	2.00

Table 10.39.: best RTD-Wrappers for SeminarAnnouncements_etime, median f1, 9 examples

algorithm	conf	ld	p	fp	rec	pre	f1	lt	tpd	r
c-BFOIL	c6	10.00	5.00	60.00	4.13	7.69	5.38	1177.20	19.00	5.00

Table 10.40.: best RTD-Wrappers for SeminarAnnouncements_location, median f1, 10 examples

algorithm	conf	ld	p	fp	rec	pre	f1	lt	tpd	r
c-BFOIL	c1	6.00	49.00	2.00	20.94	100.00	34.15	338.59	0.35	3.00

Table 10.41.: best RTD-Wrappers for SeminarAnnouncements_stime, median f1, 6 examples

algorithm	conf	ld	p	fp	rec	pre	f1	lt	tpd	r
c-BFOIL	c3	12.00	23.00	210.00	21.50	9.87	13.53	1911.13	10.55	2.00

Table 10.42.: best RTD-Wrappers for SeminarAnnouncements_speaker, median f1, 12 examples

Zagats

Similar to *LA-Weekly* the *Zagats* extraction task also contains a multi valued slot (*MV*, i.e. list). But in contrary to the list of credit cards, here a list of tuples, namely *address* and *phone number* are to be extracted. Figure 10.9 illustrates the original posted problem description as found in the *RISE repository* and the actually slightly simplified task investigated by *LIPX*. The best learned multi-slot wrapper is a *RTD-wrapper* learned by *threshold-based BFOIL* from 20 examples taken from 19 documents with *precision*: 98.3%, *recall*: 97.5% and *F1*: 97.9% using configuration *c8*. The best learned single-slot wrapper for *Adr* is a *RTD-wrapper* learned by *cluster-BFOIL* from 17 examples taken from 16 documents with *precision*: 98.4%, *recall*: 98.4% and *F1*: 98.4% using configuration *c10*.

Multi-Slot Results

algorithm	conf	ld	p	fp	rec	pre	f1	lt	tpd	r
ϵ-OSL	c1	19.00	86.00	5.00	71.67	97.26	81.55	19.39	44.05	1.00
BFOIL	c1	19.00	0.00	0.00	0.00	0.00	0.00	1856.43	0.29	0.00

Table 10.43.: best AV-Wrappers for Zagats, median f1, 20 examples

algorithm	conf	ld	p	fp	rec	pre	f1	lt	tpd	r
ϵ-OSL	c1	19.00	92.00	2.00	76.67	97.62	85.58	21.21	22.56	1.00
BFOIL	c1	19.00	0.00	0.00	0.00	0.00	0.00	1893.41	0.40	0.00

Table 10.44.: best CD-Wrappers for Zagats, median f1, 20 examples

algorithm	conf	ld	p	fp	rec	pre	f1	lt	tpd	r
c-BFOIL	c1	19.00	108.00	2.00	90.00	98.18	93.91	145.77	0.80	1.00
BFOIL	c5	19.00	105.00	2.00	87.50	98.13	92.51	860.92	0.80	1.00
ϵ-OSL	c2	19.00	107.00	6.00	89.17	95.16	91.85	57.84	4.04	1.00

Table 10.45.: best RTD-Wrappers for Zagats, median f1, 20 examples

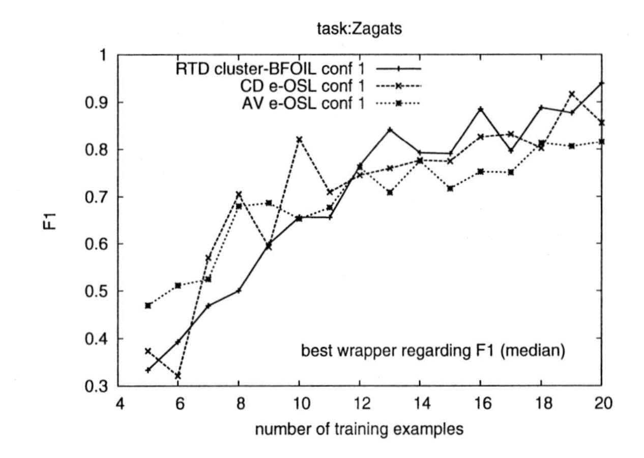

Figure 10.22.: best median f1 wrapper models for Zagats

Single-Slot Results: Adr

algorithm	conf	ld	p	fp	rec	pre	f1	lt	tpd	r
ϵ-OSL	c4	19.00	114.00	2.00	95.00	98.21	96.30	21.36	0.87	1.00
BFOIL	c1	19.00	80.00	2.00	66.67	97.40	78.26	150.94	0.47	4.00

Table 10.46.: best AV-Wrappers for Zagats_Adr, median f1, 20 examples

algorithm	conf	ld	p	fp	rec	pre	f1	lt	tpd	r
BFOIL	c14	20.00	115.00	2.00	95.83	98.20	95.87	1161.60	0.75	1.00
ϵ-OSL	c2	19.00	104.00	2.00	86.67	98.15	92.04	21.72	0.68	1.00

Table 10.47.: best CD-Wrappers for Zagats_Adr, median f1, 20 examples

algorithm	conf	ld	p	fp	rec	pre	f1	lt	tpd	r
ϵ-OSL	c4	19.00	112.00	6.00	93.33	95.16	94.92	26.01	1.00	1.00
c-BFOIL	c9	19.00	108.00	2.00	90.00	98.02	93.91	141.68	1.26	1.00
BFOIL	c4	20.00	96.00	55.00	80.00	65.19	72.97	407.05	1.52	3.00

Table 10.48.: best RTD-Wrappers for Zagats_Adr, median f1, 20 examples

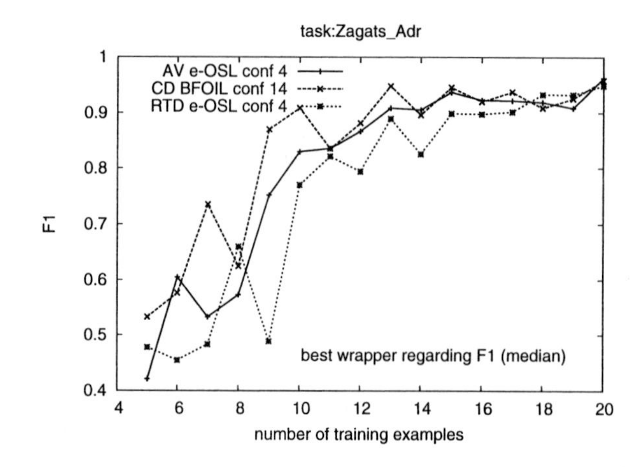

Figure 10.23.: best median f1 wrapper models for Zagats_Adr

10.4. Result Comparison and Critical Discussion

In the following experimental results of the here presented wrapper learning techniques are compared to state-of-the-art machine learning based information extraction systems. For comparison the best median results presented in Section 10.3 are taken. Note, that the presented score for precision, recall and F1 are median scores of ten test-runs as discussed in Section 10.2. In contrast to this, Table 10.24 shows the results of the best learned multi-slot wrappers by *LIPX*.

task	rec	pre	F1	model	method	examples	conf
BigBook	99.5	100	99.7	CD	bfoil	10 (4 docs)	6
IAF	62.5	90.9	74.1	CD	ϵ-*OSL*	20 (5 docs)	3
LA-Weekly	64.7	94.5	76.8	RTD	cluster	17 (9 docs)	5
Okra	99.5	100	99.8	AV	ϵ-*OSL*	19 (17 docs)	1
QuoteServer	75.0	100	85.7	CD	bfoil	13 (5 docs)	1
Zagats	97.5	98.3	97.9	RTD	bfoil	20 (19 docs)	8

Figure 10.24.: LIPX best learned multi-slot wrappers

Although there are quite a number of systems only a few publications on them report useful experimental results regarding standard evaluation metrics like *precision*, *recall*, *accuracy* or *coverage*. A further obstacle to present a comprehensive comparison is the fact that results are given regarding different metrics obtained from different test settings. For instance, *STALKER* [Muslea *et al.*, 1999] results are given in *accuracy* [2] scores, whereas they define *accuracy* as the percentage of learned rules that extracted all relevant data correctly. For instance, they report an accuracy score of 97% if out of 500 learned rules 485 were capable of extracting correctly all relevant data, while the other 15 were erroneous. It remains a bit unclear if a rule that extracts correctly all relevant data is allowed to provide false extractions, because their notion of erroneous only stats that not all relevant data was extracted correctly[3].

In contrary, results analysis and comparison for the *SoftMealy* system [Hsu and Chang, 1999] is also given in accuracy, but they define accuracy as the ratio of correctly extracted data to all data to extract. The latter definition is equivalent to the notion of *recall* whereas the definition in [Muslea *et al.*, 1999] is not directly comparable to *recall* or *precision*. For *SoftMealy* there are no precision scores given in [Hsu and Chang, 1999], hence only the recall scores are used. In [Hsu and Chang, 1999] *STALKER* results are presented and compared to their *recall* values. For the WL^2 system results are presented in a similar sketchy manner without stating precise values for precision or recall.

Other systems like *BWI*, *k-testable* and LP^2 use almost the same test settings as those used for the evaluation of the introduced learning techniques. They also provide results as precision and recall scores, and despite the fact that some report median other average results, at least a uniform platform for comparison is given.

[2] Usually accuracy is defined as $acc = \frac{tp+tn}{tp+tn+fp+fn}$ with tp and tn the true positive and true negative predicted examples and fp and fn the false positive and false negative predictions.

[3] In the case that correctly means no false positives are covered by a learned rule the used notion of accuracy is identical to the percentage of learned wrappers with $F1 = 100\%$. But since [Hsu and Chang, 1999] compares their recall results with STALKER's accuracy values this interpretation is very unlikely.

system	*BigBook*				*IAF*				*LA-Weekly*			
	e	pre	rec	F1	e	pre	rec	F1	e	pre	rec	F1
LIPX												
ϵ-OSL	20	100	92.4	96.0	14	88.6	40.0	56.6	7	39.6	14.3	24.7
BFOIL	93	100	97.3	98.6	42	77.8	20.0	32.2	time limit			
c-BFOIL	93	100	87.4	93.3	42	100	11.4	20.5	114	94.8	32.5	48.8
STALKER	8		97		10		92.5					
SoftMealy	6		100		10		57					
WIEN	274	100	100	100	too hard							
WHISK	40	100	100	100								
WL^2	6		100		1		100					
	OKRA				*QuoteServer*				*Zagats*			
LIPX												
ϵ-OSL	18	100	92.6	96.2	15	100	50.0	66.7	20	95.2	89.2	91.9
BFOIL	237	100	88.6	94.0	15	100	70	82.4	28	98.1	87.5	92.5
c-BFOIL	224	100	91.7	95.7	15	100	50.0	66.7	28	98.9	90.0	93.9
STALKER	1		97		10		79					
SoftMealy	1		100		10		85					
WIEN	46	100	100	100	too hard							
WL^2	1		100		4		100					

Table 10.49.: LIPX median results comparison with other multi-slot systems

Table 10.49 shows a comparison of system results on multi-slot extraction tasks (see Chapter 10). Due to the described problems of missing test results or not directly comparable results some of the entries are empty. The system results for *STALKER*, *WIEN* and *SoftMealy* are taken from [Hsu and Chang, 1999]. The results for *WHISK* from [Soderland, 1999] and for WL^2 from [Cohen *et al.*, 2002]. The table shows recall, precision and F1 scores and the number of examples needed to achieve a reasonable good wrapper. Those systems that do not need completely labeled documents have an apparent advantage concerning the tedious manual pre-processing of labeling examples. *SoftMealy*, *STALKER*, and WL^2 belong to this class of systems. *LIPX* and *WIEN* use completely labeled documents (i.e. exhaustively enumerated example extractions), whereas *WHISK* uses an iterative process suggesting the user some positive examples from an untagged corpus to be added to the training set. So, for *WHISK* it is a bit tricky to measure the exact set of provided learning examples, especially because in [Soderland, 1999] only the number of documents used to learn the *BigBook* wrapper is presented.

For *LIPX* it is also not quite clear how to count the number of training examples. Because the learned wrapper is generalized from a subset of the examples that have to be completely labeled in the training documents. But for evaluation during learning the complete set is needed. So, one major improvement of the presented methods in this thesis is to modify the *BFOIL* based learning algorithms such that they are capable to learn from not completely labeled documents. According to this, Table 10.49 shows the size of the exhaustive example set as the number of training examples for *BFOIL* based methods. The actual number of examples from which the wrapper is generalized is given in Section 10.3.

Two major conclusion are inferable regarding the presented scores. Due to *LIPX's* rule quality estimation based on the exhaustive example assumption, the amount of needed training data is comparable to *WIEN*. Basically, this is a real weak spot for practical applications.

Astonishingly, the *ϵ-OSL* method performs quite good on the investigated extraction tasks.

Which is of great interest, because though the underlying theory is strongly connected with *logic programming* and *ILP* this technique can be easily implemented with standard programming languages to gain higher performance.

task	rec	pre	F1	model	method	examples	conf
CS Name	66.1	86.3	74.9	AV	bfoil	18 (10 docs)	3
IAF Alt	100	87.5	93.3	CD	bfoil	4 (2 docs)	13
IAF Org	93.3	84.8	88.9	CD	bfoil	14 (5 docs)	18
LA CC	63.6	86.2	73.2	AV	bfoil	19 (10 docs)	5
QS date	100	100	100	all	all	5 (4 docs)	-
QS vol	100	100	100	all	bfoil	9 (5 docs)	-
SA etime	41.7	100	58.8	RTD	cluster	8 (7 docs)	6
SA stime	45.8	72.0	56.0	RTD	cluster	13 (13 docs)	3
SA speaker	16.2	23.0	19.0	RTD	cluster	10 (10 docs)	14
SA location	9.4	47.8	15.7	RTD	cluster	12 (12 docs)	4
Zagats Adr	98.4	98.4	98.4	RTD	cluster	17 (16 docs)	10

Figure 10.25.: LIPX best learned single-slot wrappers

A second major outcome of this comparison is, that *BFOIL* based learning yields high precision wrappers. In comparison to the systems that uses sophisticated methods either by learning wrappers based on "mixed" wrapper languages (WL^2) or by wrappers build of rules which fulfill different tasks (*STALKER*) the "one rule extracts all slots" method by *LIPX* seems to be very competitive. Once again it has to be pointed out, that *LIPX* is implemented in *Prolog* using the presented algorithms without any optimization techniques. This fact is important, because expectable better results with higher number of learning examples regarding recall scores were in some cases for *BFOIL* not realizable due to the time restrictions. Another possibility to increase the recall rate while keeping the high precision rates is to modify the *basic-BFOIL* algorithm such that it outputs a best hypothesis by testing all possible partitions of the example set. For the discussed reasons this is not really feasible in a practical application context, but a reasonable improvement can be to use some heuristic search methods e.g. $A*$.

Table 10.50 summarizes single-slot results of *LIPX* and some of the discussed systems. Results for *BWI*, *HMM*, *RAPIER*, *SRV* and *STALKER* are taken from [Freitag and Kushmerick, 2000]. *WHISK* results are taken from [Soderland, 1999], LP^2 from [Ciravegna, 2000] and results for *k-testable* from [Kosala *et al.*, 2003]. Unlike as for the multi-slot results a specification of the number of training examples is not given in the publications for most of the systems, hence it is omitted. Some of the entries are empty due to missing or not directly comparable results.

It is not too surprising that *LIPX* shows very good results for single slot extraction tasks on semi-structured documents, since it was developed with the intent to perform multi-slot extractions which is obviously a more complex task and technically spoken a superset. Flashy is the fact, that it behaves poorly on *free text* extraction challenges, like the *Semniar Announcement* problems. An explanation for this is given in the next paragraph.

system	*CS name*			*IAF alt*			*IAF org*			*QS date*			*QS vol*		
	pre	rec	F1	pre	rec	F1	pre	rec	F1	pre	rec	F1	pre	rec	F1
BWI	77.1	31.4	44.6	90.9	43.5	58.8	77.5	45.9	57.7	100	100	**100**	100	61.9	76.5
HMM	41.3	65.0	50.5	1.7	90.0	3.4	16.8	89.7	28.4	36.3	100	53.3	18.4	96.2	30.9
k-testable				100	73.9	85	100	57.9	73.3	100	60.5	75.4	100	73.6	84.8
LIPX															
ϵ-OSL	64.1	62.8	**56.3**	15	71.4	25.5	80.8	77.4	75.9	100	100	100	72.2	100	83.9
BFOIL	88.2	41.1	56.0	100	85.7	**92.3**	82.8	80.0	**81.4**	100	100	100	100	100	96.3
cluster-BFOIL	82.0	9.0	16.3	100	42.9	60	82.6	72.4	77.2	100	100	**100**	100	100	**100**
STALKER				100			48.0			0			0		
	SA speaker			*SA stime*			*SA etime*			*LA CC*			*Zagats adr*		
BWI	79.1	59.2	67.7	99.6	99.6	99.6	94.4	94.4	93.9	99.6	100	99.8	100	93.7	96.7
HMM	77.9	75.2	76.6	98.5	98.5	98.5	45.7	97.0	62.1	98.5	100	99.3	97.7	99.5	98.6
LIPX															
ϵ-OSL	time limit			time limit			time limit			79.7	44.0	55.6	98.2	95.0	96.3
BFOIL	time limit			time limit			time limit			96.3	41.7	58.4	98.2	95.8	95.9
cluster-BFOIL	9.9	21.5	13.5	100	21.0	34.2	78.6	27.6	42.7	time limit			98.0	90.0	93.9
LP^2	87.0	70.0	77.6	99.0	99.0	99.0	97.0	94.0	95.5						
RAPIER	80.9	39.4	53.0	93.3	92.9	93.4	95.8	96.6	96.2						
SRV	54.4	58.4	56.3	98.6	98.4	98.5	67.3	92.6	77.9						
STALKER										100			100		
WHISK	52.6	11.1	18.3	86.2	100.0	92.6	85.0	87.2	86.1						

Table 10.50.: LIPX median results comparison with other single-slot systems

Critical Discussion

A handful of weak points become apparent when *LIPX* results are compared to other systems results. These points are briefly discussed in the following and possible solutions to overcome these deficiencies are sketched.

LIPX performs extremely bad on *free text* extractions tasks like the *SeminarAnnouncements* because of one main reason. Since these documents are represented as a sequence of tokens, the chosen feature structure representation (*AV document representation*) is to rough in that the chosen features and their values are simply not adequate enough. It is quite obvious that a document and pattern representation incorporating morphological and semantical analysis (like *RAPIER* uses) offers much more subtle properties to describe words and natural text than a simple token representation plus one or two features representing the length of a word. Based on such a weak representation for *free text* the generalization of only two examples already yields an overly general delimiter or predicate. Which in turn leads to overly general wrappers and long and probably exceeding extraction times. Or in the case of *BFOIL* based learning to long and probably exceeding learning times, resulting in no wrapper at all, or in many extraction rules building a too specific wrapper. So it is much more reasonable for *free text* extraction tasks either to use a mixture of character and token level representation or to switch to a complex representation level using linguistic and additional semantic information. Consequently, the level of document and thus connected wrapper language representation is one major factor of a wrapper learning system.

LIPX is a multi-slot extraction system that induces rules where the application of a single rule yields all slot fillers. Comparing multi-slot extraction results of other systems with *LIPX* this feature seems to be not a real advantage. The explanation is manifold. First, it is observable that a system like WL^2 learns binary wrappers, and thus has to have a technique at hand to group extracted fields [Jensen and Cohen, 2001]. *STALKER* learns different types of rules (automata) to detect starting positions in the tree, to iterate on lists, and one extraction rule for each slot. So it seems that the quite ambitious attempt by *LIPX* to use only a single rule for extracting all slot fillers requires additional research effort, because the single-slot results showed very promising results. One starting point to improve *LIPX* multi-slot extraction quality is to introduce more sophisticated clause templates than those of *AV* and *CD-models*. As a basis for further refinements of clause templates the formal wrapper models introduced in Section 3.1 can be used to extend the presented learning techniques in that they learn from ground instantiated rule sets. More detailed, currently each example is described by one ground instantiated wrapper clause template. Obviously for more complex wrapper tasks *AV* and *CD-models* are quite limited. Instead of using solely one clause as example description a partial logic program, implementing a more complex wrapper model might yield better results. Actually, the basic learning algorithms require no or almost no modifications. Only the *lgg* calculation of example description clauses has to be adapted.

Independently of *LIPX's* results the question has to be discussed if it is really sensible to learn rules with 40 to 100 literals? This can happen if for each slot one hypothesis predicate is applicable and used for the example description clause construction. For instance, the QuoteServer wrapper has 18 slots. Hence, roughly 18 times the number of hypothesis language literals is the maximal number of literals one extraction rule can have. In fact, all tests have been performed with almost all *TDOM predicates* presented in Section 3.3.3. So, it is reasonable to conclude that smaller hypothesis languages conduct to faster learning times, because of less effort for the computation of lggs. An interesting point and future work, is to investigate for which hypothesis language size and literal selections the quality of learned wrappers especially the precision rates are significantly influenced.

Another topic not mentioned so far is the unification and anti-unification (lgg) of tokens (feature terms) with different number of feature value pairs. There are several publications on this topic [Carpenter, 1991; Smolka, 1988; Aït-Kaci *et al.*, 1997] and we already discussed this issues in Section 3.4.1. Hence, from a theoretical perspective it is straightforward to incorporate this into the *LIPX* system. For the presented methods and theoretical results in this thesis it is not directly relevant, since we can simply assume that each token of a specific type always has a fixed and ordered number of feature value pairs. But, because in the current application of the system the number of features is not fixed and thus the occurrence order differs several generalizations on token terms become overly general. So, one improvement in *LIPX's* results can be expected by the use of a correct *feature term unification* and generalization.

It is worth mentioning again that the bottleneck of the overall framework is situated in the way extractions are computed, namely by answer computation based on a logical calculus. Many of the presented results do not really state the theoretical possible quality of induced wrappers under the presented methods. Because due to runtime limits the learning or extraction process was terminated. So, now that a framework for integrating information extraction and wrapper induction into the ILP framework has been stated and basic methods have been presented an optimization phase has to take place. Such optimizations may be developed for logical calculi tailored for this specific domain or more sophisticated methods for testing the *posterior sufficiency* and *satisfiability* have to be researched. Evidently, this is subject of future work.

10.5. Empirical Learnability Analysis

This section summarizes observations made from the experiments regarding the best learning configurations and empirical learnability results.

Best Empirical Models and Methods

Figure 10.26 depicts the distribution of best wrappers and models among the learning algorithms.

As the figure shows *RTD-wrappers* yield the best results among the three presented models. In 25 test runs using ϵ-*OSL* and *threshold-based BFOIL* 11 times the best learned wrapper was a *RTD wrapper model.* The weakest of all three models is, as expected, the *AV wrapper model.* The presented results regarding *cluster-BFOIL* are a bit misleading, because *cluster-BFOIL* was only used to learn *RTD wrappers.*

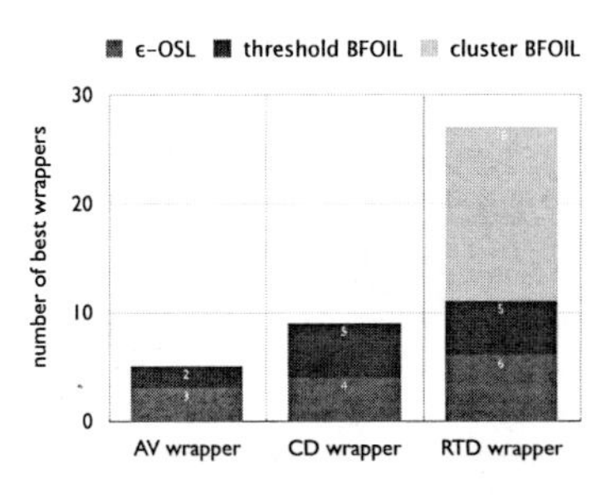

Figure 10.26.: best methods and models

To decide the best learning method we omit the *free text* task *Seminar Announcements*, because its single-slot tasks were solely investigated using *cluster-BFOIL*. According to the results presented in Section 10.3, no method dominates any other: ϵ-*OSL* (4 best wrappers), *threshold-based BFOIL* (5 best wrappers), and *cluster-BFOIL* (4 best wrappers). Obviously the decision to chose a specific learning method is problem domain dependent, but policies for an automatic selection of a best learning method are beyond the topic of this thesis.

Best Empirical ϵ-OSLConfigurations

Figure 10.27 provides an overview about the best test configurations for ϵ-*OSL*. Surprisingly, multi-slot wrappers prefer shorter delimiter and context lengths than single-slot wrappers. Seemingly, the introduced wrapper models imply strong enough constraints for multi-slot wrappers in that they are precise enough and that longer delimiters or contexts reduce the recall score. One reason for this might be the short distance between slot fillers. Thus, shorter lengths are preferred. It is the opposite with single-slot wrappers. There a short length reduces the precision rate and thus a longer length is preferred to better constrain the extraction rules.

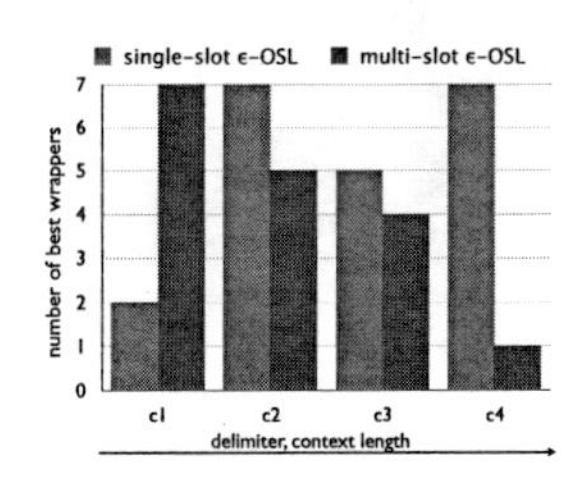

Figure 10.27.: best ϵ-*OSL* configurations

Best Empirical threshold-based BFOILConfigurations

From a theoretical perspective the proposed refinements of the *basic-BFOIL* method to use *rule quality thresholds* or dropping the *consistency check* is a reasonable method to increase the recall rate. The experiments showed, as illustrated in Figure 10.28, that the proposed refinements contribute to learning better wrappers. 76% of the best wrappers learned by *threshold-based BFOIL* use one of the proposed refinements. The largest number of best learned multi-slot wrappers were reached using refiment $c11$ and $c10$. Approximately 66% of the best learned single-slot wrappers and 58% of the best multi-slot wrappers are learned with consistency check. This underlines that for wrapper learning the consistency property as proven in Sections 6.4 and 7.2 is an important property.

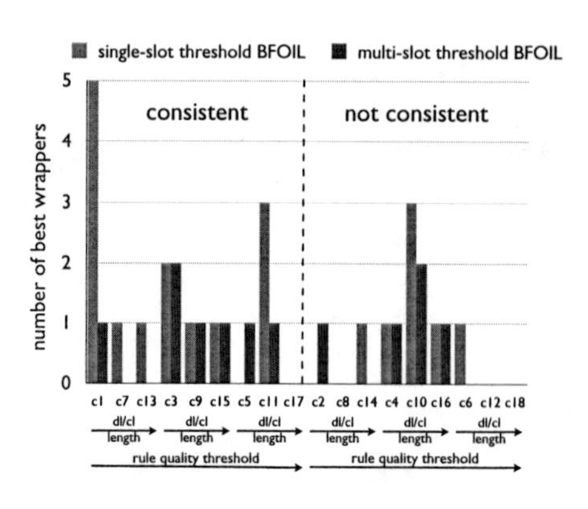

Figure 10.28.: best *threshold-based BFOIL* configurations

Best Empirical cluster-BFOILConfigurations

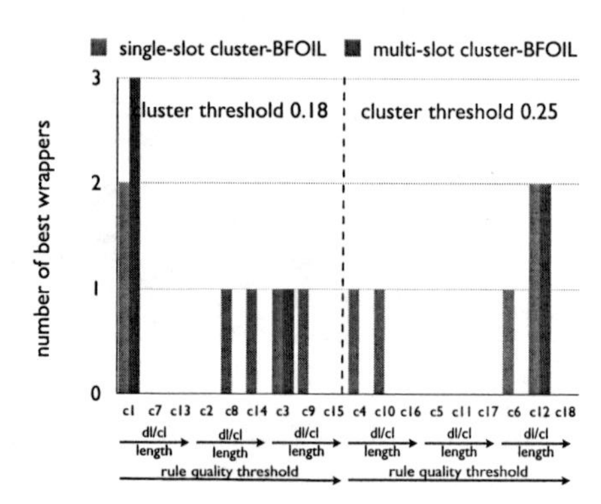

Figure 10.29.: best *cluster-BFOIL* configurations

From discussions in Section 8.5 and 8.5 we expect that wrappers learned by consistent, not false positive accepting *threshold-based BFOIL* to be less general than wrappers learned by ϵ-*OSL* and *threshold-based BFOIL*, because some example description clauses that might yield a good *lgg* are probably not contained together in one cluster. So, in general *cluster-BFOIL* will yield lower recall rates. Hence, it is reasonable that *cluster-BFOIL* configurations perform best that use a higher rule quality threshold to increase the recall rate of the learned wrapper. The distribution of best configurations shown in Figure 10.29 substantiate this conjecture. 59% of the best learned wrappers by *cluster-BFOIL* use refinements increasing the recall rate. It seems that increasing the cluster threshold, which results in larger clusters, yields only a minor improvement, because *cluster-BFOIL* degenerates more and more to *threshold-based BFOIL* which results in longer learning times exceeding the time limits. In the case of learning single-slot wrappers *cluster-BFOIL*'s behavior is almost identical to the best multi-slot configuration distribution.

learnability	wrapper class	problem, model, method
good	linear-$\wedge$	BigBook,*CD,threshold-based BFOIL*
		Okra,*RTD*,ϵ-*OSL,cluster-BFOIL*
	linear-$\wedge$-MV	Zagats,*RTD,cluster-BFOIL*
precise	linear-x-ϵ^+-$\vee$	QuoteServer,*CD,threshold-based BFOIL*
acceptable	non-linear-ϵ^+-$\vee$	InternetAddressFinder,*AV*,ϵ-*OSL*
borderline	linear-ϵ^+-MV-$\vee$	LA-Weekly,*RTD,cluster-BFOIL*
not learnable	non-linear-$\vee$-ϵ^+	Seminar Announcements (free text)

Table 10.51.: learnability of multi-slot extraction tasks

Empirical Wrapper Class Learnability Results

In Part I Section 3.6 we discussed class properties of the three wrapper models. We informally illustrated how the various wrapper classes can be modeled by appropriate rules for each wrapper model. The results presented in Section 3.6 have to be reconsidered, because several restrictions have been defined in Section 5.4. These modifications include: a strong language bias in form of clause templates and a modified predicate representation of token-patterns.

As briefly discussed in Section 3.3.4 the overall aim is to learn a partial wrapper model with a sufficiently good quality regarding precision and recall rates. Hence from a practical point of view a wrapper is set to be learnable if its quality regarding these two measurements suffice some intuitive threshold. In general such thresholds are very problem domain dependent, as for example in some application domains only 100 % precise wrappers are acceptable in others the emphasize is set on higher recall values being sensible of requiring some post-processing methods to filter out false positives. According to these considerations the practical learnability of wrappers regarding a certain wrapper class can be determined in terms of precision and recall scores, as shown in Table 10.30.

	precision		recall
perfect	100%	$\wedge$	100%
good	$> 95\%$	$\wedge$	$> 90\%$
precise	100 %	$\wedge$	$> 40\%$
acceptable	$> 80\%$	$\wedge$	$> 40\%$
borderline	$\leq 80\%$	$\vee$	$\leq 40\%$
unacceptable	$< 50\%$	$\vee$	$< 20\%$

Figure 10.30.: practical learnability scale

Following this scheme under the discussed test configurations for the 7 investigated multi-slot extraction tasks no wrapper is perfect learnable, three are good learnable, one is precise learnable, one is acceptable learnable, one is at the borderline and one is not learnable (due to time limits) using any of the presented algorithms and models. Table 10.51 summarizes these learnability observations.

For the 11 investigated single-slot extraction tasks two wrapper are perfect learnable, is good learnable, one is precise learnable, three are acceptable and four are unacceptable regarding their quality. Table 10.52 shows an overview of this results. Figures 10.31 and 10.32 depict the learnability observations with respect to the best observed median precision and recall rates from Section 10.3.

Especially from Figure 10.31 a basic characteristic is observable that the presented learning algorithms tend to learn wrapper models with higher precision and lower recall rates. We have to emphasize that the classification of the extraction tasks as shown in these two figures are based on the best results for each extraction task, wrapper model and learning method. For instance, from the multi-slot figure we cannot draw the conclusion that never a multi-slot

learnability	wrapper class	problem, model, method
perfect	∨-class	QuoteServer Vol, *RTD*, *cluster-BFOIL*
		QuoteServer Date, *AV*, *CD*, *RTD*, ϵ-*OSL*, *BFOIL*, *cluster-BFOIL*
good	∧-class	Zagats Adr,*AV*,ϵ-*OSL*,*cluster-BFOIL*
precise	∧-class	IAF Altname, *RTD*,*threshold-based BFOIL*
acceptable	∧-class	IAF Organization, *CD*,*threshold-based BFOIL*
	∧-class	LA-Weekly Credit Cards, *CD*, *threshold-based BFOIL*
	∨-class	CS Name,*CD*,*threshold-based BFOIL*
unacceptable	∨-class	SA (speaker,stime,etime,location), *RTD*, *cluster-BFOIL*

Table 10.52.: learnability of single-slot extraction tasks

wrapper with 100 % recall rate has been learned. Nevertheless it is surprising that no wrapper among the best learned ones has a 100 % recall score. This is explained by the fact that the *BFOIL*-based learning approaches learn *cautious* wrappers, because every extraction rule covering at least one false positive is rejected during learning. Even by relaxing this condition (*threshold-based BFOIL*) there was still a quite cautious strategy used for experiments (small threshold).

The assumption that more training examples are needed to gain greater recall scores is justified by the observation that for all extraction tasks the $F1$ scores of wrappers learned by *BFOIL*-based approaches increase while the precision rates remain almost unchanged. So one conclusion is that the presented *BFOIL* approaches need larger number of examples to gain high recall rates. Extensive studies of all test runs and all learning methods show that only in a few cases overly general wrappers have been learned independent of the used learning method and wrapper model. Hence its seems that also for *OSL*-based approaches one reason for this can be found in the representation of documents and wrapper languages. Most of the extraction tasks are so regular in its structural representation of extractions that under the chosen representations (document model, wrapper language) simply no over generalization takes place or a overly general wrapper does not provide any results due to terminated learning or testing times.

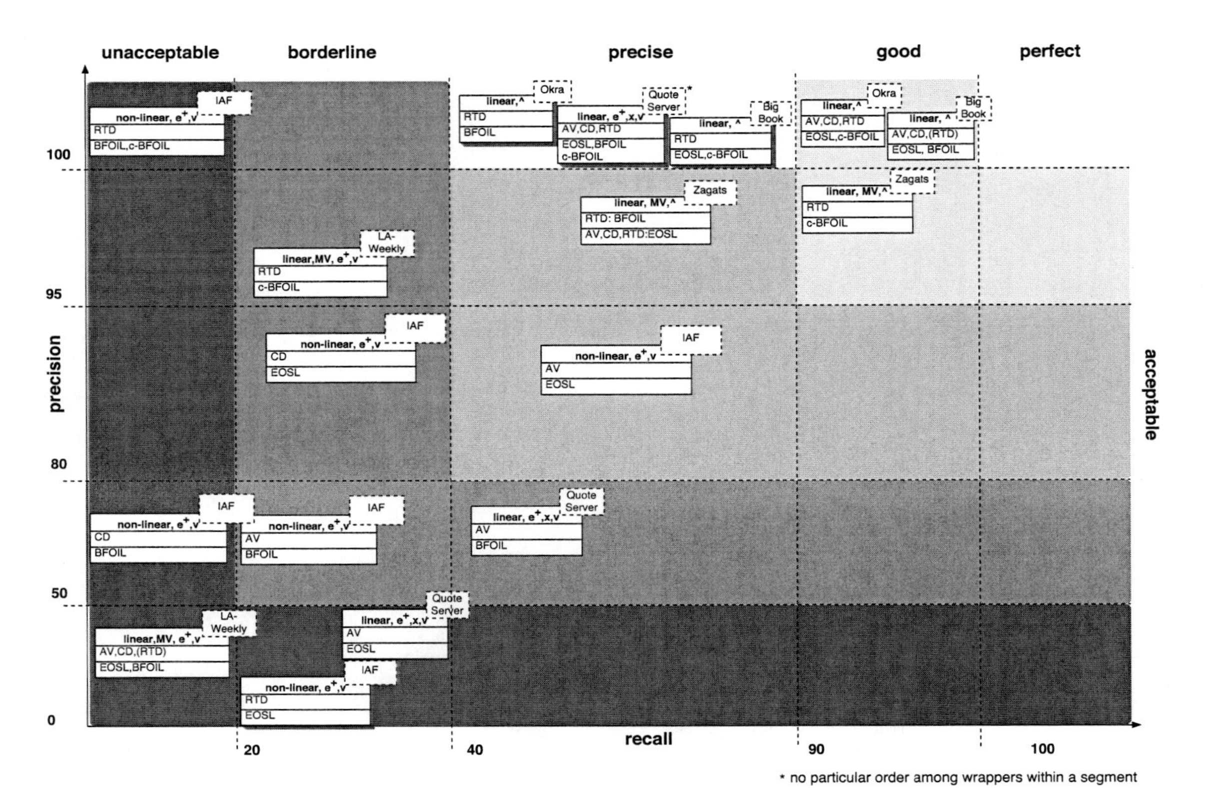

Figure 10.31.: empirical learnability observations multi slot extraction tasks

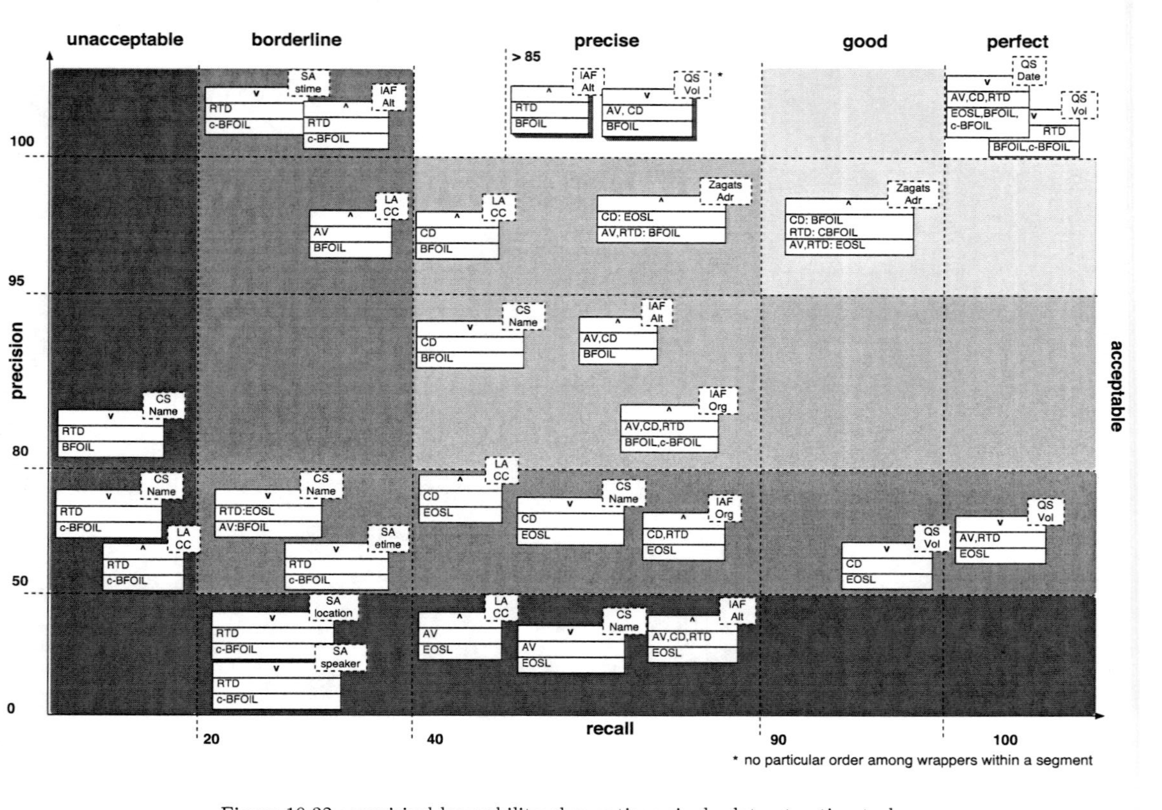

Figure 10.32.: empirical learnability observations single slot extraction tasks

11. Related Work

Some related work on *Inductive Logic Programming* already has been mentioned whilst introducing the presented methods and techniques in Chapter 5. Also an extensive overview of related *Machine Learning based Information Extraction* approaches and systems has been given in Section 10.1. Thus Section 11.1 of this chapter concludes the overview of related work with summarizing and discussing the most important adjacencies with *ILP* related approaches.

Other topics as representational issues covering the use of *XML* like document representations or feature value based document processing and representation related in a broader sense to techniques commonly used in the field of *computational linguistics (CL)* [Carpenter, 1991; Carpenter, 1992; Carpenter, 1993; Smolka and Treinen, 1994] and the therewith related *CL* approaches of logic programming based natural language analysis [Gazdar and Mellish, 1989; Fong, 1991; Johnson, 1992; Pereira and Warren, 1983] are not discussed. For the same reasons, an extensive coverage of related work to clustering methods and techniques is not given. This is explained by the fact, that the used techniques and methods from these areas are mainly understood as tools and were not in the main focus whilst developing and investigating techniques for automatic wrapper construction by combining *ILP* based techniques and information extraction techniques.

This chapter concludes with a brief presentation of a practical application, a multi-agent information system, that uses some of the introduced wrapper learning approaches.

11.1. Related ILP Approaches

The presented learning algorithms *OSL*, *ϵ-OSL*, *basic-BFOIL*, *threshold-based BFOIL* and *cluster-BFOIL* all follow a *bottom-up* learning approach using exclusively the *least general generalization* as learning operator. Compared to other sophisticated ILP systems the proposed algorithms appear to be very simplified approaches for learning clause theories. But, de facto the aim of this thesis is to present a basic uniform framework for automatic construction of wrappers in logic. Therefore it is quite reasonable to chose a simplified ILP approach nevertheless producing good results and offering a large variety for modifications and adaption possibilities to existing more sophisticated approaches. Although, the presented learning methods are developed with the intent to investigate how practical synthesized logic programs for information extraction are if learned by this basic generalization technique some additional fundamental refinements have been discussed. Consequently, the basic learning techniques, the constituted refinements and used settings like the use of clause templates or learning from positive examples only, affords a brief discussion on related ILP work.

LGG Based Learning: The origin of the *basic-BFOIL* algorithm goes back to the algorithm "*a logic of suggestion*" by [Plotkin, 1971b]. There are two points that distinguish Plotkin's algorithm from the *basic-BFOIL* approach. First, his algorithm finds a heuristic approximation to what he calls *irredundant* hypothesis (set of clauses), whereby a set of *irredundant* clauses is the minimal set of least general generalizations of clauses with respect to a given set C (e.g. the examples). The algorithm controls the selection of a hypothesis clause for further generalization by the number of failures and successes of attempts regarding previously missed consistency tests. This is different to *basic-BFOIL*'s technique where regardlessly of the number of successes and failures of evaluation checks the current hypothesis clause is tried to be generalized with all other remaining example clauses. Thus *basic-BFOIL* really partitions the set of clauses regarding the least general generalization, whereas the algorithm in [Plotkin, 1971b] might result in a hypothesis consisting of a set of generalized clauses containing two clauses subsuming each other. Secondly, Plotkin checks for consistency in a logical sense, i.e. does $lgg(C_1, C_2)$ models the observations (example clauses), whereby this consistency test is reduced to a subsumption test, justified by the subsumption theorem (Theorem 2.1.1). *basic-BFOIL* also checks hypothesis clauses if they explain the examples, but unfortunately the more efficient evaluation of potential hypothesis clauses based on subsumption checking can not be applied for several discussed reasons. *basic-BFOIL* learns from positive examples only, hence the proposed consistency test by Plotkin to check if the current clause subsumes some of the negative examples is not applicable. Additionally, in the context of wrapper learning it is not sufficient to simply test if the generalized clause allows the derivation of the positive evidence, since this is trivially the case, because the hypothesis clauses are generalizations of the examples. Hence, *basic-BFOIL* requires answer computations for the target predicate to evaluate the extraction quality of the current hypothesis. Consequently, *basic-BFOIL* in conjunction with the introduced *IE-ILP setting* is a refinement of Plotkin's original algorithm to learn from positive examples only and especially tailored for the application domain of logical wrapper induction.

Closely Related ILP Systems: Closely related to the main idea of using the least general generalization operations for learning is the *GOLEM* system [Muggleton and Feng, 1992]. *GOLEM* uses a modified *lgg* operation the *relative least general generalization.* Roughly speaking, the *rlgg* is the *lgg* of two clauses c_1 and c_2 relative to a background knowledge B. *GOLEM* expects B and the positive examples to be ground units. From this it computes the *rlgg* of two examples e_1 and e_2 by $rlgg(e_1, e_2) = lgg(e_1 \leftarrow B, e_2 \leftarrow B)$. Basically *GOLEM* can also use clausal theories as background knowledge, but therefore [Muggleton and Feng, 1992] suggest to compute a partial Herbrand Model so called *h-easy model* given by ground atoms derivable in at most h resolution steps from a theory B. Besides the fact that *GOLEM* restricts the background knowledge and examples to be units it also uses negative examples to compute hypotheses. For these reasons it is not straightforward how to adapt this system for the proposed wrapper induction tasks. Especially the possible size of a reasonable *h-easy model* for a few documents and a background knowledge implementing the hypothesis language predicates for describing *TDOM* properties seems to become fairly large. In fact, deriving only those background atoms relevant to the examples is a more effective procedure instead of giving the complete background extensionally. Systems like *CLAUDIEN* [Raedt and Bruynooghe, 1993], *TILDE* [Blockeel and Raedt, 1998] and *Progol* [Muggleton, 1995] are also based on the theory of *θ-subsumption.* But they search the θ-subsumption lattice for suitable refinements and thus do not use the *lgg* operator in the sense as the algroithms presented in this thesis or like *GOLEM.*

Strong Language Bias: Basically, the idea of using *clause templates* (Section 5.4) to strongly bias the structure of hypothesis clauses (i.e. wrappers) is related to the idea of *rule schemata* as used in systems like *ML-SMART* [Bergadano *et al.*, 1989] and *MOBAL* [Kietz and Wrobel, 1991]. Opposed to these systems the presented approaches are much more restrictive. One learning step of these systems consists in finding reasonable predicates from the hypothesis language for instantiating the second order rules (i.e rule schemata). This derivation process step is skipped at least for the learning of *AV* and *CD-wrappers* where the corresponding clause templates are fixed depending on the defined wrapper models (Section 3.3 5.4). For *RTD-wrappers* the initial *example description clause* construction can be interpreted as a simple rule schemata instantiation.

Learning Incomplete Hypothesis: In general pure bottom-up ILP systems try to invert deductive operators like unification, resolution or implication with the intent to provide methods based on a clear theoretical basis and with the advantage to find complete and consistent hypothesis. Top-down ILP methods in contrary follow in general a *generate-and-test* method and use specification heuristics to learn from general to specific. It is quite common for these approaches to measure the quality of a rule by different metrics (i.e. information gain [Shannon, 1948] as in *FOIL* [Quinlan, 1990]) for deciding what refinement operator to apply next or to decide if a stopping criterion is reached. Thus applying some of these methods to a bottom-up ILP system relaxes the condition to learn a complete and consistent hypothesis with respect to the given examples. As we have illustrated in Section 7.3 it is reasonable to renounce from this strong quality criteria to gain better predictive results (i.e. recall scores). So, *threshold-based BFOIL* is related in some degree to such top-down ILP techniques regarding the evaluation of hypothesis clauses.

Positive Examples Only: Evidently, one separability of the class of presented learning algorithms regarding standard ILP algorithms is that they learn from positive examples only. Negative examples are assumed to be given implicitly by all those atoms not contained in the exhaustive example set with respect to the target predicate to be learned. Basically, from a theoretical viewpoint this is closely related to the *non-monotonic ILP setting* [Helft, 1989; Flach, 1992] discussed in Section 5.1 and 5.2.1 and the *CLAUDIEN* system [Raedt and Bruynooghe, 1993] also based on this semantics. But since the *non-monotonic ILP setting* is not very well suited for the reasons given in Section 5.2.1 and the fact that the *CLAUDIEN* system is a top down learner this is not a real alternative for replacing the proposed techniques by this system. The *FOIDL* [Mooney and Califf, 1995] system is another closely related approach regarding the treatment of absent negative examples. By so called *mode declarations* for specific literals, the system determines if unintended implicitly given negative examples are covered or not. This is discussed in more detail in Section 5.2.1 on page 98.

Mixing Clustering and ILP: *cluster-BFOIL* uses standard document clustering methods and slightly modified metrics for clustering Horn clauses. Obviously, this approach is related to standard work on clustering techniques for documents [Miyamoto, 1990; Rasmussen, 1992; van Rijsbergen, 1980; Steinbach *et al.*, 2000]. But it is also related to research fields trying to apply non standard clustering methods either using different clustering ideas and concepts or methods to be used in a logical framework. Both fields are of interest since they provide reasonable refinements not tackled in this thesis. Firstly, in a broader sense the clustering of example clauses is related to the area of *Conceptual Clustering* [Michalski and Stepp, 1984], [Fisher, 1987]. But these approaches use valued propositional logic. As soon as the problem of clustering is investigated in a first-order logical context the problem becomes intrinsic more complicated. One reason is the problem of defining suitable distance respectively similarity metrics for atoms or clauses. There are several different approaches for defining distance metrics for first-order logical terms given by [Nienhuys-Cheng, 1997; Hutchinson, 1997; Ramon and

Bruynooghe, 1998; Ramon *et al.*, 1998; Ramon and Raedt, 1999] with which *cluster-BFOIL* can be extended. Additionally there has been research done on combining *lgg* based generalizations with clustering techniques (distance metrics) such that they define *semantic distance measures* [Markov and Pelov, 1998; Markov and Marinchev, 2000]. Not directly related to *cluster-BFOIL* but also work that brings together clustering with ILP methods, is the use of ILP techniques to cluster relational data by [Kirsten and Wrobel, 1998; Blockeel *et al.*, 1998]. An overview of this active research field is found in [Lavrac and Dzeroski, 2001].

11.2. MIA-The Mobile Information Agent

ϵ-*OSL* and a modified version of it for learning extended *CD-wrapper* and *RTD-wrappers* have been successfully used in the mobile information system *MIA*[1] [Beuster *et al.*, 2000b; Beuster *et al.*, 2000a]. *MIA* is a multi-agent based information system [Weiss, 1999; Wooldridge, 2002] focusing on the retrieval of short and precise facts from the World Wide Web, as shown in Figure 11.1.

MIA's aim is to provide location based information for a mobile user equipped with a WAP capable mobile phone or *PDA*. Therefore it identifies the users interests by means of a user profile and is capable to track the users position by using wireless ubiquitous computing and communication technologies. But also for the stationary user *MIA* provides a standard web browser access, which offers additional features like linking extraction results with other online services as illustrated in Figure 11.2.

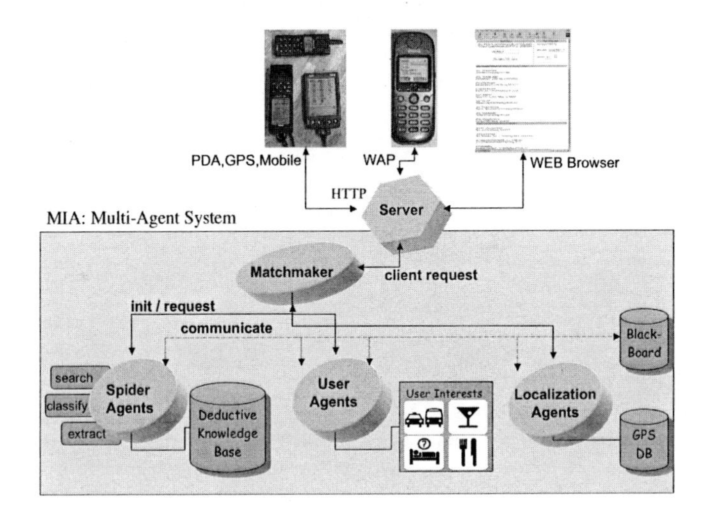

Figure 11.1.: multi-agent architecture of the MIA system

Instead of overwhelming the mobile user with documents found on the web, the MIA system offers the user a short precise piece of information he is really interested in. *MIA* monitors the position of the mobile users and autonomously updates the subject of search whenever necessary. Such changes may occur when the user travels to a different location, or when she changes her search interests (profile). Whenever this happens the search agents are updated and they adapt their current information acquisition task to the new information.

One strength of the *MIA* system is, that it is not restricted to a special search domain,

[1]MIA is part of the *IWIA - Intelligent Web Information Agents* project funded by *Stiftung Innovation* of the state Rhineland-Palatinate.

thus the user is not restricted to some predefined search domains or collection of preselected information databases as offered by some of the major telecommunication providers. So far, *MIA* is capable of automatic (multi slot) address extraction.

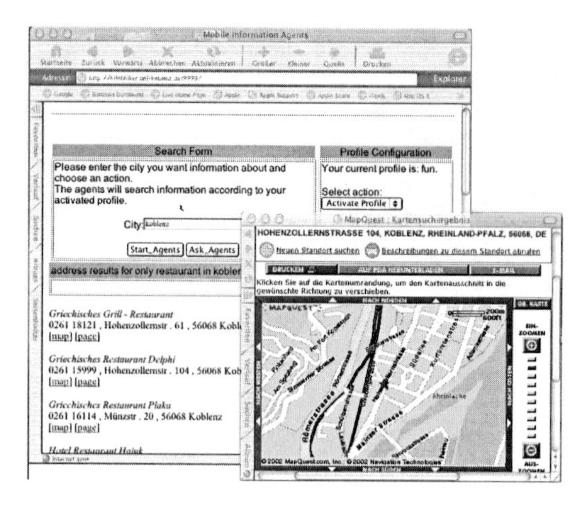

Figure 11.2.: coupling extraction results with queries to other online services

IE and Information Agents

One arduousness in building an information system offering a great degree of freedom regarding freely selectable search domains is to provide reasonable good procedures for extracting information from completely unknown web pages. To point this out clearly, in the so far discussed chapters it is implicitly assumed that a wrapper is learned for a certain class of documents. But generally in the *MIA* scenario it is not known in advance what class of documents are to be processed by the information extraction component.

Actually, *MIA* uses two methods to extract information, after the spider agent and classification agents have successfully found a potential web page of interest. Firstly, for a hand selected set of information web sites a set of wrappers are learned offline. Secondly, wrappers are learned online from automatically constructed positive examples, based on some strong assumptions.

For learning wrappers offline the MIA administrator uses *MIA's* wrapper toolkit to learn wrappers for certain domains. Whenever the extraction agents visit one of these domains during their search they use these pre-learned wrappers to extract information from one of the web pages. The wrapper toolkit uses the presented ϵ-*OSL* method and learns *RTD* based wrapper models. The online learning methods used by *MIA* to construct wrappers is explained very briefly in the following.

Learning Wrappers While Searching

The major problem someone is confronted with in the context of an autonomous multiagent system like *MIA* is the lack of available examples for learning wrappers. Because *MIA*'s web page classifier provides unknown web pages to the system, no training data is available. On the one hand, *MIA's* web page classifier is good enough to determine if an address is contained on a web page; on the other hand, arbitrary web pages vary too much for a single general purpose address wrapper to be effective.

Since it is exactly known what is to be learned, namely address wrappers, a set of restrictive *CD-wrappers* with most general delimiters but with restrictively defined slot filler patterns are used to detect some addresses in a web page. Therefore recall the idea of *clause templates* as introduced in Section 5.4. In the same manner as example extractions are represented by *example description clauses* a set of pre-defined non ground *example description clauses* can be used to detect addresses. Therefore the slot filler patterns are pre-defined, the delimiter patterns are most general and only the gap between two slot fillers is constrained.

Figure 11.3.: *MIA* results

Instead of defining the set of possible slot filler patterns extensionally they are defined intensionally as a knowledge base by means of knowledge representation techniques. During the retrieval process different address templates (clause templates) are derived from the knowledge base. As soon as two of these address templates successfully match on one web page and consequently also instantiations for delimiter patterns are obtained, the two grounded templates are stored as new *example description clauses*. This process is repeated until no more automatically constructed address templates match anymore. Before the so obtained example description clauses are passed to the learning algorithm as training examples, the slot filler patterns are replaced by most general patterns. This last step ensures that the pattern is capable of detecting other address formats than those described by the address templates. The resulting generalized clause template of all matched address templates is then applied to the web page and further addresses are obtained.

This method allows *MIA* to learn address wrappers even for unknown pages on basis of address concept patterns. This approach showed very promising results for the automatic construction of address wrappers and was tested with ϵ-*OSL* and *CD-wrapper models*.

12. Conclusion

In the remainder of this chapter we discuss if the two major claims of this thesis stated in Section 1.2 are verified by the contributions presented in the previous chapters. The chapter concludes with a brief discussion on discovered topics which are subject of future work.

12.1. Contributions

In the following the contributions of this thesis in the context of the two major claims are discussed

Claim No.1: The presented ILP *learning framework is independent of the document view and wrapper model. Different document views and wrapper models can be conveniently represented by means of logic programming and integrated into the wrapper learning process.*

In the first part of this thesis in Section 2.3, the *AV* and *TD* document representations have been introduced. They are differing in that *AV* has a sequential token view on documents and *TD* interprets documents as hierarchical structures. Used with the appropriate tokenization and pre-processing step *AV* can easily be used with natural language text. *TD* is bound to *SGML* based markup languages like *HTML* and *XML*. By these two different document views and the general method to represent documents as *unit clause sets* it has been exemplified that without modification of the learning algorithms different document representations are usable within the presented framework. Therefore the only requested property by the *IE-ILP* framework regarding documents is that they are representable as *unit clause sets.* Although, not discussed in this thesis basically a document representation can be given by an arbitrary *logic program* and does not has to be restricted to *unit clauses.* Obviously, the only requirement is that the representation has to be conform with the also freely definable logical wrapper language.

In Chapter 3 three different wrapper models with respect to the chosen document representation have been defined. The *AV* and *CD-wrapper models* follow the conventional wrapper idea of matching left and right slot delimiters. The third wrapper model *RTD-wrapper model* describes relevant slot filler text by relational properties between nodes and subtrees in a modified *document object model (DOM)* tree. In Chapter 5.4 it is shown how strongly differing wrapper models can be represented in a uniform way as *clause templates.* This enables the learning algorithm to successfully induce wrappers regarding a desired wrapper model.

Besides the logical representation of wrapper models as *clause templates* the wrapper language (hypothesis language) has to be given as a set of predicates together with a logic program implementing the predicates. But this is no limitation, because for every wrapper model and its used concepts an equivalent logic program can be constructed.

Similar to the logical representation of different document models the only requirement concerning wrapper models is that a wrapper is representable by one clause template. This clearly omits the case of very sophisticated wrapper models that require more than one clause template as in the case of recursive models. In general, the presented framework is easily extendable to learn from logical wrapper representations consisting of more than one clause template. Only the *clause set lgg* calculation has to be slightly modified. Consequently, this thesis presented a wrapper induction method independent of the used document view and wrapper model. For these reasons it offers a framework to be applied to different application domains.

Claim No.2: A pure *LP* and *bottom-up ILP* framework based on the *least general generalization of clauses* represents an adequate learning technique for single and multi-slot wrappers from positive examples only.

The use of the *clause set lgg* operation for learning wrappers has been investigated and discussed in its most simple form in Chapter 6. Regardlessly of the complexity and variance of the extractions task a *one step learning* method (*OSL*) consisting of building the *clause set lgg* of all *example description clauses* was presented. From the expectable weak points of this learning method a refinement solely based on observable properties of the given examples was introduced (ϵ-*OSL*), which yields acceptable results. Although, ϵ-*OSL* is attractive because of its fast learning time it lacks some hypothesis evaluation component to provide more precise wrappers. The *BFOIL* algorithm presented in Chapter 7 learns 100 % precise wrappers with respect to the training set. But contrary to ϵ-*OSL* its iterative technique to find partitions of the example description clause set such that each *lgg* of a partition yields a 100 % wrapper rule, takes very long computation time. The bottleneck in this approach is the number of query computations needed to estimate the current hypothesis quality and the involved long query computation times. Finding other methods for rule evaluation not based on query computations but using the introduced *IE-ILP setting* (Section 5.2), would be a major step towards high precision wrapper induction. The major problem of *BFOIL*'s long learning times is based on the evaluation test to determine if the current rule extracts some negative examples. Because negative examples are defined only implicitly under the assumption that positive examples are exhaustively enumerated, it is necessary to test if in the set of all computed extractions a false extraction is contained. Testing if a learned rule is range restricted (see Section 7.2) helps to decrease the number of needed query computations. Several refinements to improve the recall behavior of *BFOIL* learned wrappers have been presented. The *cluster-BFOIL* algorithm presented in Chapter 8 aims at speeding up the learning of *basic-BFOIL* and *threshold-based BFOIL* by using clustering techniques for clause sets. This idea was inspired by observations drawn from the ϵ-*OSL* algorithm. Like *basic-BFOIL* also the *cluster-BFOIL* algorithm learns 100 % wrappers with respect to the training set. The presented algorithms can be characterized as follows:

ϵ-*OSL* learns from a few positive examples only, does not need completely labeled documents, shows fast learning times, yields in general average precision rates.

basic-BFOIL and *threshold-based BFOIL* learn from a few positive examples only, require completely labeled documents, show long learning times, yield highly precise and medium recall wrappers.

cluster-BFOIL learns from a few positive examples only, requires completely labeled documents, shows average learning times, yields highly precise and lower recall wrappers than *basic-BFOIL*.

All *BFOIL* based algorithms and therefore also *cluster-BFOIL* does not compute the best hypothesis regarding the training set under *clause set lgg* operation. Roughly speaking, in general not the optimal partition of the example description clauses is found that result in generalized rules such that they yield the maximal number of correct extractions from the training documents. From a theoretical point of view it is easy to modify the presented *BFOIL* algorithms to output the best hypothesis wrt. the training documents. But from a practical point of view this results in higher learning times which as mentioned is the crucial problem of *BFOIL* based algorithms. Nevertheless, the thesis contributes to the *ML based IE* community a class of *ILP* learning algorithms, which offer a broad variety for improvements and adaptability to specific domains. This is underlined by the experimental results presented in Chapter 10. The three presented algorithms demonstrate that a very basic learning operator the *lgg of clause sets* can be successfully used to induce logic programs for information extraction. Therefore, this work contributes to both the area of *ILP* and *IE* and serves a starting point for more elaborated methods based on *least general generalization of logic program clauses* for wrapper induction.

Comparing the quality of existing *ML based IE* systems is not a straightforward task. Either one concentrates solely on the experimental results or one investigates how adaptable the approaches are to different domains regarding document types and extraction tasks. Chapter 10 gives an overview about the property of existing systems concerning both comparison methods. Regarding the task of learning single-slot wrappers from semi-structured documents the presented methods are highly competitive to leading systems. In fact they provide best results for 5 out of 10 problem tasks. In one case the median results are very close to the best results. In one case wrappers with average quality and in three cases no acceptable wrappers are learnable.

The problem tasks which are hard to solve by the presented methods implemented in the *LIPX* system are extraction tasks from natural text. Apparently, for natural text the used token representation (document representation) is not granular enough in that either more linguistic features have to be added to a token or that a character view and not a word view of documents has to be chosen. So, it is quite expectable that much better results are obtained using a different document representation. This is discussed in detail in Section 10.4.

Excluding the results of extraction tasks from natural text the presented *ILP* framework is indeed very competitive to state-of-the-art single-slot extraction systems if restricted to semi-structured documents. In this case, in 5 out of 7 of the standard test cases *LIPX* shows better or equal results than existing systems and for the other other two extraction tasks very closely results to the best scores of leading systems.

As discussed in Section 10.4 the reported results for existing multi-slot extraction systems differ in the used metrics and aggravate a neutral comparison of results. But the general observation made is that *LIPX* performs very good in comparison to other leading systems. Although, no best result is attained in 3 out of the 6 investigated problem tasks, wrappers with *F1* scores greater than 92 % are learned. Two of the other 3 problem tasks at least provide learned wrappers with *F1* scores larger than 55 % and the worst result is an *F1* score of 43 %. Because of incomplete reported results regarding precision and recall scores or results using different evaluation metrics the comparison with other systems can only be carried out on a quite vague level.

Comparing solely the reported recall or similar values shows that *LIPX* recall scores vary up to 60 % to the best results. But it has to be noted, that the presented *LIPX* results were selected with the intent to show the best *F1* scores hence there are better recall values. Secondly, a good recall value is obviously only one side of the coin, since it gives no hints about the number of false extractions. Nevertheless in 2 out of 4 comparable multi-slot extraction tasks *LIPX* showed almost identical recall rates compared to the leading systems. For the other two problem tasks the recall scores are between 37 % and 60 % lower than the scores of the best systems. Once again, this can be only interpreted to be a trend.

Although for two problem tasks the results are not very promising due to the low recall rates one basic advantage of the presented framework is that the learned wrappers showed very high precision scores greater than 95 % in almost all cases. This property makes the developed approach very attractive to be used within autonomous information acquisition systems (e.g. information agents) as illustrated in Section 11.2.

Due to the idea to use knowledge representation techniques to model different document views and wrapper models in logic the presented approach is one of the only approaches for multi and single slot wrapper induction that is easily applicable to different extraction domains using varying wrapper models, wrapper languages and document representations. Only the WL^2 system by [Cohen *et al.*, 2002] also aims at this goal, but does not use a pure *LP* and *ILP* framework.

The fact that the presented *ILP* framework is able to synthesize very good single-slot wrappers, high precision multi-slot wrappers yielding good overall F1 scores and that it is not bound to specific wrapper and document models distinguishes itself as an competitive approach. Hence the overall purpose of this thesis "to show that a pure *Logic Programming* and *Inductive Logic Programming* framework based on a very basic bottom-up learning technique, namely the *least general generalization* of *logic program clauses* [Plotkin, 1970], can be successfully used to learn multi-slot wrappers from positive examples only" has been successfully demonstrated.

12.2. Future Work

Several questions and new topics worth to be investigated appeared while developing the presented framework. Those topics shall be discussed in the following.

OSL was improved significantly by a simple pre-processing step, namely a syntactic based partitioning of the example description clause set (resulting in ϵ-*OSL*), there may be other similar easy to implement but very efficient pre-processing methods to be researched.

cluster-BFOIL's clustering methods are mainly based on techniques adopted from the area of document clustering. Its vector representation and similarity measurements are probably not best suited for clustering logic program clauses. Thus it is interesting to investigate how good other existing similarity metrics especially developed for clauses ([Ramon and Bruynooghe, 1998; Ramon *et al.*, 1998; Nienhuys-Cheng, 1997]) perform in this context. Also of interest is a detailed investigation how the general concept of *cluster-BFOIL* is applicable to other problem domains and ILP based learning in general.

Also closely connected with questions concerning the structure of clause templates and therefore with the used wrapper models are properties regarding the used wrapper language. For *AV* and *CD-wrapper models* this questions seems to be irrelevant, since they are based on the very strong structural idea of solely learning patterns for the left and right delimiter text fragments of each slot filler. Hence, they use only some word or token sequence predicates. But for the elaborated *RTD-wrapper models* that can use up to 21 different predicates for describing the property of one slot it is unclear to find good limits regarding the size of the hypothesis language. Too less predicates may result in fast over generalization and too many predicates may result in too specific rules or too long learning times due to the increased search space for suitable rule instantiations. Additional research has to be put into these topics.

One reason why *LIPX* showed no best results on multi-slot extraction tasks, but performed very well on the single-slot extraction tasks may be that contrary to other multi-slot systems it tries to induce only one rule for the extraction of all slot fillers, whereas other systems learn several rules each for one slot. It is worth to investigate what happens if the concept of *clause templates* is extended to consist of *partial logic programs*. Instead of describing a text fragment by a grounded clause it is described by a set of clauses with the intent to use one or more clauses to describe properties of one slot. This would require only some small modifications of the *clause set lgg* algorithm and some clustering modifications, but the general bottom-up learning idea can be still applied.

Although the different learning techniques have been evaluated in many test runs as presented in Chapter 10 some questions remain open and have to be answered by future work. For example, additional results are desirable for *BFOIL* based learning with a significantly reduced hypothesis language, e.g. only containing three or four predicates like `xpath`, `xspan` and `xbrother`. Especially for *nested slot* and *slot inclusion* wrapper classes some test cases have to be compiled and evaluated in the future. Related to the issue of testing and experimental results is the task of a more elaborated analysis of the results obtained by the different proposed refinements. Although, an overview and a brief discussion is given in Section 10.5 a deeper examination is of interest to move further improvements into a specific direction. And for the theorist a study regarding the learnability of the proposed wrapper models under the presented learning algorithms is another item on the to-do list.

One claim of this thesis is to learn from positive examples only. The presented *IE-ILP setting* was especially tailored to cover this intent in the considered information extraction domain. Although the algorithms *basic-BFOIL*, *threshold-based BFOIL* and *cluster-BFOIL* construct reasonably good wrappers from only a handful of examples they nevertheless need completely labeled documents. From a user perspective the presented algorithms still require too much hand labeling of learning examples although the actual hypothesis is only constructed from a few examples. As discussed this circumstance is based on the way how *BFOIL* based algorithms evaluate their rules whilst learning. A great spectrum of heuristic methods can be applied to automatically construct negative examples. For instance, mixing positive examples, taking surrounding text fragments as slot fillers and so on. But negative example construction in this way can only enumerate a subset of the complete set of negative examples regarding a document, and the learned rule may extract all the other negative examples not determined by such heuristics. So, both problems using not completely labeled documents and the implicit derivation (construction) of negative examples strongly influence the quality of the learned wrapper. If most of the negative examples are not present the rule evaluation yields to many erroneous rules. And if not all relevant text fragments are labeled in a document the automatic construction of negative examples may yield false negative examples, which also results in erroneous rules. The weak point of the presented approach of too long learning times due to the rule evaluation tests, can be abrogated if a more efficient method is found to prove that an extraction rule covers a negative example without computing all extractions. The ideal case would be to reduce this test solely to subsumption tests to further minimize the computational complexity caused by the answer computations. So, in general a desirable improvement and subject of future work is to modify the presented algorithms and *IE-ILP setting* such that good wrappers can be learned from not completely labeled documents without loosing the property to construct 100 % precise wrappers regarding the training corpus.

basic-BFOIL, *threshold-based BFOIL* and *cluster-BFOIL* follow the heuristic approach to find partitions that yield at least precise rules obtained under *clause set lgg*. But there are other closely related methods worth to be studied. For instance, instead of searching a partitioning one can also try to find subsets such that the sum of all extractions provided by the learned rules of these subsets is maximal regarding the validation set and still yields 100 % precise wrappers. This will probably provide less specific wrappers than those constructed by the discussed techniques, but on the other hand it would require more rule evaluation tests to find the best subsets. This problem is related to the discussion on finding the best hypothesis in Section 12.1 and under claim number two. But there is still a difference, because a wrapper obtained from the generalization of the partitions yielding the maximum number of extractions and a wrapper obtained from an arbitrary subset can significantly differ even if both extract the same number of text fragments from the training corpus. Even more important, one wrapper can be much more general than the other and therefore could provide better overall recall rates on the testing corpus. But finding such subsets or best partitions also involves more rule evaluation tests, which in turn increases the learning time. Hence, an efficient rule evaluation method in the context of learning from positive examples only is the crucial problem to be solved by future work.

Bibliography

[Aha *et al.*, 1994] D. W. Aha, S. Lapointe, C.X.Ling, and S. Matwin. Learning recursive relations with randomly selected small training sets. In W. Cohen, editor, *Proceedings of International Conference on Machine Learning*, pages 12–18. Morgan Kaufmann, 1994. New Brunswick, NJ.

[Aho, 1990] Alfred Aho. Algorithms for finding patterns in strings. In J. van Leeuwen, editor, *Handbook of Theoretical Computer Science*, pages 255–300. Elsevier, Berlin, Heidelberg, 1990.

[Aït-Kaci *et al.*, 1997] Hassan Aït-Kaci, Andreas Podelski, and Seth C. Goldstein. Order-sorted feature theory unification. *Journal of Logic Programming*, 30(2):99–124, 1997.

[Aldenderfer and Blashfield, 1984] Mark S. Aldenderfer and Roger K. Blashfield. Cluster analysis. In *Sage University Paper Series on Quantitative Applications in the Social Sciences*, pages 7–44. Sage Publications, Beverly Hills, London, New Delhi, 1984.

[Arikawa *et al.*, 1992] S. Arikawa, S. Miyano, A. Shinohara, T. Shinohara, and A. Yamamoto. Algorithmic learning theory with elementary formal systems. *IEICE Transactions on Information & Systems*, E75(4):405–414, 1992.

[Atkeson *et al.*, 1997] C.G. Atkeson, S.A. Schaal, and A.W. Moore. Locally weighted learning. *Artificial Intelligence Review*, 11(1-5):11–37, 1997.

[Baader and Schulz, 1998] Baader and Schulz. Unification theory. In Wolfgang Bibel and Peter H. Schmitt, editors, *Automated Deduction — A Basis for Applications*, pages 225–262. Kluwer Academic Publishers, 1998.

[Baeza-Yates and Ribiero-Neto, 1999] Ricardo Baeza-Yates and Berthier Ribiero-Neto. *Modern Information Retrieval.* Addison-Wesley Pub Co, 1999. ISBN-020139829X.

[Baxter, 2000] Jonathan Baxter. A model of inductive bias learning. *Journal of Artificial Intelligence Research*, 12:149–198, 2000.

[Benanav *et al.*, 1985] D. Benanav, D. Kapur, and P. Narendran. Complexity of matching problems. In J.-P. Jouannaud, editor, *Proceedings of the 1st International Conference on Rewriting Techniques and Applications*, volume 202 of *Lecture Notes in Computer Science*, pages 417–429, Berlin, 1985. Springer. Dijon, France.

[Bergadano and Giordana, 1988] F. Bergadano and A. Giordana. A knowledge intensive approach to concept induction. In J. Laird, editor, *In Proceedings of 5th International Conference on Machine Learning*, pages 305–317. Morgan Kaufmann, 1988. Ann Arbor, MI.

[Bergadano and Gunetti, 1996] Francesco Bergadano and Daniele Gunetti. *Inductive Logic Programming - From Machine Learning to Software Engineering.* MIT Press, 1996.

[Bergadano *et al.*, 1989] F. Bergadano, A. Giordana, and S. Ponsero. Deduction in top-down inductive learning. In *In Proceedings of 6th International Conference on Machine Learning*, pages 23–25. Morgan Kaufmann, 1989. Ithaca, NY.

[Beuster *et al.*, 2000a] Gerd Beuster, Bernd Thomas, and Christian Wolff. Mia - a ubiquitous multi-agent web information system. In *Proceedings of International ICSC Symposium on Multi-Agents and Mobile Agents in Virtual Organizations and E-Commerce (MAMA'2000)*, December 2000.

[Beuster *et al.*, 2000b] Gerd Beuster, Bernd Thomas, and Christian Wolff. Ubiquitous web information agents. In *Proceedings of Workshop on Artificial Intelligence In Mobile Systems*, August 2000. 14th European Conference on Aritifical Intelligence (ECAI).

[Bibel and Schmitt, 1998] Wolfgang Bibel and Peter H. Schmitt, editors. *Automated Deduction - A Basis for Applications*, volume 8/9/10 of *Applied Logic Series.* Kluwer Academic Publishers, Dordrecht, 1998.

[Blockeel and Raedt, 1998] H. Blockeel and L. De Raedt. Top-down induction of first order logical decision trees. *Artificial Intelligence*, 101(1-2):285–297, June 1998.

[Blockeel *et al.*, 1998] Hendrik Blockeel, Luc De Raedt, and Jan Ramon. Top-down induction of clustering trees. In J. Shavlik, editor, *Proceedings of the 15th International Conference on Machine Learning*, pages 55–63. Morgan Kaufmann, 1998.

[Blockeel *et al.*, 2000] H. Blockeel, L. De Raedt, N. Jacobs, and B. Demoen. Scaling up inductive logic programming by learning from interpretations. Technical Report Report CW 297, Katholieke Universiteit Leuven, Department of Computer Science, August 2000.

[Blockeel *et al.*, 2002] Hendrik Blockeel, Luc Dehaspe, Bart Demoen, Gerda Janssens, Jan Ramon, and Henk Vandecasteele. Improving the efficiency of inductive logic programming through the use of query packs. *Journal of Artificial Intelligence Research*, 16:135–166, 2002.

[Boström, 1998] H. Boström. Predicate invention and learning from positive examples only. In *Proceedings of the Tenth European Conference on Machine Learning*, 1998.

[Bratko, 1989] Ivan Bratko. Machine learning. In K. Gilhooly, editor, *Human and Machine Problem Solving.* Plenum Press, New York, 1989.

[Brewka, 1996] Gerhard Brewka, editor. *Principles of Knowledge Representation.* CSLI Publications, 1996.

[Califf, 1998] Mary Elaine Califf. *Relational Learning Techniques for Natural Language Information Extraction.* PhD thesis, University of Texas at Austin, August 1998.

[Carpenter, 1991] Bob Carpenter. Typed feature structures: an extension of first-order terms. In *Proceedings of the International Symposium on Logic Programming*, 1991. San Diego.

[Carpenter, 1992] Bob Carpenter. *The Logic of Typed Feature Structures: With Applications to Unification Grammars, Logic Programs and Constraint Resolution.* Cambridge University Press, Cambridge, New York, Melbourne, 1992.

[Carpenter, 1993] Bob Carpenter. Compilation of typed attribute-value logic grammars for parsing. In *Proceedings of the 3rd International Workshop on Parsing Technologies*, 1993. Tilburg, Netherlands and Durbuy, Belgium.

[Carraso and Oncina, 1994] R.C. Carraso and J. Oncina. Learning stochastic regular grammars by means of a state merging method. In R.C. Carrasco and J. Oncina, editors, *Grammatical Inference and Applications: Second International Colloquium, ICGI-94*. Springer, Verlag, 1994.

[Chidlovskii *et al.*, 2000] Boris Chidlovskii, Jon Ragetli, and Martin de Rijke. Wrapper generation via grammar induction. In *11th European Conference on Machine Learning, ECML'00*, volume 1810 of *Lect.Notes Comp. Science*, May 2000. Barcelona, Spain.

[Chinchor, 1992] Nancy Chinchor. Muc-4 evaluation metrics. In *Fourth Message Understanding Conference*, pages 22–29. Morgan Kaufmann, 1992.

[Ciravegna, 2000] Fabio Ciravegna. Learning to tag for information extraction from text. In *Workshop Machine Learning for Information Extraction, European Conference on Artifical Intelligence ECCAI*, August 2000. Berlin, Germany.

[Cohen *et al.*, 2002] William Cohen, Matthew Hurst, and Lee S. Jensen. A flexible learning system for wrapping tables and lists in html documents. In *The Eleventh International World Wide Web Conference WWW-2002*, 2002.

[Cowie and Lehnert, 1996] J. Cowie and W. Lehnert. Information extraction. *Communications of the ACM*, 39(1):80–91, 1996.

[Craven *et al.*, 2000] Mark Craven, Dan DiPasquo, Dayne Freitag, Andrew K. McCallum, Tom M. Mitchell, Kamal Nigam, and Seán Slattery. Learning to construct knowledge bases from the World Wide Web. *Artificial Intelligence*, 118(1/2):69–113, 2000.

[Crescenzi *et al.*, 2001] Valter Crescenzi, Giansalvatore Mecca, and Paolo Merialdo. Roadrunner: Towards automatic data extraction from large web sites. In *The VLDB Journal*, pages 109–118, 2001.

[Cutting *et al.*, 1992] Douglas Cutting, David Karger, Jan Pedersen, and John Tukey. Scatter/gather: A cluster-based approach to browsing large document collections. In *Proc. ACM SIGIR*, pages 318–329, 1992.

[de Raedt *et al.*, 1998] L. de Raedt, H. Blockeel, L. Dehaspe, and W. Van Laer. *Inductive Logic Programming for Knowledge Discovery in Databases*, chapter Three companions for first order data mining. Lecture Notes in Artificial Intelligence. Springer, 1998.

[de Raedt, 1996] Luc de Raedt, editor. *Advances in Inductive Logic Programming.* IOS Press, Amsterdam, 1996.

[Def, 1992] Defense Advanced Research Projects Agency. *Proceeding of the Fourth Message Understanding Conference (MUC-4).* Morgan Kaufmann, June 1992. McLean, Virginia.

[Def, 1993] Defense Advanced Research Projects Agency. *Proceeding of the Fifth Message Understanding Conference (MUC-5).* Morgan Kaufmann, August 1993. Baltimore, Maryland.

[Def, 1996] Defense Advanced Research Projects Agency. *Proceeding of the Fifth Message Understanding Conference (MUC-6).* Morgan Kaufmann, November 1996.

[Dom, 2000] W3C, document object model (dom) level 2 core specification, 2000. Version 1.0 `http://www.w3.org/TR/DOM-Level-2-Core/`.

[Doorenbos *et al.*, 1997] Robert B. Doorenbos, Oren Etzioni, and Daniel S. Weld. A scalable comparison-shopping agent for the world-wide web. In W. Lewis Johnson and Barbara Hayes-Roth, editors, *Proceedings of the First International Conference on Autonomous Agents (Agents'97)*, pages 39–48, Marina del Rey, CA, USA, 1997. ACM Press.

[Dzeroski *et al.*, 1992] Saso Dzeroski, Stephen Muggleton, and Stuart J. Rusell. Pac-learnability of determinate logic programs. In *Proceedings of the Fifth Annual ACM Conference on Computational Learning Theory (COLT 1992)*, pages 128–135. ACM, 1992. July 27-29, Pittsburgh, PA, USA.

[Dzeroski, 1995] Saso Dzeroski. *Numerical Constraints and Learnability in Inductive Logic Programming.* PhD thesis, Faculty of Electrical Engineering and Computer Science, University of Ljubljana, Ljubljana, Slovenia, 1995.

[ECLiPSe, 2004] *The ECLiPSe Constraint Logic Programming System*, 2004. `http://www.icparc.ic.ac.uk/eclipse/`.

[Fellbaum, 1998] Christiane Fellbaum, editor. *WordNet: An Electronic Lexical Database.* MIT Press, May 1998.

[Feng and Muggleton, 1990] Cao Feng and Stephen Muggleton. Efficient induction of logic programs. In Setsuo Arikawa, Setsuo Ohsuga S. Goto, and Takashi Yokomori, editors, *Algorithmic Learning Theory, First International Workshop, ALT '90*, pages 128–135. Springer/Ohmsha, 1990. Tokyo, Japan, October 8-10.

[Fisher, 1987] D.H. Fisher. Knowledge acquisition via incremental conceptual clustering. *Machine Learning*, 2:139–172, 1987.

[Flach, 1992] P.A. Flach. A framework for inductive logic programming. In S. Muggleton, editor, *Inductive Logic Programming*, pages 193–212. Academic Press, 1992.

[Fong, 1991] Sandiway Fong. *Natural Language Understanding and Logic Programming III*, chapter Principle-Based Parsing and Type Inference, pages 43–60. Charles G. Brown and Gregers Koch, North-Holland, 1991.

[Freitag and Kushmerick, 2000] Dayne Freitag and Nicholas Kushmerick. Boosted wrapper induction. In *Proceedings of the Seventh National Conference on Artificial*, pages 577–583, July 30 - August 3 2000. Austin, Texas.

[Freitag and McCallum, 2000] Dayne Freitag and Andrew McCallum. Information extraction with hmm structures learned by stochastic optimization. In *Proc. National Conference on Artificial Intelligence.* MIT Press, 2000. Austin, TX, USA.

[Freitag, 1998] Dayne Freitag. *Machine Learning for Information Extraction in Informal Domains.* PhD thesis, Computer Science Department, Carnegie Mellon University, Pittsburgh, PA, November 1998.

[Fürnkranz, 1999] J. Fürnkranz. Separate-and-conquer rule learning. *Artificial Intelligence Review*, 13(1):3–54, 1999.

[Gazdar and Mellish, 1989] G. Gazdar and C. Mellish. *Natural language processing in Prolog: an introduction to computational linguistics.* Addison-Wesley, Workingham, 1989.

[GNU, 1995] GNU. Flex - fast lexical analyser, 1995. `http://www.gnu.org/software/flex/`.

[GNU, 1997] GNU. Bison, 1997. `http://www.gnu.org/software/bison/`.

[Gold, 1967] E.M. Gold. Language identification in the limit. *Information and Control*, 10:447–474, 1967.

[Goldfarb, 1994] Charles F. Goldfarb. *The SGML handbook.* Clarendon Press, 1994. ISBN 0-19-853737-9.

[Gottlob and Fermüller, 1983] G. Gottlob and C.G. Fermüller. Removing redundancy from a clause. *Artifical Intelligence*, 61(2):263–289, 1983.

[Greenwald and Kephart, 1999] Amy R. Greenwald and Jeffrey O. Kephart. Shopbots and pricebots. In *Agent Mediated Electronic Commerce (IJCAI Workshop)*, pages 1–23, 1999.

[Grieser *et al.*, 2000] Gunter Grieser, Klaus P. Jantke, Steffen Lange, and Bernd Thomas. A unifying approach to html wrapper representation and learning. In *Proceedings of the Third International Conference on Discovery Science*, December 2000. Kyoto, Japan.

[Hatzivassiloglou *et al.*, 2000] Vasileios Hatzivassiloglou, Luis Gravano, and Ankineedu Maganti. An investigation of linguistic features and clustering algorithms for topical document clustering. In *SIGIR 2000*, pages 224–231, 2000.

[Haussler, 1988] David Haussler. Quantifying inductive bias: Ai learning algorithms and valiants learning framework. *Artificial Intelligence*, 36:177–221, 1988.

[Helft, 1989] Nicolas Helft. Induction as nonmonotonic inference. In *Proceedings of the 1st International Conference on Knowledge Representation and Reasoning*, pages 149–156. Morgan Kaufman, 1989. San Mateo, CA.

[Hopcroft and Ullman, 1979] John E. Hopcroft and Jeffrey D. Ullman. *Introduction to Automata Theory, Languages and Computation.* Addison-Wesley, 1979.

[Hsu and Chang, 1999] Chun-Nan Hsu and Chien-Chi Chang. Finite-state transducers for semi-structured text mining. In *Workshop on Text Mining IJCAI 99*, 1999.

[Hsu and Dung, 1998] C.-N. Hsu and M.-T. Dung. Generating finit-state transducers for semistructured data extraction from the web. *Intelligent Systems*, 23(8), 1998. Special Issue on Semistructured Data.

[Huffman, 1996] S. B. Huffman. Learning information extraction patterns from examples. *Lecture Notes in Computer Science*, 1040, 1996.

[Hutchinson, 1997] A. Hutchinson. Metrics on terms and clauses. In *Proceedings of the 9th European Conference on Machine Learning*, pages 138–145. Springer-Verlag, 1997.

[ILP2Net, 2000] ILP2Net. Ilp network of excellence: Ilp systems, 2000. `http://www-ai.ijs.si/~ilpnet2/systems/`.

[Jensen and Cohen, 2001] L. Jensen and W. Cohen. Grouping extracted fields. In *Proc. IJCAI-01 Workshop on Adaptive Text Extraction and Mining*, 2001.

[Johnson, 1992] Mark Johnson. Deductive parsing: The use of knowledge of language. In Robert C. Berwick, Steven P.Abney, and Carol Tenny, editors, *Principle-Based Parsing: Computation and Psycholinguistics*, pages 39–64. Kluwer, 1992.

[Jones *et al.*, 1999] Rosie Jones, Andrew McCallum, Kamal Nigam, and Ellen Riloff. Bootstrapping for text learning. In *Proceedings of Workshop on Text Mining: Foundations, Techniques and Applications*, pages 52–63, August 1999. International Joint Conference on Aritifical Intelligence (IJCAI).

[Junker *et al.*, 1999] Markus Junker, Michael Sintek, and Matthias Rinck. Learning for text categorization and information extraction with ilp. In *Proc. Workshop on Learning Language in Logic*, June 1999. Bled, Slovenia.

[Kietz and Lübbe, 1994] J-U. Kietz and M. Lübbe. An efficient subsumption algorithm for inductive logic programming. In S. Wrobel, editor, *Proceedings of the 4th International Workshop on Inductive Logic Programming*, volume 237, pages 97–106. Gesellschaft für Mathematik und Datenverarbeitung MBH, 1994.

[Kietz and Wrobel, 1991] J.U. Kietz and S. Wrobel. A knowledge intensive approach to concept induction. In Stephen Muggleton, editor, *In Proceedings of the 1st International Workshop on Inductive Logic Programming*, pages 107–126. Academic Press, 1991. London.

[Kim and Moldovan, 1995] J. Kim and D. Moldovan. Acqusition of linguistic patterns for knowledge-based information extraction. *IEEE Transactions on Knowledge and Data Engineering*, 7(5):713–724, 1995.

[Kirsten and Wrobel, 1998] Mathias Kirsten and Stefan Wrobel. Relational distance-based clustering. In Fritz Wysotzki, Peter Geibel, and Christina Schadler, editors, *Proc. Fachgruppentreffen Maschinelles Lernen (FGML-98)*, pages 119 – 124, 10587 Berlin, 1998. Techn. Univ. Berlin, Technischer Bericht 98/11.

[Klusch *et al.*, 2003] Matthias Klusch, S. Bergamaschi, Petta P., Edwards P., Pete Edwards, and Paolo Petta, editors. *Intelligent Information Agents: The Agentlink Perspective.* Springer, Berlin, 2003. ISBN-3540007598.

[Knight, 1989] Kevin Knight. Unification: A multidisciplinary survey. *ACM Computing Surveys*, 21(1):93–124, March 1989.

[Kosala and Blockeel, 2000] Kosala and Blockeel. Web mining research: A survey. *SIGKDD: SIGKDD Explorations: Newsletter of the Special Interest Group (SIG) on Knowledge Discovery & Data Mining, ACM*, 2, 2000.

[Kosala *et al.*, 2003] Raymondus Kosala, Maurice Bruynooghe, Hendrik Blockeel, and Jan Van den Bussche. Information extraction from web documents based on local unranked tree automaton inference. In *Proceedings of the Eighteenth International Joint Conference on Artificial Intelligence*, pages 403–408. Morgan Kaufmann, 2003.

[Koscielski and Pacholski, 1990] A. Koscielski and L. Pacholski. Complexity of unification in free groups and free semigroups. In *Proceedings of the 31st Annual IEEE Symposium on Foundations of Computer Science*, pages 824–829, Berlin, 1990. Springer. Los Alamitos.

[Kowalski and Kuehner, 1971] R. Kowalski and D. Kuehner. Linear resolution with selection function. *Artificial Intelligence*, 2:227–260, 1971.

[Kowalski, 1970] R. Kowalski. *Studies in the completeness and efficiency of theorem-proving by Resolution.* PhD thesis, Metamathematics Unit, University of Edinburgh, 1970.

[Kowalski, 1974] R. A. Kowalski. Predicate logic as a programming language. In *Proceedings IFIPS*, pages 569–574, 1974.

[Krupka, 1995] G. Krupka. Description of the sra system as used for muc-6. In *Proceedings of the Sixth Message Understanding Conference*, pages 221–235, 1995.

[Kushmerick and Thomas, 2003] Nicholas Kushmerick and Bernd Thomas. *Intelligent Information Agents - The AgentLink perspective*, volume 2586 of *Lecture Notes in Computer Science*, chapter Adaptive Information Extraction: A Core Technology for Information Agents, pages 79–103. Springer, 2003.

[Kushmerick *et al.*, 1997] Nicholas Kushmerick, Daniel S. Weld, and Robert Doorenbos. Wrapper induction for information extraction. In Martha E. Pollack, editor, *Fifteenth International Joint Conference on Artificial Intelligence*, volume 1, pages 729–735, August 1997. Japan.

[Kushmerick, 1997] Nicholas Kushmerick. *Wrapper Induction for Information Extraction.* PhD thesis, University of Washington, 1997.

[Kushmerick, 2000] N. Kushmerick. Wrapper induction: Efficiency and expressiveness. *Artificial Intelligence*, 118(1–2):15–68, 2000.

[Lange *et al.*, 2003] Steffen Lange, Gunter Grieser, and Klaus P. Jantke. Advanced elementary formal systems. *Theoretical Computer Science*, 298:51–70, 2003.

[Lapointe and Matwin, 1992] S. Lapointe and S. Matwin. Sub-unification: A tool for efficient induction of recursive programs. In D. Sleeman and P. Edwards, editors, *Proceedings of International Conference on Machine Learning*, pages 273–281. Morgan Kaufmann, 1992. Aberdeen, Scotland.

[Lavrac and Dzeroski, 1994] Nada Lavrac and Saso Dzeroski. *Inductive Logic Programming: Techniques and Applications.* Ellis Horwood, New York, 1994.

[Lavrac and Dzeroski, 2001] Nada Lavrac and Saso Dzeroski, editors. *Relational Data Mining.* Springer, Berlin, 2001.

[Lewis, 1992] David D. Lewis. *Representation and Learning in Information Retrieval.* PhD thesis, Department of Computer Science, University of Massachusetts, 1992.

[LExIKON, 2000] LExIKON, 2000. `http://lexikon.dfki.de`.

[Lloyd, 1987] J.W. Lloyd. *Foundations of Logic Programming.* Springer-Verlag, 2 edition, 1987.

[Manandhar *et al.*, 1998] Suresh Manandhar, Saso Dzeroski, and Tomaz Erjavec. Learning multilingual morphology with clog. In *The Eight Internation Conference on Inductive Logic Programming (ILP'98)*, 1998. Madison, Wisconsin, USA.

[Marcinkowski and Pacholski, 1992] J. Marcinkowski and L. Pacholski. Undecidability of the horn clause implication problem. In *33rd IEEE Symposium on Foundations of Computer Science*, pages 354–362. IEEE Computer Society Press, 1992. Pittsburgh, PA.

[Markov and Marinchev, 2000] Zdravko Markov and Ivo Marinchev. Metric-based inductive learning using semantic height functions. In *Machine Learning: ECML 2000, 11th European Conference on Machine Learning, Barcelona, Catalonia, Spain, May 31 - June 2, 2000, Proceedings*, volume 1810, pages 254–262. Springer, Berlin, 2000.

[Markov and Pelov, 1998] Zdravko Markov and Nikolay Pelov. A framework for inductive learning based on subsumption lattices. *Lecture Notes in Computer Science*, 1480, 1998.

[McCallum *et al.*, 1999] Andrew McCallum, Kamal Nigam, Jason Rennie, and Kristie Seymore. A machine learning approach to building domain-specific search engines. In *The Sixteenth International Joint Conference on Artificial Intelligence (IJCAI-99)*, 1999.

[McCallum, 1999] Dayne Freitag Andrew McCallum. Information extraction using hmms and shrinkage. In *Proc. AAAI-99 Workshop on Machine Learning for Information Extraction*, 1999.

[Michalski and Stepp, 1984] R. S. Michalski and R. E. Stepp. Learning from observation: Conceptual clustering. In R. S. Michalski, J. G. Carbonell, and T. M. Mitchell, editors, *Machine Learning: An Artificial Intelligence Approach*, pages 331–363. Springer, Berlin, Heidelberg, 1984.

[Mitchell, 1977] Tom M. Mitchell. Version spaces: A candidate elimination approach to rule learning. In *Fifth International Joint Conference on Artificial Intelligence*, pages 305–310. MIT Press, 1977. Cambridge, MA.

[Mitchell, 1982] Tom M. Mitchell. Generalization as search. *Artifical Intelligence*, 18(2):517–542, 1982.

[Mitchell, 1990] T.M. Mitchell. The need for biases in learning generalizations. In J. Shavlik and T. Dietterich, editors, *Readings in Machine Learning.* Morgan Kaufmann, 1990.

[Mitchell, 1997] Tom M. Mitchell. *Machine Learning.* McGraw-Hill, 1997. ISBN 0-07-042807-7.

[Miyamoto, 1990] Saadaki Miyamoto. *Fuzzy sets in information retrieval and cluster analysis.* Kluwer Academic, Netherlands, 1990.

[Miyano *et al.*, 2000] S. Miyano, A. Shinohara, and T. Shinohara. Polynomial-time learning of elementary formal systems. *New Generation Computing*, 18:217–242, 2000.

[Mooney and Califf, 1995] R.J. Mooney and M.-E. Califf. Induction of first-order decision lists: Results on learning the past tense of english verbs. *Journal of Artificial Intelligence Research*, 3:1–24, 1995. Morgan Kaufmann.

[Mooney *et al.*, 1994] Raymond J. Mooney, Joshua B. Konvisser, and John M. Zelle. Combining top-down and bottom-up techniques in inductive logic programming. In William W. Cohen and Haym Hirsh, editors, *In Proceedings of 11th International Conference on Machine Learning*, pages 343–351. Morgan Kaufmann, 1994. New Brunswick, NJ, USA, July 10-13.

[Muggleton and Buntine, 1988] Stephen Muggleton and W. Buntine. Machine invention of first order predicates by inverting resolution. In J. Laird, editor, *Proceedings of 5th International Conference on Machine Learning*, pages 339–352. Morgan Kaufmann, 1988. Ann Arbor, MI.

[Muggleton and Feng, 1992] S. Muggleton and C. Feng. Efficient induction of logic programs. In S. Muggleton, editor, *Inductive Logic Programming*, pages 281–298. Academic Press, 1992.

[Muggleton and Raedt, 1994] Stephen Muggleton and Luc De Raedt. Inductive logic programming: Theory and methods. *Journal of Logic Programming*, 19(20):629–679, 1994.

[Muggleton, 1991] Stephen Muggleton. Inductive logic programming. *New Generation Computing*, 20(4):295–318, 1991.

[Muggleton, 1992] Stephen Muggleton. Inverting implication. In S. Muggleton and K. Furukawa, editors, *Proceedings of the Second International Workshop on Inductive Logic Programming (ILP92)*. ICOT Technical Memorandum, 1992. TM-1182.

[Muggleton, 1995] Stephen Muggleton. Inverse entailment and progol. *New Generation Computing, Special Issue on Inductive Logic Programming*, 13(3-4):245–286, 1995.

[Muggleton, 1996] Stephen Muggleton. Learning from positive data. In S. Muggleton, editor, *Proceedings of the 6th International Workshop on Inductive Logic Programming*, pages 225–244, 1996. Stockholm University, Royal Institute of Technology.

[Murphy, 1996] K. Murphy. Learning finite automata. Technical report, Santa Fe Institute, 1996. TR 96-04017.

[Muslea *et al.*, 1999] Ion Muslea, Steve Minton, and Craig Knoblock. A hierarchical approach to wrapper induction. In Oren Etzioni, Jörg P. Müller, and Jeffrey M. Bradshaw, editors, *Proceedings of the Third International Conference on Autonomous Agents (Agents'99)*, pages 190–197, Seattle, WA, USA, 1999. ACM Press.

[Muslea *et al.*, 2000] I. Muslea, S. Minton, and C. Knoblock. Selective sampling with redundant views. In *Proc. National Conference on Artificial Intelligence*, 2000.

[Muslea *et al.*, 2001] Ion Muslea, Steven Minton, and Craig A. Knoblock. Selective sampling with naive co-testing: Preliminary results. In *Proceedings of Workshop Machine Learning for Information Extraction*, August 2001. 14th European Conference on Aritifical Intelligence (ECAI).

[Muslea, 1998] Ion Muslea. A repository of online information sources used in information extraction tasks, 1998. `http://www.isi.edu/info-agents/RISE/index.html`.

[Muslea, 1999] I. Muslea. Extraction patterns for information extraction tasks: A survey. In *Proc. AAAI-99 Workshop on Machine Learning for Information Extraction*, 1999.

[Nienhuys-Cheng, 1997] S-H. Nienhuys-Cheng. Distance between herbrand interpretations: A measure for approximations to a target concept. In S. Džeroski and N. Lavrač, editors, *Proceedings of the 7th International Workshop on Inductive Logic Programming*, volume 1297, pages 213–226. Springer-Verlag, 1997.

[Parekh and Honavar, 1998] R. Parekh and V. Honavar. *Handbook of Natural Language Processing*, chapter Automata Induction, Grammar Inference, and Language Acquisition. Marcel Dekker, New York, 1998.

[Pereira and Warren, 1983] F. C. N. Pereira and D. H. D. Warren. Parsing as deduction. In *Proceedings of 21st Annual Meeting (ACL'83)*, pages 137–144. Association of Computational Linguistics, 1983.

[Plotkin, 1970] G. Plotkin. A note on inductive generalization. *Machine Intelligence*, 5:153–163, 1970. Edinburgh Univ. Press.

[Plotkin, 1971a] G. Plotkin. *Automatic Methods of Inductive Inference.* PhD thesis, University of Edinburgh, 1971.

[Plotkin, 1971b] G. Plotkin. A further note on inductive generalization. *Machine Intelligence*, 6:101–124, 1971. Edinburgh Univ. Press.

[Quinlan, 1990] J. R. Quinlan. Learning logical definitions from relations. *Machine Learning*, 5:239–266, 1990.

[Raedt and Bruynooghe, 1993] L. De Raedt and M. Bruynooghe. A theory of clausal discovery. In *Thirteenth International Joint Conference on Artificial Intelligence*, pages 1058–1063. Morgan Kaufmann, 1993. San Mateo, CA.

[Raedt, 1997] Luc De Raedt. Logical settings for concept-learning. *Artifical Intelligence*, 95(20):187–201, 1997.

[Ramon and Bruynooghe, 1998] J. Ramon and M. Bruynooghe. A framework for defining distances between first-order logic objects. In D. Page, editor, *Proceedings of the 8th International Conference on Inductive Logic Programming*, volume 1446, pages 271–280. Springer-Verlag, 1998.

[Ramon and Raedt, 1999] Jan Ramon and Luc De Raedt. Instance based function learning. *Lecture Notes in Computer Science*, 1634, 1999.

[Ramon *et al.*, 1998] Jan Ramon, Maurice Bruynooghe, and Wim Van Laer. Distance measures between atoms. In *CompulogNet Area Meeting on Computational Logic and Machine Learing*, pages 35–41. University of Manchester, UK, 1998.

[Rasmussen, 1992] Edie Rasmussen. *Information retrieval: data structures and algorithms*, chapter Clustering algorithms, pages 419–442. Prentice-Hall, Inc., 1992.

[Reiter, 1978] R. Reiter. On closed world data bases. In H. Gallaire and J. Minker, editors, *Logic and Data Bases*, volume 961, pages 55–76. Plenum Press, New York, 1978.

[Rico-Juan *et al.*, 2000] J.R. Rico-Juan, J. Calera-Rubio, and R.C. Carrasco. Probabilistic testable tree-languages. In *Grammatical Inference: Algorithms and Applications ICGI 2000*, pages 221–228, 2000.

[Riloff, 1994] Ellen M. Riloff. *Information Extraction as a Basis for Portable Text Classification Systems.* PhD thesis, University of Massachusetts Amherst, 1994.

[Robinson, 1965] J.A. Robinson. A machine oriented logic based on the resolution principle. *ACM*, 12(1):23–41, 1965.

[Rulot and Vidal, 1988] H. Rulot and E. Vidal. An efficient algorithm for the inference of circuit-free automata. In G.A. Ferrate, editor, *Syntactic and Structural Pattern Recognition.* Springer, 1988.

[Sable and Church, 2001] Carl Sable and Ken Church. Using bins to empirically estimate term weights for text categorization. In Lillian Lee and Donna Harman, editors, *Proceedings of EMNLP-01, 6th Conference on Empirical Methods in Natural Language Processing*, pages 58–66, Pittsburgh, US, 2001. Association for Computational Linguistics, Morristown, US.

[Salton and Buckley, 1988] G. Salton and C. Buckley. Term weighting approaches in automatic text retrieval. *Information Processing and Management*, 24(5):513–523, 1988.

[Salton, 1989] Gerard Salton. *Automatic Text Processing – The Transformation, Analysis, and Retrieval of Information by Computer.* Addison–Wesley, 1989.

[Schmidt-Schauss, 1988] M. Schmidt-Schauss. Implication of clauses is undecidable. *Theoretical Computer Science*, 59:287–296, 1988.

[Scott, 1998] Mona L. Scott. *Dewey Decimal Classification: A Study Manual and Number Building Guide.* Libraries Unlimited, 1998.

[Shannon, 1948] C.E. Shannon. A mathematical theory of communication. *Bell System Technical Journal*, 27:379–423, 1948.

[Shapire and Singer, 1998] R. Shapire and Y. Singer. Improved boosting algorithms using confidence-rated predicitions. In *Proceedings of the Eleventh Annual Anunal Conference on Computational Learning Theory*, 1998.

[Shapiro, 1981] E. Y. Shapiro. An algorithm that infers theories from facts. Technical Report 192, Yale University, Department of Computer Science, 1981.

[Smolka and Treinen, 1994] Gert Smolka and Ralf Treinen. Records for logic programming. *Journal of Logic Programming*, 18:229–258, 1994.

[Smolka, 1988] Gert Smolka. A feature logic with subsorts. Technical Report LILOG Report 33, IBM Deutschland GmbH, WT LILOG/Dept. 3508, May 1988.

[Smullyan, 1961] R.M. Smullyan. Theory of formal systems. *Annals of Mathematical Studies*, 47, 1961.

[Soderland, 1997] Stephen G. Soderland. *Learning Text Analysis Rules for Domain-Specific Natural Language Processing.* PhD thesis, University of Massachusetts Amherst, 1997.

[Soderland, 1999] Stephen Soderland. Learning information extraction rules for semi-structured and free text. *Machine Learning*, 34(1-3):233–272, 1999.

[Steinbach *et al.*, 2000] M. Steinbach, G. Karypis, and V. Kumar. A comparison of document clustering techniques, 2000.

[Taniguchi *et al.*, 2001] Katsuaki Taniguchi, Hiroshi Sakamoto, Hiroki Arimura, Shinichi Shimozono, and Setsuo Arikawa. Mining semi-structured data by path expressions. In Klaus P. Jantke and Ayumi Shinohara, editors, *4th International Conference on Discovery Science*, volume 2226 of *Lecture Notes in Computer Science*, pages 378–388. Springer, November 2001. Washington DC, USA.

[Thomas, 1999a] Bernd Thomas. Anti-unification based learning of T-Wrappers for information extraction. In *Proc. AAAI-99 Workshop on Machine Learning for Information Extraction*, 1999.

[Thomas, 1999b] Bernd Thomas. Learning T-Wrappers for information extraction. In *Workshop on Machine Learning in Human Language Technology*, July 1999. ACAI'99 Advanced Course on Artificial Intelligence.

[Thomas, 2000] Bernd Thomas. Token-Templates and Logic Programs for intelligent web search. *Intelligent Information Systems*, 14(2/3):241–261, March-June 2000. Special Issue: Methodologies for Intelligent Information Systems.

[Thomas, 2003] Bernd Thomas. Bottom-up learning of logic programs for information extraction from hypertext documents. In *European Conference on Machine Learning / Principles and Practice of Knowledge Discovery in Databases ECML/PKDD 2003, Cavtat-Dubrovnik, Kroatia*, Berlin, 2003. Springer.

[Utgoff, 1986] Paul E. Utgoff. *Machine Learning of Inductive Bias.* Kluwer Academic Publishers, 1986.

[Valiant, 1984] L. Valiant. A theory of the learnable. *Communications of the ACM*, 27(11):1134–1142, 1984.

[van Rijsbergen, 1980] C. J. van Rijsbergen. *Information Retrieval.* Butterworths, 2nd edition, 1980.

[W3C, 2004] W3C. Semantic web, 2004. `http://www.w3.org/2001/sw/`.

[Weiss, 1999] Gerhard Weiss. *Multiagent Systems.* MIT Press, 1999. ISBN-0262731312.

[Wooldridge, 2002] Michael J. Wooldridge. *An Introduction to MultiAgent Systems.* John Wiley & Sons, 2002. ISBN-047149691X.

[Wrobel and Dzeroski, 1995] Stefan Wrobel and Saso Dzeroski. The ilp description learning problem: Towards a general model-level definition of data mining in ilp. In K. Morik and J. Herrmann, editors, *FGML-95 Annual Workshop of the GI Special Interest Group Machine Learning (GI FG 1.1.3).* University Dortmund, 1995. Research Report.

[Wrobel, 1996] Stefan Wrobel. Inductive logic programming. In Gerhard Brewka, editor, *Principles of Knowledge Representation*, pages 153–189. CSLI Publications, Stanford, California, 1996.

[Xerces, 2003] Xerces: Xml parsers in java and c++ (plus perl and com), 2003. `http://xml.apache.org/`.

[XPa, 1999] W3C, *xpath specification*, 1999. `http://www.w3.org/TR/xpath`.

[Zelle and Mooney, 1996] J.M. Zelle and R.J. Mooney. Learning to parse database queries using inductive logic programming. In *Proceedings of the 14th National Conference on Artificial Intelligence*. AAAI Press/MIT Press, 1996.

[Zeugmann and Lange, 1995] Thomas Zeugmann and Steffen Lange. A guided tour across the boundaries of learning recursive languages. In Klaus P. Jantke and Steffen Lange, editors, *Algorithmic Learning for Knowledge-Based Systems*, volume 961, pages 190–258. Springer-Verlag, 1995.

List of Algorithms

List of Figures

List of Tables

Index

Curriculum Vitae

1.1.1969	Geboren in Hagen, Nordrhein-Westfalen, Deutschland
1975-1979	Grundschule, Parkschule in Hagen
1979-1989	Gymnasium, Theodor-Heuss-Gymnasium in Hagen Abschluss: Abitur (22.5.1989)
1989-1990	Grundwehrdienstzeit, 3. Fallschirmjäger Bataillon, Iserlohn, Nordrhein-Westfalen
1990-1998	Studium der Informatik, Anwendungsschwerpunkt Computerlinguistik, Universität Koblenz-Landau, Koblenz, Rheinland-Pfalz, Deutschland Abschluss: Diplom Informatik (31.3.1998)
1998-2003	Wissenschaftlicher Mitarbeiter, Institut für Informatik, Arbeitsgruppe Künstliche Intelligenz, Universität Koblenz-Landau, Koblenz